PALEOECOLOGY
Concepts and Applications

SCOTT W. STARRATT

PALEOECOLOGY
Concepts and Applications

Second Edition

J. Robert Dodd
Professor of Geology
Department of Geology
Indiana University

Robert J. Stanton, Jr.
Professor of Geology
Department of Geology
Texas A & M University

WILEY

A WILEY-INTERSCIENCE PUBLICATION

JOHN WILEY & SONS

New York / Chichester / Brisbane / Toronto / Singapore

Library of Congress Cataloging in Publication Data:

Dodd, J. Robert (James Robert), 1934-
    Paleoecology, concepts and applications / J. Robert Dodd, Robert
J. Stanton. Jr.
        p.   cm.

    "A Wiley-Interscience publication."
    Includes bibliographical references.
        1. Paleoecology.   I. Stanton, Robert J. (Robert James), 1931-
II. Title.

QE720.D62   1990
560'.45—dc20   89-22502
ISBN 0-471-85711-4   CIP

Printed in the United States of America

10 9 8 7 6 5 4 3 2 1

To our wives, Joann Dodd and Patricia Stanton

Preface to Second Edition

As might have been expected, many new developments have occurred in the field of paleoecology since publication of the first edition of this book. However, the developments have not always occurred in the areas we would have predicted. A virtual flood of papers has appeared on the topic of taphonomy—how it affects the fossil record and how taphonomic information itself can be used to study ancient depositional environments and processes affecting those environments. Studies in paleobiogeography have appeared in profusion. Biogeographers have developed and expanded quantitative techniques of recognizing biotic provinces and the history of their development. The concepts of vicariance and cladistic approaches have been especially well studied. Paleobiogeographic methods have been used effectively in conjunction with studies of plate motions, especially documentation of accretion of exotic terrains. Some areas that were flourishing in the late 1970s are no longer as popular—population dynamics, shell chemistry and structure, and functional morphologic studies have not received the attention in recent years that they once did.

Because of these developments, the first edition of our book has become increasingly dated and less useful as a textbook and reference volume. We have attempted to correct this with the new edition, which reflects recent developments and changing emphasis in the field. Although the basic organization of the book remains the same, we have made several changes that we hope will make the book more useful. Some of these changes have developed from our own experience in using the book as a text in our courses for the last several years. We have also incorporated suggestions of several colleagues who have used the book.

Major changes include:

1. Extensive reorganization and shortening of Chapter 2 (Taxonomic Uniformitarianism), with a changed emphasis to environmental parameters rather than individual taxonomic groups.

2. Greater use of tables, with references to pertinent literature.
3. Inclusion of a new chapter on taphonomy.
4. Elimination of chapter on skeletons as sedimentary particles; some material from this chapter is now included in the chapter on taphonomy.
5. Removal of many of the recurring examples from the Neogene of the Kettlemen Hills.
6. Inclusion of new references on all topics.

At the conclusion of our rewriting, we were surprised at the length of the reference list. This is the result of holding on to the old references as new ones are added. Consequently, the reference list serves as an entry to the literature of the fascinating field of paleoecology by containing a blend of both historical references to early work in the field, important "milestones" in the development of paleoecology, and references to the most current research.

We would like to thank the many people who have helped at various stages in the development of this edition. We want to thank those colleagues who took the time to write to us with suggestions of how the book could be improved: Dick Alexander, Bill Ausich, Karl Flessa, Bob Frey, Ethan Grossman, Roger Kaesler, Jim Valentine, and Art Boucot. We are especially grateful to the students at Indiana University and Texas A & M University who have taken our courses in paleoecology; they have been an essential part in the development of the ideas and approaches that we have used. Finally, special thanks go to Joann Dodd, who spent hundreds of hours with editorial details and preparation of the reference list.

J. ROBERT DODD
ROBERT J. STANTON

Bloomington, Indiana
College Station, Texas

December 1989

Preface to
First Edition

This book is intended to serve two major functions: to be a textbook for an advanced undergraduate- or graduate-level course in paleoecology and to be a reference work and review for the professional paleontologist. Palcoecology is both specialized and broad. It is specialized in the sense of being only one aspect in the field of paleontology, and it is broad in being concerned with many aspects of biology, sedimentology, and geochemistry, in addition to paleontology, as concepts from all these fields are applied to the study of ancient organisms and their environments. Obviously, one book cannot cover such a broad area comprehensively. We do hope that this book can serve as an introduction to major aspects of paleoecology and to the voluminous literature in the field.

We assume the reader to have at least a minimal knowledge of the various fields that contribute to paleoecology. We particularly depend on the reader's having a basic knowledge of the structure and function of the organisms with important fossil records. We also assume a basic understanding of fundamental geologic concepts and terminology as they impinge on the field of paleoecology.

Paleoecology is not a very well-defined subject. It has no basis in neatly structured series of principles that make a natural organization for the subject. In fact, each person working in the field probably has his or her own ideas as to what constitutes the coverage of the subject. Consequently, the method of approach and the organization of a book on paleoecology is not well established. Two major viewpoints or approaches to the subject are perhaps most often taken: theory and application. A good deal of the early interest and research in paleoecology was on the application of fossils and their paleoecological relationships to the determination of ancient environments. Relationships observed between modern organisms and their environment were applied to paleoenvironmental interpretation without really understanding in many cases the reason for the relationships. In recent years, much paleoecological research has been on theory, especially the application to the fossil record of modern ecological concepts (such as those that encompass population and community ecological relationships). This recent trend has been toward understanding the functioning of ancient ecosystems and their evolution rather than using the relationships to determine paleoenviron- ments. Clearly, both aspects are important and supplement each other. We can better determine ancient environmental conditions if we understand why the environment affects the organisms present. Also, we can better under- stand the reason for a certain relationship if we know the details of the ancient environment. Thus in this book we try to show a balance between theory and application, but emphasize application. We stress application because other recent books and reviews have concentrated on the theoretical aspects. No comparable recent treatment has been given to the applied approach.

In recent years much progress has been made in interpreting depositional environments on the basis of sedimentological features. Paleontologists have in some cases deferred to sedimentologists to determine the environmental conditions under which the fossils lived, but paleoecology clearly has much to contribute to environmental interpretation. This was forcibly brought home to us when we were recently in the field with a sedimentologist. After spend- ing some time looking at a section of strata, we asked him if he could identify the depositional environment. He replied that the observed features could be produced either in deep water or in a restricted shallow-water setting and that if we could tell him which of these the fossils indicated, he could identify the sedimentary environment! The greatest precision of environmental interpreta- tion comes from an integration of interpretations based on paleoecology *and* sedimentology as well as on other approaches.

Many people have helped us in the preparation of this book. Perhaps our greatest dept of gratitude should go to the many students at Indiana Univer- sity and Texas A & M University who have taken our courses in paleoecology and have helped to stimulate and then sharpen many of the ideas presented here. Special thanks are due to colleagues with whom we have discussed our ideas and some of whom have read all or parts of the manuscript. Included in this group are Alan Horowitz, Richard Alexander, Gary Lane, Don Hattin, David Kersey, Eric Powell, and Stefan Gartner. A number of people have generously allowed us to use published and unpublished figures and photo- graphs. We especially wish to thank Ken Towe, J. D. Hudson, George Clark, Copeland MacClintock, and Allen Archer for contributing photographs.

J. ROBERT DODD
ROBERT J. STANTON

Bloomington, Indiana
College Station, Texas

January 1981

Contents

SCOTT W. STARRATT

PALEOECOLOGY
Concepts and Applications

1

Introduction

DEFINITIONS AND SUBJECT MATTER

Paleoecology has grown during the past several decades to become a major component of paleontology. In this period of time, activities in paleoecology have expanded from an initial strong focus on the reconstruction of the ancient physical environment to a wide range of topics of both biological and geological emphasis.

A definition of paleoecology can be approached by first defining ecology, but we are confronted immediately with diverse definitions of ecology that have been the result of different perspectives and objectives of ecologists. We favor the simple and concise statement that *ecology* is the *study of interactions of organisms with one another and with their environment.* This definition encompasses all of the physical, chemical, and biological materials, processes, and responses and thus all of the aspects of ecosystem. Correspondingly, then, paleoecology is the *study of interactions of organisms with one another and with their environment in the geologic past.*

The emphasis in paleoecology is, however, distinctly different from that in ecology, for several reasons. One reason is that ecologists study the biota within the framework of an environment that they can describe in as much detail as they wish. In contrast, if the paleoenvironment is known in advance, it is only in very general and limited terms, and more commonly, paleoecologists work in the opposite direction, making inferences about the environment from the paleontologic evidence. A second reason is that the fossil record commonly consists of such a small part of the original biota that paleoecologists are precluded from studying many topics that are a basic part of ecology. A third reason is the great difference in the time scale of ecologic and paleoecologic events and data. Whereas ecologists deal with processes taking

1

place during intervals measured in years, and commonly have data from a single sampling or, rarely, from samples collected during a year or two, paleoecologists have difficulty recognizing phenomena of such short duration in the geologic record. Instead, they generally work within a framework of thousands or millions of years. Consequently, long-term environmental, even evolutionary, changes are an integral part of their perspective.

Ecologists are potentially able to examine directly all aspects of the ecosystem under study, to determine the life histories and interactions of all organisms present, and to relate these biologic data to instantaneous characteristics of the environment. Thus they may be able to develop multivariate and quantitative models that are both precise and realistic. In the fossil record, however, most of the organisms, even those that were most abundant, are not preserved, and short-term phenomena cannot be distinguished. Consequently, the inadequacies of the data base in paleoecology may appear insurmountable, particularly for questions that demand a complete and detailed record of the ancient biota. For two reasons, however, the capabilities of paleoecology are more encouraging than this comparison with ecology and the imperfections of the data base would suggest.

1. Ecologic models based on the available detailed data may be precise and real, but their generality can be tested only by observing essentially contemporaneous geographic gradients in environment and biota; they cannot be tested through time because of the short span of scientific observation. Time is, on the other hand, a parameter readily available to paleoecologists. This permits them to study phenomena that exceed the observational possibilities of ecologists, even of the historical baseline.
2. Although ecologists may, if they wish, examine simultaneously many aspects of the ecosystem, commonly they do not. Much of ecologic study does not depend on the potentially available comprehensive data base, but relies instead on limited components of the biota and on samples gathered during a short time span. Thus the data base available to the paleoecologist is, in reality, similar in quantity and quality to that on which much of ecology is based.

The definition of paleoecology we have adopted emphasizes two major subject areas in paleoecology: One deals with organism–environment interactions; the other, with the more strictly biological attributes of the organisms—their individual life histories, their interactions with one another, and their integration into communities. Our present level of understanding in paleoecology, and the relative amount of interest and research activity in the various aspects of paleoecology, reflect the historical development of the subject and the blending of distinctive orientations of workers in different countries. As one example, the study from a geologic/paleontologic perspective of organisms in their modern environment (actuopaleontology) was originally carried out much more vigorously in Germany than elsewhere (Richter, 1929; Abel, 1935; Schäfer, 1962). This approach was little used in the United States until the years shortly after World War II, and even until the 1960s, paleoecology in the United States was carried out largely with a stratigraphic/paleontologic orientation, with relatively little modern ecologic data and theories to support it. This is evident from a review of the *Treatise on Marine Ecology and Paleoecology* (Hedgpeth, 1957a; Ladd, 1957), in which the majority of ecologic contributions focused on environmental tolerances of organisms, and paleoecologic papers concentrated on the determination of physical aspects of the environment.

Regional differences in emphasis have tended to blur in recent years, but a major historical trend has been increasing interest in the more purely biological aspects of paleoecology. Increasing effort is being expended in trying to understand fossils as once-living organisms, in trying to understand lists of fossils as once-integrated communities, and in trying to understand range charts as they represent biogeographies changing through geologic time. Consequently, paleoecology has increasingly looked to ecology and biology for new approaches to be adapted and applied to the fossil record. Examples would be the analysis of fossils in terms of diversity, community ecology, evolution, and trophic structure in the ecosystem, and the use of electrophoretic techniques to characterize population variability. Of course, because paleoecology is not a derivative science based on ecology, information flow has not been in just one direction: Paleoecologists have made significant contributions to these topics as well as to others, such as biomineralization. Today, paleoecology is progressing on a broad front as an increasingly full and diverse base of fossil evidence is being used, as new analytical and computer-based techniques are being applied, and as new ecologic concepts are being incorporated.

Paleoecology is an important factor in the evolutionary process and as a determinant of broad patterns through geologic time in the history of life, but its role in the process of evolution has yet to be investigated systematically. Ecologic factors, however, may play a major role in the discussion of extinction events, as reviewed, for example, by Hsu (1986).

Application of paleontologic data in reconstructing ancient environments has remained a strong force in paleoecologic work, but many of the interpretive criteria are still not rigorously established. In fact, many paleoenvironmental studies in the literature seem to result in possible but not necessarily probable explanations of the available data—ad hoc explanations that are based neither on sound general principles nor on the application of the method of multiple working hypotheses, and thus do not contribute to the establishment of predictive general theories in paleoecology. Thus the development of interpretive methodology continues to be a vital intellectual challenge.

Improvement in our ability to determine ancient environmental conditions is essential in order to provide a framework for describing earth history and for understanding geologic processes that have been active. In addition, the application of paleoecology in reconstruction of ancient environments has been essential in exploration for and development of many earth resources, and is the justification for much of the paleoecologic work that has been done.

OBJECTIVES AND ORGANIZATION OF THIS BOOK

The emphasis in this book is on organism–environment interactions and on paleontologic techniques by which ancient depositional environments can be determined. Our objective is to provide a concise review of current paleontologic approaches in paleoenvironmental analysis. We have attempted to provide a balanced and comprehensive discussion of techniques and examples relevant to all groups of fossils. However, our own research has been largely with macroinvertebrates, and the paleoecologic literature dealing with these fossils is much larger than that with vertebrates, plants, or microinvertebrates. Consequently, the preponderance of examples we use deal with macroinvertebrates. By discussing the theoretical basis for each paleoecologic approach, however, we hope that the material will be in a form applicable to the full spectrum of taxonomic groups.

In order both to describe present capabilities and to evaluate the potential for future development, we have emphasized the biological, chemical, or physical principals that form the theoretical basis of each topic discussed. From this, underlying assumptions and logic of the method can be evaluated, and aspects most in need of further study can be identified. The theoretical potential of each approach must be evaluated in light of both the assumptions and the nature of the fossil record—the fossil data that are required, compared to those that are likely to be available. In the course of discussing each approach, its application to paleoenvironmental analysis is illustrated by examples representing a range of geologic ages, locations, and kinds of fossils.

Underlying principles, as far as they exist, should be emphasized for another more important reason. As noted earlier, much of the literature in paleoecology consists of descriptive examples and ad hoc explanations, but little in the way of testable models of general and predictive value. Paleoecology will mature as it develops explanatory hypotheses from this descriptive data base.

The application of paleoecology to better understand evolution, the broad patterns of the history of life, and biologic interactions are peripheral to the major thrust of the book, but the principles and phenomena laid out here should be directly applicable to these topics. Autecologic topics, those dealing with fossils and their characteristics at the level of the individual specimen or taxon, are discussed in Chapters 2 to 7. Synecologic topics, those dealing with assemblages or with a more inclusive grouping as the unit of analysis, are discussed in the remaining chapters.

THE DATA BASE IN PALEOECOLOGY

Paleoenvironmental reconstruction depends on three ingredients: a well-established stratigraphic framework, good taxonomy, and a comprehensive ecologic background. The stratigraphic setting provides the spatial and temporal relationships for comparison of fossils within contemporaneous environmental gradients and within geologic history. The basic data of paleoecology are the fossils, adequately identified and correctly positioned within the stratigraphic framework. The necessary ecology consists of an understanding of the ways in which living organisms function within their ecosystems: how their morphology and physiology are adaptive to their conditions of life, the ways in which they interact with one another, and the ways in which they may modify their life history to fit the environment.

The ecologic information required by the paleoecologist is largely at the level of natural history—a field of biology that is relatively inactive in this present era of emphasis on biochemistry, cell biology, and medically oriented topics. Consequently, paleoecologists are commonly confronted with the task of gathering for themselves the ecologic information necessary to interpret the fossils. Ecologic data are necessary ingredients in paleoecology because they usually provide the best basis for developing a sense of the possible ways in which fossils could have interacted with one another and have coped with their environment. Information garnered from ecology is applied to paleoecology in many ways. At one end of the spectrum general ecologic "laws" developed inductively from the living world are applied deductively to the fossil record. These might be general relationships of diversity with environmental resources or stability, for example. At the other end of the spectrum the present-day significance of a particular species or morphologic feature is applied empirically, without benefit of any general, theoretical basis to the same species or biotic characteristic in the fossil record.

Use of ecologic information in paleoecology involves uniformitarianism, analogy, and simplicity. The concept of uniformitarianism has generated much discussion in geology. This has been in part because of the initial broad range of meanings assigned to it by Lyell and other early geologists as they sought to establish geology as a valid science. Uniformitarianism can be classified as either substantive or methodological (Gould, 1965). *Substantive uniformitarianism* implies that the materials, conditions, and rates of processes have remained constant during earth history. *Methodological uniformitarianism* implies that the laws of nature (such as gravity, the properties of fluid flow, and thermodynamics) have been constant in their operation through earth history.

Rigorously defined, substantive uniformitarianism has been largely abandoned because earth materials have not remained constant in composition and proportions during geologic time and because rates of some processes have fluctuated widely beyond the ranges presently observed. Consequently, the present may not always be a good key to the past. In addition, and perhaps more important, substantive uniformitarianism constrains thought and speculation: By specifying dogmatically the limits within which geologic explanations must fall, it inhibits the generation of "outrageous hypotheses" (Davis, 1926) that lead geologic thought into novel and stimulating avenues. Methodological uniformitarianism, in contrast, is a statement of the inductive–deductive logical processes that are inherent in science in general. Consequently, Gould (1965) has argued that the term *uniformitarianism* is not necessary—that geology has already abandoned the substantive variety and that the methodological variety is merely the normal scientific mode of thought, for which geology does not require a special name. Nevertheless, the term is widespread in the geologic literature, and the use of both substantive and methodological uniformitarianism is prevalent in paleoecology. Thus we do well to keep in mind the diverse connotations of uniformitarianism as we study the fossil record.

In many fields of geology, fundamental laws of nature are available and readily applicable. For example, thermodynamic principles can be used in experimental petrology to generate phase diagrams that specify with a high degree of confidence the physical and chemical conditions of igneous and metamorphic phenomena. Similarly, laws of hydrodynamics can be used to understand fluid characteristics resulting in the transportation and deposition of a sediment with specific textural features and sedimentary structures. In paleoecology, however, fundamental laws are not readily established. For example, the apparently inherent and constant characteristics of the organic chemical system and of the established reproductive methods do not provide much guidance for paleoenvironmental reconstruction because they are so general that they prove to be trivial. Consequently, in the absence of fundamental principals to guide their methodological uniformitarian reasoning, paleoecologists commonly rely on substantive uniformitarianism. At the most elementary and simplistic level, of taxonomic uniformitarianism, environmental interpretation of a fossil is based on the habitat characteristics of the most closely related living taxon. When a fossil oyster is found, the paleoecologist may determine the environmental tolerances of the living equivalent taxon and then infer that the fossil also lived within this range of conditions, even though the modern organism may be in a different species, genus, or family. The implicit assumption is made that the evolutionary history that separates the two organisms was neither a response to nor a cause for changing environmental requirements. The confidence with which this assumption can be used clearly decreases as the temporal and taxonomic gaps between the fossil and the extant analog increases. The validity of such a substantive uniformitarian interpretation is strengthened, however, if the additional assumption is made that the name of the taxon serves as a transfer unit for morphologic and physiologic attributes, which are determined by external environmental conditions. Then we are arguing by process-determined *analogy*. Much of paleoecologic reasoning is of this sort, intermediate in rigor and validity between substantive and methodological approaches.

Analogy, whether, for example, in morphology of the individual organism, in community structure, or in population dynamics, is inferred to be valid because environmental forces are time independent. In the study of the morphology of fossils, for instance, we recognize that most living animals moving in water or the air are streamlined. By analogy we can recognize streamlining in fossils and infer from it the causative paleoenvironmental condition. Lorenz (1974), in an article entitled "Analogy as a source of knowledge," has described the value of analogy:

> Whenever we find, in two forms of life that are unrelated to each other, a similarity of form or of behavior patterns which relates to more than a few minor details, we assume it to be caused by parallel adaptation to the same life-preserving function. The improbability of coincidental similarity is proportional to the number of independent traits of similarity, and is, for n such characters, equal to $2n - 1$. If we find, in a swift and in an airplane, or in a shark and a dolphin, and in a torpedo the striking resemblances illustrated in Figure 1.1, we can safely assume that in the organisms as well as in the manmade machines, the need to reduce friction has led to parallel adaptations. Though the independent points of similarity are in these cases not very many, it is still a safe guess that any organism or vehicle possessing them is adapted to fast motion.
>
> There are conformities which concern an incomparably greater number of independent details. Figure 1.2 shows cross sections through the eyes of a vertebrate and a cephalopod. In both cases there is a lens, a retina connected

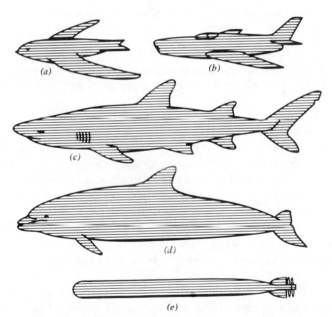

Figure 1.1 Analogy of streamlined form attributed to functional requirement for speed in each case: (*a*) swift; (*b*) airplane; (*c*) shark: (*d*) dolphin; (*e*) torpedo. After Lorenz (1974), copyright The Nobel Foundation.

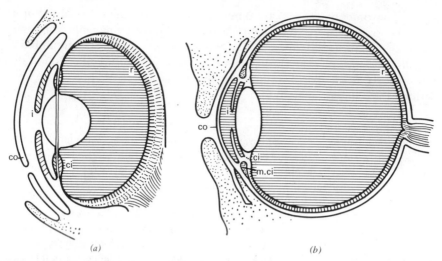

(a)   *(b)*

Figure 1.2 Eye in (*a*) invertebrate (octopus) and (*b*) vertebrate (man). Although independently evolved, striking similarity in detail is determined by identical function. co, Cornea; ci, corpus ciliare; m.ci., musculus ciliris; i, iris; r, retina. After Lorenz (1974), copyright The Nobel Foundation.

by nerves with the brain, a muscle moving the lens in order to focus, a contractile iris acting as a diaphragm, a diaphanous cornea in front of the camera, a layer of pigmented cells shielding it from behind as well as many other matching details. If a zoologist who knew nothing whatever of the existence of cephalopods were examining such an eye for the very first time, he would conclude without further ado that it was indeed a light-perceiving organ. He would not even need to observe a live octopus to know this much with certainty. (Lorenz, 1974, p. 230)

In paleoecology, the essential assumption is that a common process results in analogous structures. However, to assume that streamlined shape can be equated with rapid motion in a fluid medium is simplistic, as Lorenz points out, for streamlining may have a variety of functional explanations. Different organisms may be more or less "able" to become streamlined because of constraints imposed by their physiology and skeletal materials, and they may solve a specific environmental problem in a number of ways. Consequently, the presence and degree of streamlining may not be strongly correlated with fast motion by the organism or by the motion of the fluid environment in which it lived.

The problem of understanding the logic of paleoecology relative to the traditional concepts of uniformitarianism and of critically considering the validity of types of paleoecologic reasoning has been philosophically and pragmatically analyzed by numerous workers. Both Scott (1963) and Lawrence (1971a) have argued that analogy contains the error of substantive uniformitarianism. This is true, but it is also true that analogy between the fossil and living world has been the primary source of paleoecologic hypotheses.

Paleoenvironmental analysis is based in general on the viewpoint that the phenomena of paleontology are *deterministic*, that is, the fossil record is the result of specific and unique causes and that the causes can be determined from its analysis. One reason the deterministic approach is characteristic in paleoecology is because of our historical viewpoint. Just as the historian may strive for a simple explanation for the cause or outcome of some historical event, so we have tended to seek simple explanations for paleontologic data. We have felt comfortable with this because biologists have given us an evolutionary rationale for it. This is exemplified by Cody's (1974) discussion of "optimization in ecology," in which he maintains that natural selection produces optimal results and maximizes fitness. In paleoecology, this has been applied most directly in the study of adaptive functional morphology, but Cody also applies it to a wide range of autecologic and synecologic phenomena. Raup (1977) and others have shown that many broad phylogenetic and diversity patterns through geologic time cannot be demonstrated to be deterministic in that they are not significantly different from *stochastic* (random chance) computer-generated patterns. This work has led to much discussion in the paleontologic literature about the deterministic and/or stochastic nature of the fossil record. We believe that much of this discussion reflects confusion caused by the scale at which observations are analyzed and interpreted. When modern phenomena are analyzed at the most detailed scale, causes can be identified for effects if the scale of analysis is sufficiently detailed—the natural world is deterministic. When modern phenomena are viewed from a more distant perspective, the immediate causes cannot be identified and the proportions of alternative effects seem to fit stochastic expectations—the natural world is not clearly deterministic. Nevertheless, we are dealing with organisms living individually in a deterministic world. Ecologic observations indicate that a deterministic viewpoint is valid in paleoecology, but that causes may often be indeterminate because the paleontologic data are not detailed enough. This topic has been discussed with clarity by Hoffman (1981).

The life histories and consequent abundances and distributions in space and time of living organisms can only be understood if the physical and biological parameters of their environment are studied in detail. Obviously, a great many parameters may be important, but some parameters are more important than others, and the explanation of observed phenomena generally involves an unstated ordering of environmental parameters by their previously established or presumed importance. Thus the explanation of a complex ecologic system is simplified by presenting the effects of only the most important parameters, although with more detailed observations, additional second- and higher-order parameters could presumably be incorporated into the explanation. Because paleoecologic data are commonly sparse and incomplete, the environmental reconstruction usually includes only the most basic parameters. Thus the adequacy of the interpretation depends on the degree to which only a few parameters are important. This simplifying procedure should be valid in paleoecology because it is exactly that used in ecology and in science in general. It saves us from the despair of attempting to derive from the limited paleontologic data an explanation encompassing the myriad of environmental parameters.

The characterization of humid and mesic (= semiarid) forests of the northern hemisphere by a few temperature parameters (Wolfe, 1978) is an example of this procedure (Fig. 1.3). In this analysis, mean temperature and temperature range are clearly considered dominant, and although many other environmental parameters could be used to separate vegetation types and would add detail to the explanation, they are either less important or are dependent and correlated with those Wolfe has used. The result is a model that is justified by the qualitative available data and is a starting point for further refinement. In this analysis, and in general, model building is a simplifying process (Levins, 1968), so the value of the result depends on correctly identifying the most essential parameters.

THE NATURE OF THE FOSSIL RECORD

The quality of the record that is available for us to study and interpret is a major concern to paleontologist. That this has been a long-standing concern is indicated by discussion such as that of Lawrence (1968) or Schopf (1978) on the completeness of the fossil record. These emphasize the small fraction of the total biota that is likely to be preserved, and the differences in preservation potential of different organisms, whether categorized by life habit or taxonomic groups, leading to strong biases in the composition of the fossil record. In addition, Durham (1967), in his paper "The incompleteness of our knowledge of the fossil record," has pointed out that because of the normally small sample size, we generally do not know very well even the small part of the original biota that is preserved.

The quality of the fossil record that is required for paleoecologic analysis depends, of course, on the objective. For example, biogeochemical analysis is not a feasible approach if the original skeletal chemistry has been altered, no matter how large the collection of fossils is. Community reconstruction, on the other hand, requires a high level of preservation of the original community, but not necessarily excellent preservation of the specimens themselves. When the analytic approach to be used has been established, the sampling procedure and sample size can be established, using basic statistical concepts and techniques. The paleontologic perspective in this has been particularly well developed by Dennison and Hay (1967) and Chang (1967).

It is clear that much of the original biota is, except in very rare instances, not preserved in the fossil record. It is also clear that it is generally difficult to establish the time span that a fossil assemblage encompasses—the extent to which an assemblage represents a single community accumulated during a short period of time, or is the residue of a number of successive communities

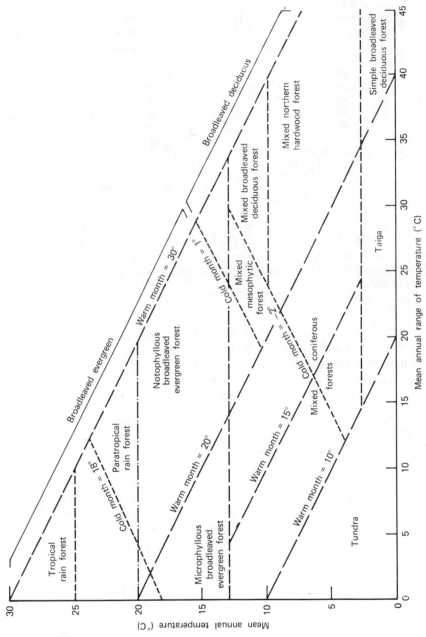

Figure 1.3 Major temperature parameters that are correlated with the different humid and mesic forests of the northern hemisphere. After Wolfe (1978), reprinted by permission of *American Scientist*, journal of Sigma Xi, The Scientific Research Society.

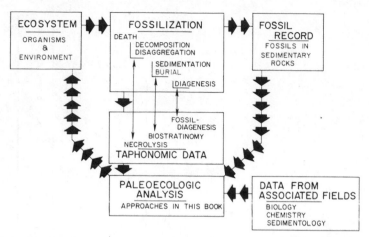

Figure 1.4   The fossil record is formed from the ecosystem by the processes of fossilization. The fossil record interpreted to reconstruct aspects of the ecosystem by incorporating into the paleoecologic analysis concepts and data from associated fields and taphonomic data.

and is a time-averaged accumulation of a longer time interval. The processes that form a fossil assemblage from the original living biota make up the subject matter of taphonomy (Fig. 1.4). This is important throughout paleoecology and will be a part of each of the following chapters. In addition, however, many of the general aspects are summarized in Chapter 7.

2

Taxonomic Uniformitarianism

The most basic method of interpreting environments from fossils is by assuming that the environmental requirements of the fossils were the same as those of the most closely related living representatives and transferring this environmental information from the modern to the fossil. This is based on a strict substantive application of the principle of uniformitarianism—that the ecology of present organisms is the key to that of past organisms. This method has been used successfully for decades and in fact was probably the first paleoecological method employed. The Greek scholar Xenophanes (ca. 540 B.C.) noted fossil seashells preserved in rocks well above the level of the present Mediterranean Sea. The shells he observed were very much like (or even identical to) living mollusks which he had observed in the Mediterranean. He made the simple uniformitarian deduction that this meant that the sea had once been more extensive and had occupied the area where he observed the fossil shells (Adams, 1938).

Paleoecologic interpretation of environments such as this can be found in many places in the ancient as well as more modern literature. Leonardo da Vinci used the presence of fossil shells miles from the sea to prove the once greater extent of the Mediterranean Sea. Edward Forbes, the so-called Father of Modern Marine Ecology and Paleoecology, suggested using uniformitarian extension of modern data on the distribution of organisms in the sea to interpret ancient environments (Forbes, 1843). He made a detailed study of the distribution of organisms in the Aegean Sea, especially in relation to depth. He noted marked differences in the organisms found in different environments and suggested that fossil organisms must have been similarly influenced by environment. M. K. Elias (1937) used data on modern organism distribution from Forbes and others in his classical paleoecologic study of the Permian Big Blue Limestone of Kansas. He studied the vertical distribution of fossils through the strata in order to determine variation in water depth during deposition. His results were rather crude by modern standards, but his study was especially important as a pioneering effort in paleoecology.

Another classical and important study using taxonomic uniformitarianism is that of M. L. Natland (1933), who investigated the distribution of modern and fossil benthic foraminifera and made paleoenvironmental interpretations of the Cenozoic sediments of the Ventura Basin in California. He recognized five foraminiferal assemblages off the modern southern California coast. The species characterizing these assemblages lived at differents depths. The distribution of fossil foraminifera in the sediments suggested marked changes in water depth, which were ultimately interpreted as due to introduction of the shallow-water forms by turbidity current flows. Although it was not appreciated for many years, Natland's paleoecologic study was a vital part of the evidence supporting the hypothesis that sediments can be transported by this process. His study was one of the earliest to describe examples of turbidites.

The taxonomic uniformitarian approach is certainly not limited to the invertebrates. Palynologists have made extensive use of this method in interpreting environments from the distribution of fossil pollen and spores. The technique has been especially effective in studying cores from Pleistocene and Holocene lake deposits (e.g., Adams and West, 1983; Baker, 1986).

Some have criticized the uniformitarian approach on the grounds that it often does not consider why a particular taxon is restricted to certain environmental settings; that is, it uses the substantive uniformitarian approach (see Chapter 1). Evolution could occur that changes the physiology of the organism and its response to environmental parameters without affecting its morphology. Apparent examples of this have been found (e.g., Stanton and Dodd, 1970). Certainly, the ideal approach would be to determine on the basis of physical laws why a given measurable feature of the fossil is related to specific environmental parameters, the methodological uniformitarian approach (see Chapter 1). (Many of the techniques discussed elsewhere in this book, such as the functional morphology approach and the oxygen isotopic paleotemperature determination method, are of the methodological uniformitarian type.) Such relationships are not as subject to evolutionary change; however, even the methodological approach may be complicated by the fact that through evolution, organisms develop increasingly superior morphologies to cope with environmental problems. Thus the relationship noted between a given morphology and environment in the Recent may not have yet evolved in fossils of an earlier age. Another problem is that a certain general morphology may be a response to more than one environmental parameter.

Despite these complicating factors, in many cases the reason for the relationships between environmental parameter and taxon can be explained. For example, echinoderms are not found in fresh or very low salinity water because their excretory systems cannot maintain a concentration gradient between internal fluids and the surrounding medium. This fundamental physiological feature of the echinoderms is not likely to have evolved independently of morphology; so we can be fairly confident that echinoderms have never lived in fresh or very brackish water. Although the blind application of the uniformitarian approach may involve risks, it has proven many times over to be a useful method of paleoenvironmental interpretation and will certainly continue to be widely used.

The taxonomic uniformitarian approach can be applied at many different taxonomic levels. As an example, a Pleistocene limestone may contain specimens of the coral *Acropora palmata* (Fig. 2.1). This species is living today in the Caribbean area, where it is restricted to very shallow (usually less than 5 m) turbulent water in the open reef environment. It is most abundant on the shallowest, most turbulent portion of reef. Thus the depositional environment of the limestone containing this species (assuming that it is in place) can be quite specifically determined by the taxonomic uniformitarian approach.

Another limestone, this one of Miocene age, may contain a now extinct species of *Acropora*. The entire genus *Acropora*, of course, has a much broader range of environmental requirements than any one of its species. The genus is widespread in reef and associated environments throughout the

Figure 2.1   *Acropora palmata* (right) and *A. cervicornis* (left). Photo taken in Florida Keys.

world. Two common species of *Acropora* occur in the Caribbean area, *A. palmata* and *A. cervicornis*. *A. cervicornis* is most abundant in moderately turbulent water in the back reef area and on patch reefs some distance from the shelf edge in less turbulent environments than *A. palmata*. The total environmental range would be further broadened if the many Indo-Pacific species were included. Thus the taxonomic uniformitarian approach used at the generic level would yield less precise information than at the species level, but would still indicate a probable reef or reef-associated environment.

The next step up the taxonomic scale to the family level would yield still less information. An early Cenozoic limestone might contain *Dendracis,* an extinct genus of the family Acroporidae. The family Acroporidae today largely consists of species of the genus *Acropora* but includes a few other less common genera. All live in tropical, shallow, full-salinity environments, but their total environmental range is greater than that for the genus *Acropora* and certainly of any of its species.

A Cretaceous limestone may contain a species of an extinct scleractinian family such as Stylinidae. The order Scleractinia includes all of the modern hexacorals, from the reef-building forms of the tropics to the relatively deep water, ahermatypic forms of the high latitudes. The total environmental range of the order is much broader than lower taxonomic levels. The majority of scleractinian species (but certainly not all) are tropical. Most (but not all) live in shallow water. All live in normal or near-normal salinities. All require well to moderately well oxygenated water that is relatively sediment free. So the presence of a fossil scleractinian gives much less specific information, although a diverse assemblage of scleractinian corals generally suggests shallow, tropical seas.

A Paleozoic limestone may contain an extinct order, Rugosa, of the class Anthozoa. Modern anthozoans include the orders Octocorallia, Scleractinia, and several other orders without skeletons. Modern anthozoans are most abundant in the shallow tropics but are common down to considerable depths. They occur at temperatures from 0 to 40°C or above. They are most common at normal marine salinities, but some forms extend into brackish or hypersaline waters. Consequently, a very limited amount of information concerning environment can be obtained by taxonomic uniformitarianism at the class level. Even less could be said at the phylum level. Coelenterates occur in practically all aquatic environments, both marine and nonmarine.

As is clear from the foregoing example, the substantive taxonomic uniformitarian approach increases in usefulness (or precision) with decreasing geologic age. The best results are obtained in rocks of late Cenozoic age which contain many extant species or at least genera. Only the most general applications can be made to rocks of Paleozoic age. However, the methodologic uniformitarian approach in which, for example, the morphology of the modern organism is causally related to the environment can be used with older as well as geologically younger species.

In our example of the corals, the methodological uniformitarian approach would tie the morphology of the coral to a function that is environmentally related. Most of the shallow-water corals contain symbiotic photosynthesizing dinoflagellates (zooxanthellae) in their tissue. Corals containing zooxanthellae are called *hermatypic* and those without are *ahermatypic* forms. The zooxanthellae utilize CO_2 and waste products produced by the coral as raw materials for photosynthesis. The coral benefits from the relationship by having a source of oxygen from photosynthesis and a sink for some of its waste products. Uptake of CO_2 also aids calcification by the coral, allowing more rapid growth. As this relationship depends on the presence of abundant light to provide the energy source for photosynthesis, hermatypic corals are restricted to shallow water.

If one could determine that the coral contained zooxanthellae in its tissues, one could be certain that the coral lived in shallow water. Zooxanthellae are, of course, never preserved as fossils, but their effects on coral growth can perhaps be observed. Today, all massive, rapidly growing scleractinians are hermatypic. Highly developed growth of the calcareous tissue between the corallites in a coral colony is also restricted to hermatypic corals (Fagerstrom, 1987). Presence of these features in fossils thus suggests growth in a shallow environment. This determination would be based on a known process and would therefore be methodological uniformitarianism rather than strict substantive uniformitarianism.

In the simplest case, the taxonomic uniformitarian approach involves transferring ecological information from a single extant species to its fossil counterpart. Greater precision might be obtained by using information for as many species as possible. Perhaps the ultimate application of the method is to recognize and utilize information from the modern community which is the counterpart of the fossil community. This approach is discussed in detail in Chapter 9. More commonly, the uniformitarian approach has been used with several species of one or a few taxonomic groups rather than an entire community.

How can information from several species or other taxonomic units be combined into a single environmental interpretation? At one end of a spectrum of approaches the investigator may simply present the data and then make a subjective interpretation based on those data. Data for all species (such as depth or temperature distribution) may be presented in a systematic form that shows overlap of ranges. Figure 2.2 is an example of this format. This method is comparable to the method of concurrent ranges or assemblage zones used in biostratigraphy. Ideally, all ranges will overlap in only one narrow band, which allows a very precise interpretation of the environmental parameter (Fig. 2.2). In practice, this ideal is seldom realized. This may be due to several factors: (1) a certain amount of postmortem mixing may have occurred (e.g., some shallow-water species may be transported by currents into deeper water; (2) data on modern environmental requirements may be incomplete; or (3) as indicated above, some species may have evolved between the time when the fossils lived and today so that the environmental requirements have changed. In the case of nonoverlapping ranges, some at-

Degrees Latitude

	20	40	60	80

Gastropods

Calliostoma

Calyptraea

Cancellaria

Jaton

Littorina

Nassarius

Neverita

Olivella

Opalia

Bivalves

Anadara

Chama

Chaceia

Chione

Chlamys

Florimetis

Glycymeris

Macoma

Modiolus

Mya

Mytilus

Ostrea

Panopea

Pecten

Protothaca

Saxidomus

Semele

Solen

Tellina

Trachycardium

Tresus

Figure 2.2 Modern latitudinal ranges of marine *Pecten* zone (Neogene of Kettleman Hills, California) molluskan genera along the west coast of North America. Data from Keen (1937).

tempt is usually made to explain the discrepancy, and the area of greatest overlap is considered to indicate the most likely environment.

Greater precision should be attainable if quantitative information on the relative abundance of species is included in the analysis rather than simple presence or absence of the species. Environmental information from an abundant species is likely to be more indicative of depositional environment than that for a rare species, which may not even be indigenous to the environment where it was buried. Relative abundance of the modern species throughout its environmental range should be taken into account. For example, the rare occurrence of a modern species in water deeper than most members of the species results in an increase in total depth range, but most members do *not* live in deep water. Abundant occurrence of that species probably means a shallower depth. Delorme (1971) uses this approach in his study of the depth distribution of fossil ostracods.

Perhaps one of the most rigorously quantitative approaches to using the uniformitarian method attempted to date is that of Imbrie and Kipp (1971) on the temperature distribution of modern planktic foraminifera. Their approach has been used extensively in connection with the CLIMAP project, which has the goal of mapping the climatic pattern for various times during the Quaternary Period (Cline and Hays, 1976).

Imbrie and Kipp have taken data on the relative abundance of modern planktic foraminifera in various temperature regimes and subjected it to a principal component analysis in order to establish assemblages characteristic of different climates. Then by viewing a sample as a mixture of these assemblages in different proportions, they are able to determine the temperature at which the specimens in the sample lived. Imbrie and Kipp claim a precision within 1°C for this method when it is applied to samples from deep-sea cores. In principle this technique should be usable with other fossil groups in other geologic settings. However, the study of foraminifera in the deep sea is an ideal setting for this quantitative approach. The environment is ecologically relatively simple, amount of mixing and diagenetic effect is minimal, and information on modern distributions comes from the same type of samples (deep-sea sediment) as those to which it is applied. A complicating factor in interpretations based on planktic organisms is the effect of currents in transporting organisms before and after death (Weyl, 1978).

Throughout this discussion we have implied that fossils are preserved in the place where they lived. As indicated in Chapter 7, this is clearly not always the case. A part of any analysis using the taxonomic uniformitarian approach (or any paleoecologic study for that matter) should be consideration of possible mixing of fossils from several environments. This mixing can be either mixing in area (such as transport by currents) or in time (such as stratigraphic mixing by biologic or physical processes). On the positive side, our own experience (Dodd and Stanton, 1975; Powell and Stanton, 1985) and that of others (e.g., Fürsich and Flessa, 1987) suggest that transport and mixing of fossils may not be as common as was once thought. Using a complete assemblage of fossils helps to alleviate the problem of transport by diluting the contribution of a few incidental transported specimens.

Differential preservation, perhaps due to differential chemical diagenesis, is also important in studies of this kind. Studies such as those of Imbrie and Kipp, which are based on the relative abundance of an entire assemblage, especially require unbiased preservation. Differential preservation may cause the final assemblage investigated by the paleoecologist to be quite different from the original living assemblage. The fossil assemblage preserved in this way may look much more like an assemblage from another environment. Chave (1964) gives an excellent example of this from the Upper Cretaceous Navesink Formation of New Jersey. The foraminifera in assemblages from the Navesink Formation consist entirely of planktic species. Modern assemblages dominated by planktic species are almost exclusively found in deep-sea sediments, although planktic species also occur in shallow-water sediments along with more abundant benthic species. The implication from the uniformitarian approach is thus that the Cretaceous Navesink assemblage was deposited in deep water. This is surprising because the formation contains sedimentary features suggesting deposition in shallow water. Its stratigraphic setting also suggests relatively shallow water. More important from the paleoecologic point of view, the total fossil assemblage also contains oysters, which by the uniformitarian approach, suggest shallow water. Chave explains the apparent dilemma as resulting from differential preservation. The assemblage contains only fossils originally composed of low-magnesium calcite. All evidence of the aragonitic, shallow-water bivalves and gastropods that were probably once in the assemblage is gone. The benthic foraminifera which were originally high-magnesium calcite also appear to have been dissolved. The final assemblage is thus highly biased.

Perhaps the safest procedure to follow is to base interpretations using the uniformitarian approach on fossils that are present and not on absence of fossils. In the New Jersey Cretaceous example, the planktic forams by themselves have no special depth implication. They can occur in sediments deposited at all depths. The absence of fossils normally expected in shallow-water sediments leads to the erroneous interpretation. In a more general sense, the absence of fossils in sedimentary rocks anywhere should not be taken to imply that they were never there. The vast volumes of unfossiliferous sedimentary rocks in the world today could not conceivably all have been deposited in environments without life. Also, that all skeletal remains ever produced would be preserved is inconceivable. Thus differential preservation of fossils either within a given assemblage or between fossiliferous and unfossiliferous rocks is a fact of life that must always be taken into account in paleoecological studies.

The literature on the relationships between modern organisms and their environment that is potentially relevant to interpretation of the geologic record is extremely large. In this chapter we only briefly summarize some of the most general relationships for large taxonomic units. The greatest use of the uniformitarian approach (especially as applied to formations of Cenozoic age) has been the substantive approach at the generic and specific level. A large collection of papers and monographs using this approach has been published. We cannot hope to review them all here.

Much information helpful in applying the taxonomic uniformitarian approach is summarized in textbooks on invertebrate zoology and on paleontology, and in compendia such as the *Treatise of Invertebrate Paleontology* and the *Treatise of Marine Ecology and Paleoecology.* Good sources of information about environmental requirements and function in living and extinct groups are the paleontology textbook by Boardman et al. (1987) and the invertebrate zoology text by Pearse et al. (1987). The series of short-course notes published by the Paleontological Society is also a good source of general information and references useful in the taxonomic uniformitarian approach. An especially useful, brief summary of the taxonomic uniformitarian approach, including a summary of environmental requirements of fossil groups, has been prepared by Heckel (1972). The reader is referred to these and more basic sources for details. However, taxonomic uniformitarianism is so fundamental that we will include some basic information on the ecologic requirements of important fossil groups.

In the remainder of this chapter we presuppose a basic knowledge on the part of the reader of the morphology and biological functioning and mode of life of the major invertebrate groups with important fossil records. We discuss the major environmental parameters of importance to the distribution and morphology of these groups (currents, temperature, salinity, nutrient elements, dissolved oxygen, depth, and substrate). We discuss examples of the relationship of distribution and morphology to environment as observed in modern and fossil representatives of each group. The latter approach impinges on functional morphology which is treated more fully in Chapter 5. In the current chapter we concentrate on relationships observed in living forms, whereas Chapter 5 contains deductions of function from the morphology of fossil forms. Table 2.1 lists important references that provide more detail on the uniformitarian approach.

The taxonomic uniformitarian approach has also been used extensively for plant and vertebrate fossils. We briefly mention these groups and give references for more extensive treatments but do not consider them in the detail we give for the invertebrate groups. We do this because plants and vertebrate fossils are not as abundant as invertebrates in most rocks and because we are not as familiar with the literature on vertebrate and plant fossils.

ENVIRONMENTAL PARAMETERS

In this section we discuss taxonomic uniformitarianism as it can be used to study several major environmental parameters of the marine environment: currents and water turbulence, temperature, salinity, nutrient elements (productivity),

TABLE 2.1   References to Literature on Ecology and Taxonomic Uniformitarian Studies of Important Groups of Fossil Organisms

Group	General References on Ecology	Studies Using Taxonomic Uniformitarianism
Coccoliths	Paasche, 1968; Haq, 1978; Buzas et al., 1987	MacIntyre, 1967; Geitzenauer, 1969; MacIntyre et al., 1970
Diatoms	Simonsen, 1972; Burckle, 1978; Brasier, 1980	Barron, 1973; Burckle, 1984; Sanchez et al., 1987
Stromatolites	Walter, 1976; Wray, 1977; Walter, 1977; Fagerstrom, 1987	Ahr, 1971; Playford et al., 1976; Peryt and Piatikowski, 1977; Dill et al., 1986
Green algae	Johnson, 1961; Wray, 1977; Fagerstrom, 1987	Conrad, 1977; Elliott, 1984; Beadle, 1988
Red algae	Adey and Macintyre, 1973; Wray, 1977; Johansen, 1981	Buchbinder, 1977; Bosence, 1985; Manker and Carter, 1987
Radiolaria	Casey, 1977; Anderson, 1983	Casey, 1972; Palmer, 1986
Foraminifera	Boltovskoy and Wright, 1976; Boersma, 1978; Haynes, 1981; Buzas and Sen Gupta, 1982	Imbrie and Kipp, 1971; Hayward and Buzas, 1979; Miller, 1982; Hallock and Glenn, 1986
Sponges	Fry, 1970; Bergquist, 1978; Rigby and Stearn, 1983; Rigby, 1987	Jackson et al., 1971; Termier and Termier, 1975
Corals	Wells, 1957; Goreau et al., 1979; Oliver and Coates, 1987; Fagerstrom, 1987	Frost and Langenheim, 1974; Frost, 1977; Rosen, 1977
Bryozoa	Ryland, 1970; Woollacott and Zimmer, 1977; Ross, 1987; Boardman and Cheetham, 1987	Lagaaij and Gautier, 1965; Brood, 1972; Lagaaij and Cook, 1973; Cuffey and McKinney, 1982
Brachiopods	Ager, 1967; Rudwick, 1970; Richardson, 1986; Rowell and Grant, 1987	Surlyk, 1972; Fürsich and Hurst, 1974; Alexander, 1975; Emig, 1981 ·
Mollusks[a]	Stanley, 1970 (B); Denton and Gilpin-Brown, 1973 (C); Yonge and Thompson, 1976; Purchon, 1977; Linsley, 1978a (G); Morton, 1979; Lehmann, 1981 (C); Bottjer et al., 1985; Pojeta et al., 1987	Trueman, 1940 (C); Valentine, 1961 (B,G); Strauch, 1968 (B); Cowen et al., 1973 (C); Mutvei and Reyment, 1973 (C); Peel, 1975 (G); Linsley et al., 1978 (G); Hickman, 1984 (B); Good, 1987 (B,G)

TABLE 2.1   (Continued)

Group	General References on Ecology	Studies Using Taxonomic Uniformitarianism
Ostracods	Benson, 1975; Loffler and Danielopol, 1977; Pokorny, 1978; Bate et al., 1982	Hazel, 1971; Cronin, 1983; Benson, 1984; Cohen and Nielsen, 1986
Crinoids	Fell, 1966; Breimer and Lane, 1978; McCurda and Meyer, 1983; Lawrence, 1987	Breimer, 1969; Meyer and Lane, 1976; Ausich, 1980
Echinoids	Nichols, 1972, Kier, 1974; Smith, 1984; Lawrence, 1987	Fell, 1954; Nichols, 1959; Kier, 1972; McKinney and Zachos, 1986

[a]Emphasis on bivalves (B), gastropods (G), and cephalopods (C).

dissolved oxygen levels, depth, and substrate type. These are certainly not the only environmental parameters of interest, but include those most routinely investigated in paleoenvironmental analysis. Throughout this book we emphasize the marine environment, but we will on occasion also refer to nonmarine environmental parameters. Our approach will be to (1) discuss briefly the parameter; (2) review the spatial variation of that parameter, including effects of that variation on organisms (especially those which are likely to be preserved in the fossil record); (3) present information, partly in tabular form, on the relationship of modern organisms to that parameter; and (4) give a few examples of paleoecologic studies using the taxonomic uniformitarianism approach with relation to this parameter.

Because of space limitations, our discussion of the nature of each parameter and its variation will necessarily be brief. The reader is encouraged to look to other sources for details. Textbooks and reviews of oceanography and marine biology and geology are convenient sources of information in this regard. Especially useful sources are *GSA Memoir 67* (Hedgpeth, 1957a), Valentine (1973a), and Kennett (1982). More detailed information is found in the various volumes of *Marine Biology* (Kinne, 1971, etc.) and *The Sea* (Hill, 1962, etc.). Other sources could also be mentioned, but those listed above should serve as an introduction to the rich literature on this topic.

CURRENTS

Currents can be discussed at two scales: (1) major, oceanic currents or global circulation, and (2) local currents such as those generated by waves and tides. In some cases the two may be one and the same, but in general they can readily be subdivided. Global circulation is very important in paleoecology,

but most of the effects are due to indirect control of some other parameter by
major current patterns. Local currents or turbulence have a more direct effect
on organisms, which must adapt to cope either with currents or their absence.
We begin with a very brief discussion of major, oceanic circulation and then
consider locally generated currents and turbulence.

Global Circulation

Global currents have a large effect on other parameters, especially temperature, salinity, and concentration of nutrients. These major currents thus indirectly have a very large effect on the distribution and abundance of organisms.
They may also directly affect distribution of biota by transporting pelagic
species, transporting the pelagic, larval stage of many benthic forms, and in
some cases even transporting shells after death of the organism (e.g., *Nautilus*). We usually think in terms of horizontal currents in the oceans, but
vertical circulation is also extremely important in its effect on organisms.

Wind is the driving force behind surface currents. The oceanic circulation
pattern is directly related to the pattern of global winds and pressure systems
(Fig. 2.3). Just as winds are strongly influenced by the Coreolis effect, so are
the major currents in the oceans. The Coreolis effect, which is due to rotation
of the earth, causes currents to be deflected to the right in the northern

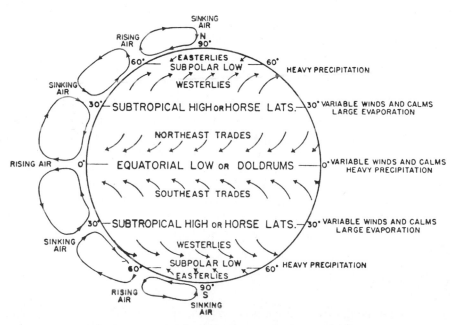

Figure 2.3 Schematic global wind pattern for hypothetical completely water-covered
earth. From Fleming (1957), Geological Society of America Memoir 67, Vol. 1, p. 89.

hemisphere and to the left in the southern hemisphere. Von Arx (1962) and Valentine (1973a) include good discussions of the Coreolis effect. If the earth were completely covered by the oceans, winds and the Coreolis effect would result in major current bands completely circling the globe. Because the oceans are broken up by continents, major currents form gyres within the ocean basins, circulating clockwise in the northern hemisphere and counter-clockwise in the southern hemisphere (Fig. 2.4). This pattern results in strong currents on the ocean margins and relatively weak circulation in the centers of the oceans. Productivity, temperature, and salinity are strongly influenced by this current pattern.

Vertical circulation in the oceans is controlled by density differences in the water. Factors controlling the density of seawater are temperature, salinity, and suspended sediment content. Temperature is by far the most important of these. The oceans are density stratified with the warmer, less dense water on top and cooler, more dense water on the bottom (Fig. 2.5). Density differences are greatest at low latitudes, where the temperature difference between the surface and bottom are greatest. In general, vertical circulation is weakest here. Temperature and thus density differences tend to disappear at high latitudes. Vertical circulation is enhanced in these areas by this lack of density stratification.

Vertical circulation of the oceans in general is driven by cooling of water near the poles. Cooling and even freezing of seawater in these areas produces especially dense water. Density is not only increased by cooling, but by an increase in salinity caused when relatively fresh water freezes out, leaving the remaining seawater with more dissolved salts. This dense water, produced at limited locations in the North Atlantic and the Wedell Sea off Antarctica, sinks to the bottom and slowly flows along the bottoms of all the oceans (Fig. 2.6). Water at intermediate depths also comes from high latitudes by slightly less cooling and sinking to lesser depths. This deep water slowly warms due to heat flow through the seafloor and gradual mixing with the less dense water above. It will eventually return to the surface. Vertical circulation is especially important in recycling nutrient elements released by decay of organic matter, which sinks to the bottom. It also brings dissolved oxygen from the surface to the depths, where it can be utilized by organisms living there.

Another important aspect of vertical circulation is the occurrence of divergent and convergent currents. When surface currents diverge from one another or move away from land areas, water must rise from below to replace the water that has moved away laterally (Fig. 2.7). This process is called upwelling. The water that "wells up" from below is cooler and richer in nutrient elements than the water it replaces. Some of the important effects of upwelling on organisms are discussed below. The biological effect of these major current and circulation patterns is immense, but for the most part the effect is indirect. We discuss these effects in relation to the various parameters controlled by currents such as temperature, nutrient element concentration, and dissolved oxygen levels.

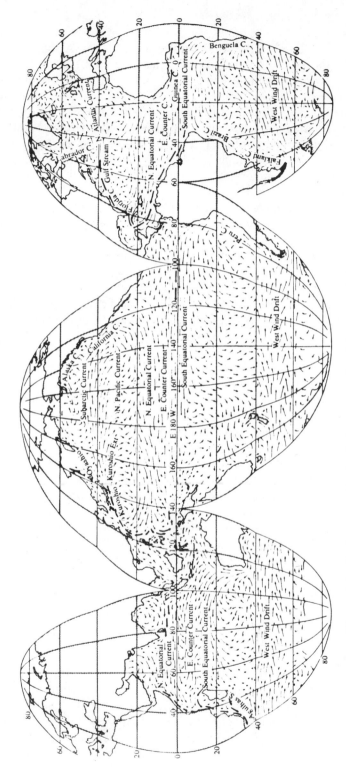

Figure 2.4 Surface oceanic circulation pattern for February and March. From Sverdrup/Johnson/Fleming, The oceans: Their physics, chemistry, and general biology, copyright 1942, Chart VII. Reprinted by permission of Prentice-Hall, Inc., Englewood Cliffs, N.J.

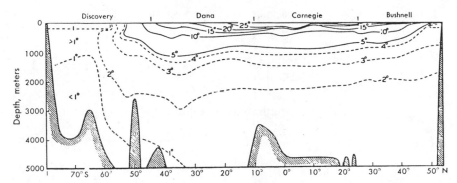

Figure 2.5 Vertical section approximately along the 170° meridian in the Pacific Ocean showing temperature distribution. From Sverdrup/Johnson/Fleming, The oceans: Their physics, chemistry, and general biology, copyright 1942, p. 753. Reprinted by permission of Prentice-Hall, Inc., Englewood Cliffs, N.J.

Local Currents

Local currents are generated by winds, tides, and local density differences (e.g., turbidity currents), or they may be local expressions of global currents. Wind-generated currents are associated primarily with waves. Orbital motion of water in waves and currents produced by breaking waves is a type of local

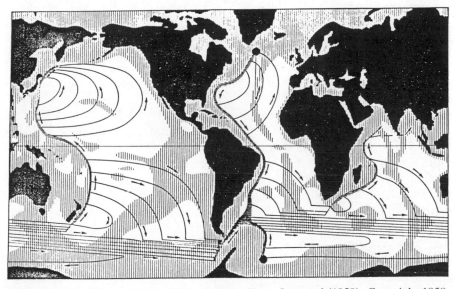

Figure 2.6 Deep oceanic circulation pattern. From Stommel (1958). Copyright 1958, Pergamon Journals Ltd.

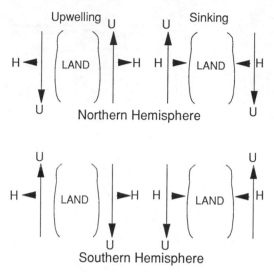

Figure 2.7 Upwelling and sinking of surface water along hypothetical coastlines. *U,* direction of current flow; *H,* Ekman transport direction due to Coreolis effect.

current. As waves pass over the surface of the ocean, the water below moves in closed (or nearly closed) orbits (Fig. 2.8). The diameter of the orbits decreases with depth until motion effectively ceases. Thus wave-related currents are effective only in shallow water. When waves break, water no longer simply moves in orbit but moves entirely toward shore. Breaking waves cause water to be piled up along the shore. If the waves do not break perfectly perpendicular to shore, they will have a component of motion parallel to shore. This results in formation of longshore currents that move parallel to the shoreline. These currents have extremely important geologic and biologic effects. Water that flows up onto the shore must return offshore by some process. Rip currents result from this return flow (Fig. 2.9).

Tides also generate currents, especially where tides pass through constrictions. Tides may either be diurnal or semidiurnal. They can vary in magnitude from a few centimeters to many meters, depending on location and configuration of the coastline. Highest tides occur at the ends of long, narrow bays, and lowest tides are measured on oceanic islands. Tides are by nature bidirectional. Tidal effects are limited almost entirely to very shallow water, but currents with tidal periodicity have been measured at depth in submarine canyons.

Density-generated currents, especially turbidity currents, are very important in their effects on organisms. Their low frequency and unpredictability makes most of their effects indirect. Their effects by burial of fossils and on changing the substrate are especially important. Density currents differ

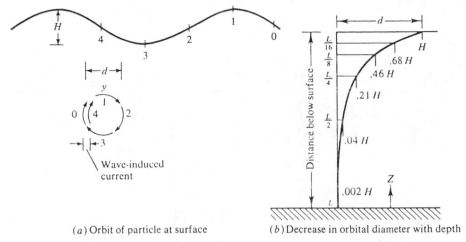

(a) Orbit of particle at surface   (b) Decrease in orbital diameter with depth

Figure 2.8 Theoretical orbital velocity of water due to wave traveling in deep water. H, wave height; d, orbital diameter; L, wavelength; Z, distance above bottom. From submarine geology, second edition, by Francis P. Shepard. Copyright 1963, 1973 by Francis P. Shepard. Reprinted by permission of Harper and Row, Publishers, Inc.

from wind- and tide-generated currents in not being confined to shallow water.

Global currents may be strong enough to have local effects in places. They are likely to be particularly strong where they are confined by various constrictions. For example, the Gulf Stream may have currents with a velocity of 2 m/s or more. Even deep-water currents may be strong enough to produce ripples and move relatively coarse sediment where they pass over topographic highs such as seamounts (Genin et al., 1986).

Biological Effects

Life in an environment with strong currents offers a number of advantages to organisms because currents (1) bring food to the organism, (2) remove waste products, (3) bring dissolved oxygen, (4) bring nutrient elements for plants or animals containing photosynthesizing symbionts, and (5) may eliminate predators that are not able to cope with life in strong currents. On the other hand, life in environments with strong currents presents some difficulties. The main problem is difficulty of "holding on" or maintaining position in strong currents. Organisms have adopted a number of strategies to cope with this problem. Sessile forms have developed strong attachment by (1) cementation, (2) byssus, (3) flexible or articulated skeletons that "give" with the current, (4) strong muscle attachment, (5) suction, or (6) holdfasts of various types. Table 2.2 gives a sample from the paleoecological literature of the many examples of the effect of water turbulence on organisms.

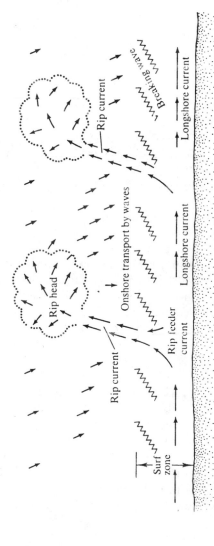

Figure 2.9 Nearshore water-transport system. From Submarine geology, Second Edition, by Francis P. Shepard. Copyright 1963, 1973 by Francis P. Shepard. Reprinted by permission by Harper and Row, Publishers, Inc.

TABLE 2.2   Examples of Effects of Water Turbulence on Common Fossil Groups

Group	Effect	Reference
Algae	Stromatolite shape varies from undulatory to club shape in quiet to turbulent water; coraline algal branching from loose to compact in quiet-to-turbulent water	Logan et al., 1974; Bosence, 1976, 1985
Foraminifera	Mainly large or encrusting types in turbulent water; larger foraminifera are more spherical in turbulent water and are thin and flat in quiet water	Boltovskoy and Wright, 1976; Leutenegger, 1984; Hallock and Glenn, 1986
Sponges	Vase shaped with small osculum in quiet water; often fan or bowl shaped in turbulent water; sclerosponges live in quiet, cryptic habitats	Bidder, 1923; Leigh, 1971; Rigby and Stearn, 1983
Corals	Many forms require turbulence; branching forms orient in current; colony shape more compact in currents; branching in quiet water	Wells, 1957; Graus et al., 1977; Chappell, 1981
Bryozoans	Encrusting, massive, and flexible forms in turbulent water; rigid, delicate branching forms in quiet water	Schopf, 1969; Brood, 1972; Harmelin, 1975
Brachiopods	Most modern forms need some current; large pedicle, heavy shell, supporting spines in turbulent water; fold and sulcus and long hinge in quiet water	Ager, 1965; Rudwick, 1970
Mollusks	Broadly adapted to turbulence from very high to very low; thick shell, strong attachment, and burrowing in turbulent water; well-developed siphons in quiet water	Purchon, 1968; Stanley, 1970; Wilbur, 1983–1985; Vogel, 1988
Anthropods	Ostracods develop shell-strengthening sculpture for turbulence	Benson, 1975
Echinoderms	Crinoids mostly require currents for feeding; heavy calyx and stem for turbulence; some echinoids strong burrowers or wedge in crevices in turbulent settings	Kier, 1974; Breimer and Lane, 1978; Smith, 1984

Vagile forms have developed strong locomotion (e.g., certain bivalves and gastropods). All forms tend to have strong skeletons, although some have opted for flexibility (e.g., articulated coraline algae). Some vagile forms live in burrows to escape the strong currents but extend siphons or other food-collecting devices into the current (e.g., burrowing bivalves). Many forms have developed streamlining or a low profile to minimize the drag in currents (e.g., limpets and chitons). Branching organisms such as some corals orient their branches in the direction of strong currents (Fig. 2.10). Such orientation, when observed in fossil assemblages, may potentially be used to study paleocurrent strength and direction (Graus et al., 1977).

Life in environments with weak currents also offers some advantages. Most important is elimination of the necessity of expending energy on the various adaptations for life in strong currents indicated above. More variability in morphology is possible. The organism can devote its energy to food gathering and other life processes rather than in maintaining position in strong currents.

Organisms that live in environments with weak currents must adopt quite different food-gathering strategies than those living in strong currents. Basically, they must produce their own currents to bring in food and dissolved oxygen and to eliminate wastes. Sponges are well adapted to this strategy (Fig. 2.11). Many other groups living in quiet water do not use food suspended in the water but actively go after food deposited on the bottom or in the sediment.

Other environmental parameters are likely to vary along with current strength. Hard or coarse substrates are more common in high-current regimes. Soft, perhaps soupy substrates are common in low-current settings. Dissolved oxygen concentration tends to correlate directly with current strength. High currents generally are confined to shallow environments, whereas quiet water may be either shallow or deep. Because of this partial correlation between parameters, separation of the environmental factor responsible for a certain adaptation may thus be difficult.

TEMPERATURE

The temperature range in the ocean is approximately -2 to $40°C$. Coldest waters are found in the polar regions, where seawater is freezing. Highest surface temperatures occur in lagoonal or bay settings marginal to tropical oceans, where the temperature may sometimes exceed $40°C$. Much higher temperatures recently have been recorded in thermal springs associated with submarine rift zones (Jannasch and Mottl, 1985). The temperature range in nonmarine environments is of course much greater, ranging from about $-50°C$ to nearly $100°C$ in some continental hot springs.

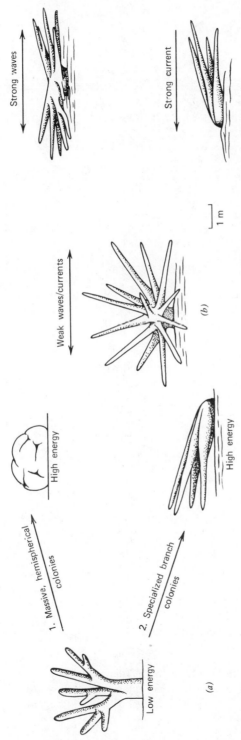

Figure 2.10 Adaptation of coral colonial form to differing conditions of current and wave energy. (*a*) Colonies that branch under low-energy conditions may either become more massive (1) or branch toward the current (2) under high-energy conditions. (*b*) In strong waves branches orient bidirectionally parallel to wave direction. They branch toward unidirectional currents. From Graus et al. (1977).

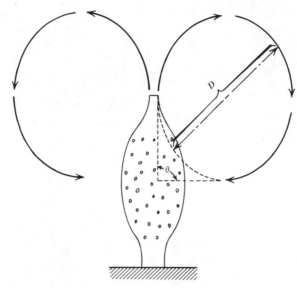

Figure 2.11   Diagrammatic quiet-water sponge. D, diameter of supply; B, angle of supply.

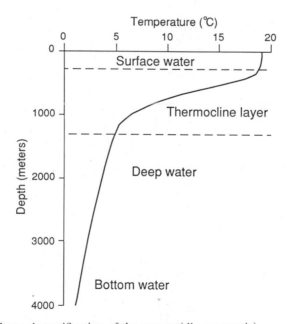

Figure 2.12   Thermal stratification of the oceans (diagrammatic).

Vertical Variation

As discussed earlier, except in polar regions, temperature decreases with depth in the oceans. Surface temperatures are generally more variable as well as warmer than deeper water. A zone of rapidly decreasing temperature (the thermocline) separates the warmer surface waters from the colder deep water (Fig. 2.12). Because bottom water forms by cooling and sinking of surface water in the polar regions, it will always have the approximate temperature of the coldest surface water in the oceans. Water above the bottom zone but below the thermocline is also cold, but not quite as cold as the bottom water. Often, zones separated by slight temperature differences can be recognized at these intermediate depths. These correspond to intermediate water masses that have a common area of origin at the surface and can be traced over large regions.

Geographic Variation

Surface temperature is basically controlled by input of solar energy. Surface temperature thus varies inversely with latitude. Oceanic circulation causes major deviations from this simple correlation between temperature and latitude (Fig. 2.13). Currents flowing from low to high latitudes (such as the Gulf Stream) cause warm waters to be displaced toward higher latitudes. Currents flowing from high latitudes to low (such as the Alaskan Current) cause cold water to be displaced toward lower latitude. Because of water flow direction in the circulation gyres, the net result is to broaden the tropical zone on the western sides of all ocean basins and narrow it on the eastern sides.

The vertical component of circulation also has a major effect on ocean surface temperature. Areas of divergence (upwelling) have lower temperatures than might otherwise be expected at the latitude where they occur. The extensive band of slightly lower temperature near the equator in the Pacific Ocean is due to this effect. Relatively low temperature along the west coast of North and South America is in large part due to upwelling of deep water to compensate for the offshore-directed surface current components.

Variation in temperature, as well as the absolute temperature, has important biological effects. Surface temperature variation is usually greater in midlatitudes and lower at both high and low latitudes (Fig. 2.14). Although temperature also correlates with latitude in nonmarine environments, it is more variable than in marine environments. Altitude and nearness to large water bodies are especially important factors causing geographic irregularities in temperature on land.

Biological Effects

The effect of temperature is obviously different on poikilotherms (cold-blooded) organisms than on homiotherms (warm-blooded) organisms. As our

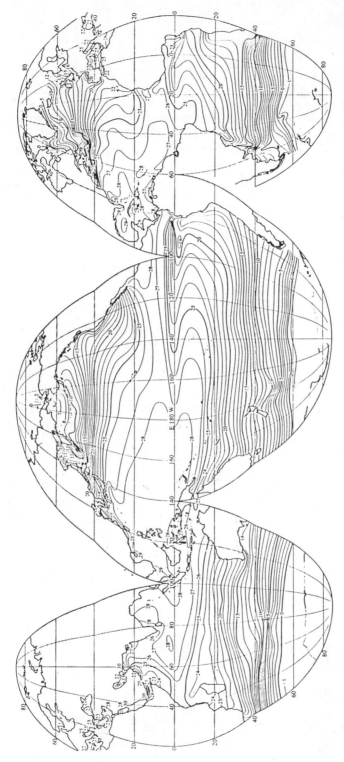

Figure 2.13 Surface temperature in the oceans during the northern summer. From Sverdrup/Johnson/Fleming, The oceans: Their physics, chemistry, and general biology, copyright 1942, Chart III. Reprinted by permission of Prentice-Hall, Inc., Englewood Cliffs, N.J.

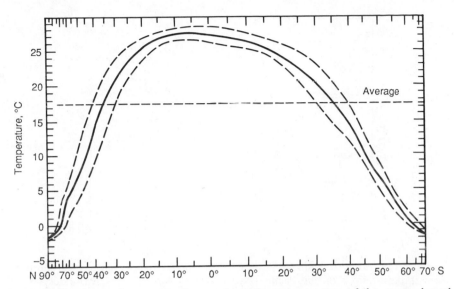

Figure 2.14 Approximate annual range in surface temperature of the oceans based on February and August temperatures. From Wüst et al. (1954), by permission of Institut für Meerskunde an der Universität Kiel, Kiel, West Germany.

emphasis here is on invertebrates, we confine our comments to poikilotherms. The rate of metabolism in poikilotherms is proportional to temperature (Fig. 2.15). Indeed, the rate approximately doubles for each 10°C rise in temperature. This relationship has been called van't Hoff's rule. Actually, this relationship is very approximate, varying between different groups of organisms. This variability has given rise to the Q-10 index, the fractional increase in metabolism caused by a 10°C rise in temperature.

The general pattern is for organisms to have a minimal metabolic rate at a particular low temperature. That rate increases with temperature to a maximum. At temperatures higher than that maximum, metabolic rate decreases abruptly and death will normally ensue (Fig. 2.16). Organisms often show adaptation of their metabolic rate in response to temperature. Different groups of organisms have different rates of metabolism at the same temperature. Indeed, experiments have shown that in some cases organisms can acclimate to temperature changes by changing their metabolic rates. In general, organisms that are adapted to life in warm environments have a slower metabolic rate at a given temperature than do organisms adapted to life in cold environments (Fig. 2.15).

Obviously, metabolic rates are difficult to measure in fossil organisms! But one major expression of metabolic rate is growth rate. Growth rate can potentially be recorded in fossilized skeletons. For example, variation in growth rate during the year may be recorded as growth rings within shells or on their surfaces (see Chapter 4). Such growth rings not only suggest seasonally vari-

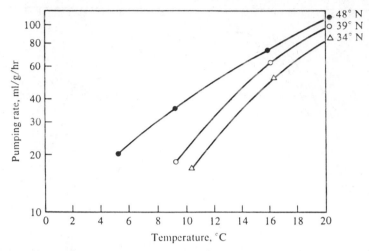

Figure 2.15 Variation with temperature of pumping rate in *Mytilus californianus* from three locations on the Pacific coast of North America. From Bullock (1955), Biological Reviews, Cambridge University Press.

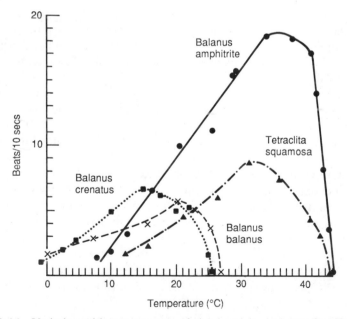

Figure 2.16 Variation with temperature of cirral activity in barnacles. From Southward (1964), from Helgolandes Wissenschaffliche Meeresuntersuchungen.

able temperature, but also allow determination of the average annual growth rate, which may be a function of average annual temperature.

Temperature often controls the time of reproduction. It is also one of the most important factors determining the geographic range of a species (due to effects on reproduction and survival). Temperature may influence the size of the organism in two ways. In general, species living in warm water that is supersaturated with calcium carbonate grow large skeletons. Thus, on balance, tropical organisms tend to be larger than temperate organisms and much larger than organisms living in high latitudes. However, within a given species an opposite trend may be observed (Fig. 2.17). A species living at the cold end of its range may grow slowly but also mature slowly. It may thus live to an old age and grow larger than individuals nearer the warm end of the species range. Many examples of the effect of temperature on distribution and morphology of the major fossil invertebrate groups are shown in Table 2.3.

SALINITY

Salinity is a measure of the concentration of dissolved salts in seawater. It is formally defined as the weight in grams of dissolved solids in 1 k of seawater.

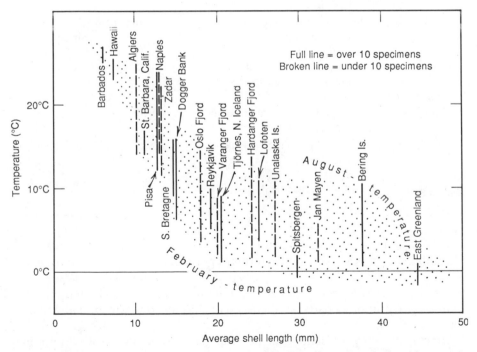

Figure 2.17 Variation in average length of adult *Hiatella arctica* shells with annual temperature range. From Strauch (1968).

TABLE 2.3   Examples of Effects of Temperature on Common Marine Invertebrate Fossil Groups

Group	Effect	References
Algae	Calcareous green algae tropical and subtropical; strong temperature control on distribution of coralline algae, coccoliths, and diatoms	Elliot, 1984; Wray, 1977; Haq, 1978; Burckle, 1978
Foraminifera	Temperature control on distribution of species; coiling direction; porosity of chambers; larger forams only in tropics	Boltovskoy and Wright, 1976; Kennett, 1976; Bé, 1968; Hallock and Glenn, 1986
Sponges	Sclerosponges tropical	Hartman, 1983
Corals	Hermatypic forms tropical; ahermatypic forms eurythermal	Wells, 1957; Fagerstrom, 1987
Bryozoans	Species ranges temperature controlled	Ryland, 1970
Brachiopods	Modern forms especially in temperate waters; punctae density positively correlates with temperature	Rudwick, 1970; Foster, 1974
Mollusks	Species ranges strongly temperature dependent; tropical forms larger, more ornate than cold-water forms; maximum size within species at cold end of geographic range	Wilbur, 1983–1985; Nicol, 1964, 1967; Strauch, 1968
Arthropods	Ostracod species ranges temperature dependent	Neale, 1969
Echinoderms	Some echinoid families and genera only in tropics; long, simple crinoid arms in cold water; shorter, branched in warm water	Fell, 1954; Breimer and Lane, 1978

Commonly, salinity is expressed in terms of per mille (‰), which is equivalent to parts per thousand or ppt (*not* parts per million) of dissolved solids in the water. The term *chlorinity* is sometimes used in the literature. This term refers to the concentration in parts per thousand of chlorine ions, the most abundant ion in seawater.

Practically every element in the periodic table occurs in dissolved form in seawater (Brewer, 1975). But relatively few ions account for most of the dissolved solids. In order of abundance, the most common cations are Na^+,

Mg^{2+}, Ca^{2+}, K^+, and Sr^{2+}. In order of abundance, the most common anions are Cl^-, SO_4^{2-}, Br^-, and F^-. These ions form relatively soluble compounds and hence do not precipitate out of solution. The relative proportions of these elements in seawater are nearly constant regardless of absolute concentration. For this reason they are called conservative elements. Other elements, such as P, N, and Si, are low and highly variable in concentration in seawater. This is largely because of their involvement in photosynthesis and biological processes. They are called nutrient or biolimiting elements. Of course, some elements behave in an intermediate way, varying in relative concentration to some extent but not so much as the nutrient elements. Broecker (1974) gives an excellent discussion of the distribution of dissolved ions in seawater.

Normal seawater has a salinity of approximately 35‰. Salinity may be much higher in evaporitic settings, such as restricted lagoons and salinas. Salinity may, of course, decrease to near zero in areas of mixing with fresh water. Open oceans have nearly constant salinity of 34.5 to 35.0‰. Surface salinity varies slightly due to climatic patterns and the oceanic circulation pattern (Fig. 2.18). Salinity tends to be highest in the centers of circulation gyres, in the "horse latitudes" (about 30°). It is lowest in the high latitudes, where evaporation is minimal and some runoff of fresh water from the continents occurs. Salinity is considerably elevated in semienclosed bodies of water, such as the Mediterranean and Red Seas, in regions of dry climate. Semi enclosed bodies of water in humid climates, such as the Baltic Sea, have low surface salinities. Smaller bays and lagoons may even have greater departures from normal marine salinity. For example, Galveston Bay in the United States and the Waddensea in northwestern Europe are examples of low-salinity bays in humid climates. Laguna Madre in the United States and the Gulf of Aquaba in the Middle East are examples of high-salinity water bodies in arid climates. These smaller, restricted bodies are frequently the sites of deposition of fossiliferous sediments. Salinity may be an especially important environmental variable.

Both the composition and absolute concentration of dissolved solids in seawater are very important to organisms living there. Nutrient elements (discussed below) may be limiting factors in the growth of populations of organisms. The more abundant, conservative elements may also be important, but due to their abundance, they do not limit productivity. Some organisms are very tolerant of salinity variation (euryhaline forms), and others have narrow tolerances (stenohaline forms).

The major effect of salinity is on osmotic pressure across cell membranes within organisms (Pearse and Gunter, 1957). Some organisms are able to regulate ionic concentration within their body (homiosmotic forms or osmoregulators), but others do not (poikilosmotic or osmoconformers). In addition to actively regulating ionic concentrations within cells, some organisms cope with variable salinity in other ways. If salinity variation is seasonal, some vagile forms migrate when local salinity becomes unfavorable. Some forms cope with short-term variation in salinity by closing themselves in their shells until favor-

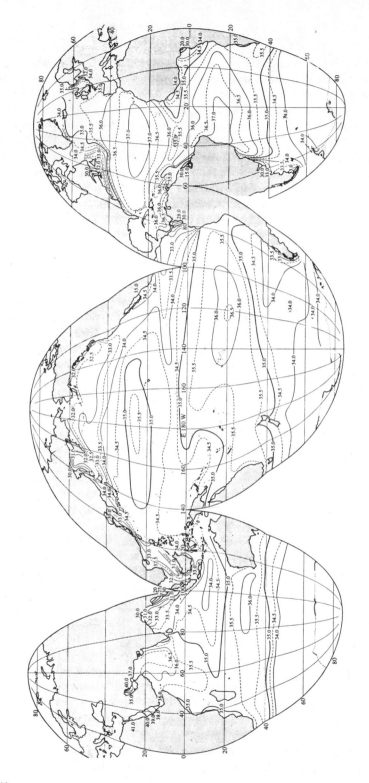

Figure 2.18 Surface salinity in the oceans during the northern summer. From Sverdrup/ Johnson/Fleming, The oceans: Their physics, chemistry, and general biology, copyright 1942, Chart VI. Reprinted by permission of Prentice-Hall, Inc., Englewood Cliffs, N.J.

able salinity conditions return. Burrowing forms may withdraw deep within their burrows during times of unfavorable salinity conditions. Some forms seem to be able to function with a variable cellular ionic concentration.

Areas with either reduced or elevated salinity are almost always also areas of variable salinity. Thus the organisms must be able to cope with variation in salinity as well as nonnormal marine salinites. Relatively few organisms are able to cope with this instability. By far the greatest diversity of marine invertebrates is in areas with normal, ultrahaline salinity. Diversity declines at both lower and higher salinity but increases somewhat in fresh water (Fig. 2.19). This is perhaps due to the stability of salinity (i.e., essentially zero) in fresh water.

Many examples of variation in morphology have been linked to salinity variation. A common effect of both decreasing and increasing salinity away from the normal marine range is a decrease in maximum size of individuals within the species; however, many exceptions to this relationship have been noted. Species that are specifically adapted to life in brackish waters (e.g., some oysters) reach their maximum size under brackish conditions. Massiveness of the skeleton or thickness of the shell may be less under brackish conditions. This condition may be related to degree of saturation of the water with $CaCO_3$. Smaller, less massive shells form in undersaturated water. Many examples of salinity effects on distribution and morphology of the major fossil groups are given in Table 2.4.

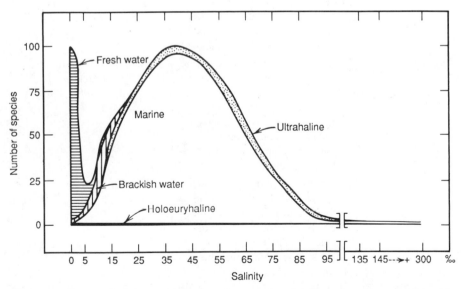

Figure 2.19 Variation in diversity of species with salinity. After Hedgpeth (1967), copyright 1967 by the American Association for the Advancement of Science, AAAS Publ. 83: 408–419, figure 5.

TABLE 2.4   Examples of the Effect of Salinity on Distribution and Morphology of the Major Fossil Groups

Group	Effect	Reference
Algae	Stromatolites (blue-green) and diatoms euryhaline, most calcareous greens, reds, and coccoliths normal marine	Wray, 1977; Haq, 1978; Burckle, 1978; Brasier, 1980
Foraminifera	Euryhaline as a group; some groups stenohaline, especially planktic forms; wall type in part salinity dependent (chapter 4)	Boltovskoy and Wright, 1976; Buzas and Sen Gupta, 1982
Sponges	Euryhaline as a group; sclerosponges only normal marine	deLaubenfels, 1957; Hartman, 1983
Corals	Very stenohaline as a group; some genera slightly euryhaline; Paleozoic forms normal marine	Wells, 1957; Fagerstrom, 1987
Bryozoans	Euryhaline as a group; greatest diversity in normal marine indicating many stenohaline genera; Paleozoic forms mainly normal marine.	Ryland, 1970; Schopf, 1969; Boardman and Cheetham, 1987
Brachiopods	Mostly stenohaline; linguloids and a few others brackish; Paleozoic forms largely normal marine but linguloids brackish	Rudwick, 1970; Emig et al., 1978
Mollusks	Euryhaline as a group; some forms (e.g., unionids) entirely fresh water; many families and all cephalopods stenohaline; other groups (oysters, mussels) euryhaline; Paleozoic forms show same pattern; many nearshore	Bretsky, 1968; Stanley, 1970; Purchon, 1977; Wilbur, 1983–1985
Arthropods	Ostracods very euryhaline; abundant freshwater group; also brackish and hypersaline; may be very abundant but not diverse in restricted environments; Paleozoic forms show same pattern; trilobites normal marine?	Benson, 1961
Echinoderms	Mostly stenohaline, especially crinoids; some echinoids more euryhaline	Binyon, 1966; Meyer, 1980

NUTRIENT ELEMENTS (PRODUCTIVITY)

To this point we have emphasized the "qualitative" effect of environmental parameters on organisms. For example, we have discussed ways in which salinity affects the types of organisms that live in a given environment or the morphological features of those organisms. In this section we discuss quantitative aspects (i.e., how many organisms live in given environments). In marine environments, this is controlled in large part by the availability of light and concentration of nutrient elements (i.e., those elements whose concentration varies in seawater due to biological processes). In terrestrial environments, temperature and availability of water are additional factors.

Productivity is the rate of production of organic material by biological processes. Productivity is sometimes divided into primary productivity, the production of organic material by photosynthesis, and secondary productivity, the production of organic material by animals utilizing preexisting plant and animal organic matter. For the most part we will be concerned with primary productivity. Related to the concept of productivity is *biomass* or *standing crop*. This is the amount of living organic material present in a given area at any one time. It is obviously related to but not equivalent to productivity. Biomass is a function of longevity as well as productivity. An area with large, long-lived plants will have a larger biomass than an equally productive area with small, short-lived plants.

How does one measure productivity or standing crop in the geologic record? Only with great difficulty and uncertainty. But knowing something about productivity is so important to a full understanding of paleoecology that it is worth an effort. Fossil biomass is difficult to measure for two main reasons: (1) very little of the original biomass is preserved in the geologic record, and (2) the concentration of fossils is controlled not only by the rate of their production but also by sedimentation rate. Strictly speaking, biomass refers to the organic portion of organisms, which is almost never preserved. The concentration of skeletal materials may be crudely correlated with biomass under certain conditions. But even the concentration of fossilized skeletal materials is subject to diagenetic factors and may not accurately reflect the original concentration. Under some conditions, sedimentation rate can be estimated, but usually such estimates are very approximate. Even if one could determine biomass, productivity would be difficult to measure accurately because longevity of the organisms involved is not known.

Vertical Variation

Light is the energy source for the formation of organic material by photosynthesis. In the shallow sea, light is always available; although its distribution may be seasonal. Temperature, salinity, and other environmental factors may affect photosynthetic rate and the nature of the plants doing the photosynthesizing, but they do not limit the total amount of organic material formed. In the long term, concentration of the nutrient elements, N, P, and Si is the limiting factor on productivity in shallow water. With increasing depth, the availability of light as an energy source becomes more important than nutrient element concentration. In shallow water, plants are photosynthesizing faster than they are respiring. Thus more organic material is being produced than is being consumed. With increasing depth, photosynthesis decreases because of decreasing light levels, but respiration does not. At the *compensation depth,* photosynthesis and respiration are equal. Below the compensation depth, respiration exceeds photosynthesis and thus plants cannot survive. (Compensation depth in this context is, of course, different from the *carbonate compensation depth,* the depth below which calcite or aragonite dissolves.) Below the compensation depth, respiration by animals, as well as inorganic oxidation, continues. Destruction of organic matter frees nutrient elements to go back into solution in the seawater. This produces a pattern of low concentration of nutrient elements at the surface with increasing concentration at depth below the compensation depth (Fig. 2.20).

Vertical circulation causes some variation in this basic pattern. In polar areas, where vertical circulation is vigorous due to minimal density difference between shallow and deep water, nutrient concentration varies less with depth than it does in tropical regions, where marked density difference results in sluggish circulation. Bottom water may have a slightly lower concentration of nutrient elements than intermediate water because it has come more recently from the surface in polar regions.

Some primary production occurs in deep water well below the compensation depth. This is productivity by organisms using chemical energy rather than light energy. The most spectacular example of this process is the bacteria, which obtain their energy by oxidizing sulfide in the thermal springs associated with spreading centers (Jannasch and Mottl, 1985). These bacteria form the basis of an entire, distinctive ecosystem associated with the springs. But the bacteria are special exceptions to the more general pattern described above.

Geographic Variation

Geographic variation in nutrient element concentration and productivity is strongly influenced by global circulation patterns (Fig. 2.21). Areas of divergent currents are usually areas of high nutrient element concentration and of high productivity. This is because water enriched in nutrient elements from below the compensation depth upwells in these areas. Areas of convergence in the center of the circulation gyres have low nutrient element concentration because the surface water has been at the surface for a long time and photosynthesis has removed practically all of its nutrient elements. Productivity is on average higher near continents than in the centers of oceans (a great convenience for those who depend on fishing for their livelihood!). Three factors are responsible for this: (1) Upwelling often occurs near coasts where surface

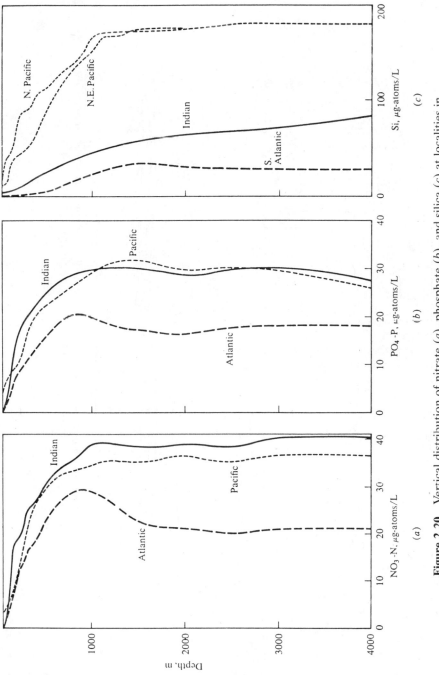

Figure 2.20 Vertical distribution of nitrate (*a*), phosphate (*b*), and silica (*c*) at localities in the Atlantic, Pacific, and Indian Oceans. From Sverdrup/Johnson/Fleming, The oceans: Their physics, chemistry, and general biology, copyright 1942, pp. 241, 242, and 245. Reprinted by permission of Prentice-Hall, Inc., Englewood Cliffs, N.J.

Figure 2.21 Estimated annual photosynthetic productivity of the oceans in grams of carbon per square meter per year. From Fleming, 1957, Geological Society of America Memoir 67, Vol. 1, p. 89.

water has a component of motion away from the continent; (2) a major ultimate source of nutrients in the oceans is material weathered from rocks on land, brought to the oceans at the continental margins by rivers; and (3) organic material produced by photosynthesis in the surface waters does not sink as deep on the relatively shallow continental shelves as it does in the ocean basins. At shelf depths, it can decay and the released nutrient elements can be returned to the surface more quickly than organic material in the ocean basins.

Surface waters in polar regions tend to be more productive than surface waters in low latitudes. This is because the lower density contrast in the high latitudes allows vertical circulation to bring nutrients from deep water more readily than the sluggish vertical circulation at low latitudes, where the density contrast is great.

Temporal Contrast

Productivity also varies with time on a number of different scales: (1) It varies daily with the intensity of sunlight, stopping at night; (2) seasonal variability in productivity is great at high latitudes, where much more solar energy is available during summer than winter; (3) cycles of several years have been observed, perhaps the best known of which are those related to the El Niño currents in the eastern Pacific (Drake et al., 1978); (4) thousand- to million-year cycles relate to major climatic variation such as those associated with continental glaciation (Berger et al., 1984); and (5) cycles of tens of millions of years associated with plate tectonics processes, of which climatic and productivity differences between early and late Cenozoic time are an example.

Biological Effects

Productivity itself is a biological effect. As indicated above, variation in surface-water productivity is largely the result of nutrient-element variation. To some extent, the nature of organisms found in a given area depends on the level of productivity. Hallock and Schlager (1986) theorized that the growth of coral reefs is strongly influenced by productivity level. Hermatypic corals are adapted for life in areas of low productivity. When productivity increases, other organisms tend to replace corals. Depth zonation may be strongly influenced by productivity because planktic plants suspended in water greatly decrease its transparency. Thus organisms that are susceptible to light levels (such as hermatypic corals and algae) will vary in their depth distributions as productivity levels vary (Fig. 2.22).

Patterns of diversity and trophic structure of ecosystems are influenced by productivity level. This is discussed in more detail in Chapter 10. Major groups of organisms such as diatoms and other groups of algae tend to be proportionally more abundant in areas of high productivity. The methods used by animals to capture food vary somewhat depending on the abundance

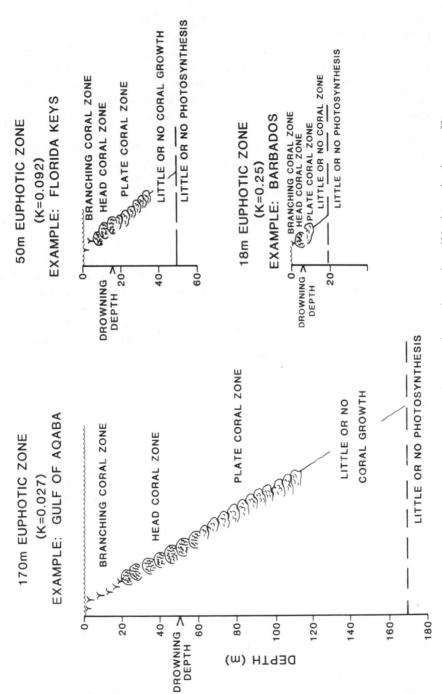

Figure 2.22 Effect of water clarity on hermatypic coral zonation. K is the extinction coefficient of light. Drowning depth is approximate submergence depth required to drown reef. From Hallock and Schlager (1986), by permission of Society of Economic Paleontologists and Mineralogists.

of food. Thus, to some extent, different groups of animals are more abundant in highly productive areas than in areas of low productivity. Little paleoecological research has been done on this relationship. Development of tools to identify areas of high productivity in the geologic past would be a valuable contribution. High organic content in sediments has sometimes been attributed to high productivity. However, organic content of sediments is a balance between productivity, redox potential, and sedimentation rate.

DEPTH

We are often interested in knowing the depth of water in which a particular sediment was deposited. Water depth is one of the most basic aspects of both paleogeography and interpretations of geologic history. Determination of water depth is often vital to study of sedimentary basins and the occurrence of hydrocarbons or other economic resources within these basins.

Water depth has an important influence on sedimentary processes and properties. For this reason water depth in ancient oceans is often most readily determined by sedimentologic methods. However, sedimentologic data are not always definitive; paleoecologic approaches may prove more useful. Very fine grained sediments (shales) may be especially difficult to interpret. They were clearly deposited under quiet-water conditions, but was the water quiet because it was deep or because it was protected by some barrier from wave action? A good example of this ambiguity is that of the so-called core shales of the Pennsylvanian cyclothems of the North American midcontinent region. The ambiguity in this case has led to sometimes heated debate (Merrill and Martin, 1976; Heckel, 1977). If the shales are relatively deep water deposits, the cyclothems can be explained by a simple transgression–regression sequence. If they are shallow-water deposits, two transgression–regression sequences are needed to explain the cyclothems. Unfortunately, often neither sedimentologic nor paleoecologic methods are infallible indicators of depositional environment!

Water depth is a complex variable. Many factors vary with depth, including hydrostatic pressure; light intensity; temperature; salinity; concentration of nutrient elements and quantity of dissolved gases, including oxygen; and water turbulence. Indeed, every environmental variable discussed in this chapter varies with depth in a more-or-less regular fashion. Not only the absolute values of these parameters, but also their variability in time, differ with depth. Hydrostatic pressure, and to a somewhat less extent, light, correlate most directly with depth and do not show significant geographic variation. The effects of hydrostatic pressure on the distribution and morphology of organisms are poorly understood, even for modern organisms, let alone fossils.

The effect of light on organisms is somewhat better understood, especially in connection with photosynthesis in plants. We often cannot identify which of the other factors is responsible for variation in organisms with depth. And if the cause of variation can be identified (e.g., water turbulence), we may not know if the parameter is varying with depth or geographically. We thus often observe variation in distribution and morphology of organisms with depth without specifying what depth-related factor is really causing the variation. Many of our data are thus purely empirical (substantive uniformitarianism), but they are the best we have at present. Such data have been used extensively and successfully in environmental reconstruction, particularly for Cenozoic and Mesozoic rocks. This approach has been especially extensively used with apparent success with the foraminifera (Fig. 2.23).

Most of the marine geologic record preserved on continents is "shallow" in the oceanographic sense. However, in recent years, increasing study of the deep ocean basins via DSDP (Deep Sea Drilling Program) cores has greatly added to our knowledge of really deep organisms, both modern and fossil. Much of the obvious variation in organisms with depth occurs in the upper few tens to hundreds of meters. This is probably because most variables change more rapidly at these relatively shallow depths. Thus the relatively shallow water sedimentary record on the continents contains much of the depth-related variability.

As indicated above, the actual cause of depth-related variation in organisms often cannot be determined. Indeed, in some cases the variation may be due to a combination of factors. Variation with light intensity perhaps offers the most direct method of determining paleodepths. Because light is needed for photosynthesis, the abundance and characteristics of in situ benthic plants

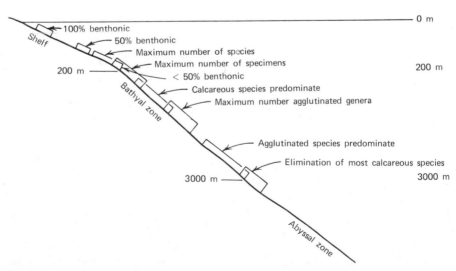

Figure 2.23 Variation with depth of proportion of benthic calcareous and agglutinated foraminifera. From Boltovskoy and Wright (1976), Dr. W. Junk B.V., Publishers.

correlates strongly with water depth. Plant abundance decreases rapidly in the lower part of the photic zone (approaching the compensation depth) and plants are absent below. The position of the compensation depth differs for each plant species, and therefore there is a zone of transition rather than a sharp horizon. Some plants, such as the red algae, are able to photosynthesize in excess of respiration at depths as great as 250 m (Wray, 1977). Other groups, such as the green algae, have a much shallower limit. Thus the gross taxonomic composition of plants may be helpful in determining depositional depth.

Certain algal groups such as the Dascycladaceae are largely restricted to very shallow depths [less than 30 m, and common only at a depth of less than 5 m (Wray, 1977)]. Many animal groups have genera that contain symbiotic algae within their tissues (e.g., corals, foraminifera, bivalves). These forms behave much like plants in that their distribution is thus strongly controlled by light intensity and depth. Swinchatt (1969) and others have suggested that in carbonate rocks, the degree of micritization of grains by boring algae can be used to determine approximate depositional depth. Because of downslope transport and micritization by boring fungae as well as algae, Friedman et al. (1971) suggest caution in using this method. Many examples of correlation of distribution and morphology of invertebrate animals with depth have been described. Some of these are listed in Table 2.5.

DISSOLVED OXYGEN

Increasing attention has been given in recent years to the effects of dissolved oxygen level on the distribution of organisms. Researchers working with trace fossils have led the way in recognizing the importance of this variable (e.g., Rhoads and Morse, 1971; Ekdale and Mason, 1988). More attention is now being given to dissolved oxygen as a controlling factor in the distribution of body fossils as well (Savrda et al., 1984). Dissolved oxygen level is very important in determining early diagenetic processes at the sediment–water interface or at shallow depths within the sediment. Preservation of organic compounds especially correlated with dissolved oxygen levels. Organic content of sediments has major economic implications in terms of hydrocarbon source beds.

Dissolved oxygen levels can range from saturation or supersaturation to complete absence of oxygen. Three categories of environments are commonly recognized on the basis of dissolved oxygen levels: aerobic (greater than 1 ml/l), dysaerobic (0.1 to 1.0 ml/l), and anaerobic (less than 0.1 ml/l).

In the open ocean, surface water is saturated or supersaturated with oxygen (Fig. 2.24). Solubility of oxygen in seawater varies inversely with temperature and salinity. Oxygen can dissolve from the atmosphere, especially in turbulent surface water. Another important source of oxygen is from photosynthesis by plants in the photic zone. Oxygen levels decline with depth, especially below

TABLE 2.5 Examples of the Effect of Depth on Distribution and Morphology of the Major Fossil Groups

Group	Effect	Reference
Algae	Stromatolites usually shallow-water indicators; algal families show depth zonation; all algae in photic zone; algal produced micrite envelopes in shallow water	Walter, 1976; Dill et al., 1986; Wray, 1977; Swinchatt, 1969
Foraminifera	Ratio of planktic to benthic increases with depth; only arenaceous types at great depth; some genera become larger with depth; crusts on deeper-water planktic genera; many empirical studies of distribution versus depth; large forams hermatypic, shallow	Boltovskoy and Wright, 1976; Bandy, 1964; Buzas and Sen Gupta, 1982; Hallock and Glenn, 1986
Sponges	Hyalosponges mostly deep water; sclerosponges (and stromatoporoids?) only in shallow water	Finks, 1970; Rigby and Stearn, 1983
Corals	Hermatypic forms only in photic zone; most diverse in water less than 10 m deep; ahermatypic in deep water; morphology varies with depth (hemispherical colonies in shallow to platelike in deep water)	Wells, 1957; Graus and Macintyre, 1976; Goreau, 1963; Chappell, 1981; Fagerstrom, 1987
Bryozoans	Rigid, branching forms in deeper water; strong correlation with turbulence which correlates with depth	Schopf, 1969; Lagaaij and Gautier, 1965
Brachiopods	Strong control by turbulence and substrate which correlate with depth; small size, thin shell, in deeper water	Ager, 1965; Fürsich and Hurst, 1974
Mollusks	Some hermatypic forms (*Tridacna*) in shallow water; small, thin-shelled, deposit-feeding bivalves in deep water; planktic gastropods (pteropods) in deep water	Wilbur, 1983–1985; Nicol, 1967; Herman, 1978
Arthropods	Ostracod size and reticulation increase and species with eye tubercles decrease with depth	Benson, 1975; Kontrovitz and Myers, 1988
Echinoderms	Modern stalked forms deep water, but not Paleozoic forms; comatulids shallow water	Fell, 1966; McCurda and Meyer, 1983

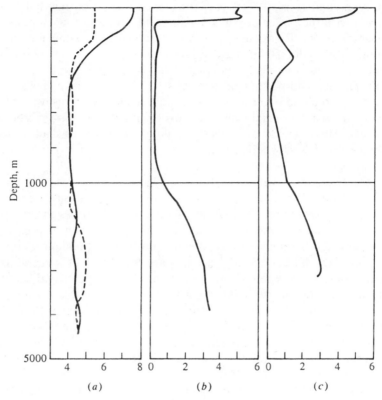

Figure 2.24 Vertical distribution of dissolved oxygen at stations in the Antarctic (*a*) and tropical Pacific [(*b*) and (*c*)]. Oxygen concentration in ml/l. From Richards, 1957, Geological Society of America Memoir 67, Vol. 1, p. 89.

the photic zone as oxygen is removed by respiration and oxidation of suspended organic matter. Oxygen reaches a minimum (*oxygen minimum layer*) at intermediate depths (usually 1 to 2 km below the surface) where most oxygen may be removed by organic and inorganic oxidation processes. Oxygen can be added at these depths only by sluggish vertical circulation of water from the photic zone. Oxygen concentration commonly increases slightly in deeper water because fewer organisms live at this depth and the concentration of suspended organic material is lower. Also, vertical circulation tends to be more vigorous in deeper water than at intermediate depths.

Lowest oxygen levels occur in silled basins and restricted bays and lagoons. Below the sill level, silled basins commonly have cold, dense water, which does not circulate because denser water is not being produced at the surface which can replace the deep water. Such stagnation is particularly pronounced in settings in which the surface water has a lower salinity and is thus much less dense than the bottom water. The Black Sea is an especially good example of this (Degens and Ross, 1974). Ancient epicontinental basins that were not connected with the larger ocean were especially prone to develop anaerobic conditions at depth. Restricted shallow seas and lakes also develop anaerobic conditions, especially if high surface productivity provides abundant organic material, the oxidation of which depletes the oxygen.

Both seasonal and permanent density stratification in lakes results in limited vertical circulation and anaerobic conditions in these freshwater bodies. Such conditions are especially likely to develop if the lake bottom is below the photic zone. But low oxygen levels can develop even at very shallow depth in swamps and marshes, where oxidation of very abundant organic matter removes the oxygen.

Oxygen levels may vary with time. Oxygen levels are balanced by the interplay of rate of production of oxygen by photosynthesis, depletion by respiration and oxidation, and vertical circulation. At times in the geologic past, anaerobic conditions have been more prevalent than at present. The Middle Cretaceous seems to have been such a time (Schlanger and Cita, 1982). This may have been at least in part the result of sluggish vertical circulation due to a low temperature gradient between poles and equator, or salinity stratification due to higher-salinity water at depth than at the surface. This resulted in minimal density contrast in seawater and sluggish circulation. Especially high productivity of the surface waters at this time has also been suggested; however, times of sluggish vertical circulation are likely to be times of low productivity, due to slow recycling of nutrient elements.

Oxygen levels in the deep Mediterranean Sea varied between aerobic and anaerobic during Quaternary time (Van Hinte et al., 1987). This also seems to have been the result of variable vertical circulation. During times of humid climates in the area, fresh water entering from the drainage basin produced low-density, brackish surface water. This resulted in strong density stratification and limited vertical circulation in the deep water. During times of arid climates, surface waters became slightly hypersaline. This dense surface water, with its dissolved oxygen, could then sink to the bottom.

The effect of oxygen levels on distribution and morphology of fossils has not been studied as extensively as that of some other environmental variables. In general, fewer and fewer species of organisms are able to survive as oxygen levels decrease. Certain groups of organisms, such as hermatypic corals and cephalopods, have long been known to require relatively high oxygen levels. Other groups (e.g., some annelids) do well in low-oxygen environments.

In general, the most tolerant organisms are soft-bodied forms that leave only a trace fossil record. Thus perhaps the best way to study oxygen levels is via trace fossils. Indeed, many researchers have emphasized the importance of oxygen levels in trace fossil distribution. The general pattern of trace fossil distribution seems to be for diverse assemblages with abundant dwelling burrows to develop under aerobic conditions. Relatively few trace fossil genera occur under dysaerobic conditions. First, locomotion traces and then deposit feeding traces become more common as oxygen levels decrease (Ekdale and

Mason, 1988). No trace fossils are found under anaerobic conditions. Certain trace fossils (e.g., *Chondrites*) are especially tolerant of low oxygen levels (Bromley and Ekdale, 1984a). This topic is discussed in greater detail in Chapter 6. In anaerobic environments, no organism that utilizes respiration as an energy source can live. Only a few species of anaerobic microorganisms, which use alternative energy sources, are able to survive under these conditions.

SUBSTRATE

Benthic organisms are strongly influenced by the nature of the substrate on or in which they live. Variation in bivalve morphology is an especially good example of this effect (Fig.2.25). The most basic subdivision of substrate type is between soft and hard substrates. But both hard and soft substates can be further subdivided. Ekdale (1985) has suggested subdividing soft substrates into *soupgrounds, softgrounds,* and *firmgrounds* on the basis of cohesiveness of the sediment. Most hard substrates would be classified as *hardgrounds,* usually consisting of lithified sediments. Bromley et al. (1984) suggests *woodgrounds* as a special type of hard substrate. Skeletons of other organisms are also important hard substrates. Perhaps they could be called *shellgrounds*.

Many factors determine the nature of the substrate: current strength, sediment source, sedimentation processes and rates, diagenetic processes, and the organisms themselves. Because fossils are usually found embedded in sediment, soft substrates are of most interest to us. Fine-grained sediments deposited in settings with weak currents are usually softer than coarse-grained sediments deposited by strong currents. Hardgrounds require lithification. Lithification may occur after burial, in which case erosion (unconformity formation) is necessary to expose the hardground later at the surface. But hardgrounds can also be produced by submarine cementation at the sediment–water interface, especially in carbonate rocks. Exoskeletons of organisms themselves may serve directly as hard substrates. In addition, filamentous algae may add firmness to the substrate by forming matlike surfaces.

Sedimentation rate, as well as softness, may be important to organisms. Organisms living in areas with rapid sedimentation rates may require special adaptations to keep from becoming buried or to keep the food-gathering structures from becoming clogged with sediment. Rapid vertical growth and flexible, articulated skeletons are adaptations for rapid sedimentation rate.

Both the distribution and morphology of organisms is strongly influenced by the nature of the substrate (Table 2.6). Researchers have described many adaptations for soft substrates in several fossil groups. One major problem with life on a soft substrate is the need to keep from sinking below the surface. Several methods have been evolved to cope with this problem. Thayer (1975a) noted development of the "snowshoe" (large, flat area of shell next to substrate) and "iceberg" (large portion of shell buried with commisure float-

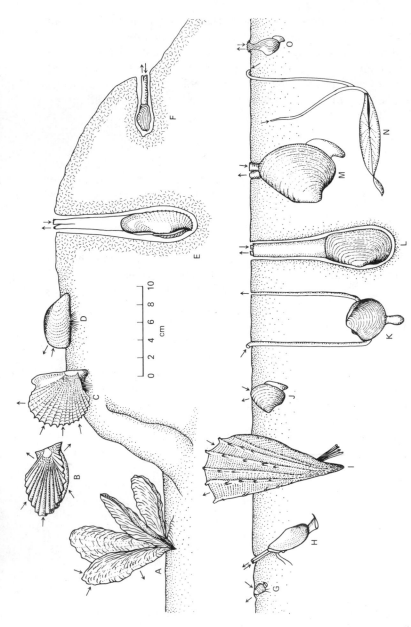

Figure 2.25 Modes of life of bivalves. A, Cemented (*Crassostrea*); B, swimmer (and recliner) (*Pecten*); C and D, attached by byssal threads (*Pinctada* and *Mytilus*); E, boring (*Pholas*); F, nestling in preexisting cavity (*Hiatella*); G, labial palp deposit feeder (*Nucula*); H, siphonate labial palp deposit feeder (*Yoldia*); I, nonsiphonate, byssal attached infaunal suspension feeder (*Pinna*); J, nonsiphonate suspension feeder (*Astarte*); K, mucus tube feeder (*Phacoides*); L and M, siphonate suspension feeders (*Mya* and *Mercenaria*); N, siphonate deposit feeder (*Tellina*); O, siphonate carnivore (*Cuspidaria*). From Stanley (1968), Journal of Paleontology, Society of Economic Paleontologists and Mineralogists.

TABLE 2.6   Examples of the Effect of Substrate on Distribution and Morphology of Organisms

Group	Effect	Reference
Algae	Stromatolites make a firm-to-hard substrate; many encrusting red algae on hard substrates	Wray, 1977; Walter, 1977
Foraminifera	Species distribution correlates with substrate type	Boltovskoy and Wright, 1976
Sponges	Many attach to hard substrate (including sclerosponges); fossil spicules common in fine, originally soft substrate	Finks, 1970; Jackson et al., 1971
Corals	Hermatypic scleractinians mostly attach to hard substrates; many *Rugosa* lived on soft substrates and grew crooked shapes to keep calyx above the sediment; some tall, slender forms where sedimentation rapid	Wells, 1957; Hubbard, 1970; Philcox, 1971; Fagerstrom, 1987
Bryozoans	Many attach to hard substrate, but they are also common on soft substrates; erect, flexible types and lunulitiform zooaria where sedimentation is rapid	Schopf, 1969; Lagaaij and Gautier, 1965; Rider and Cowan, 1977
Brachiopods	Most modern forms attach to hard substrate; many fossil types on soft substrates; snowshoe and iceberg strategy and thin, small shells to avoid sinking; zigzag commissure to exclude sediment grains?	Rudwick, 1970; Thayer, 1975a and b; Rudwick, 1964a; Alexander, 1975
Mollusks	Bivalves and gastropods live on both soft and hard substrates; many adaptations for substrate have evolved in these groups; small, thin, flat shells to avoid sinking; cementation, byssus, nestling, etc., on hard substrates; streamlining and shell sculpture to aid burrowing	Stanley, 1970; Stanley, 1972; Seilacher, 1984; Lindsley, 1978a and b
Arthropods	Shell thick, large, sculptured ostracods in coarse sediment; smooth, elongate, streamlined shells in soft substrate	Benson, 1961; Pokorny, 1978

TABLE 2.6   (Continued)

Group	Effect	Reference
Echinoderms	Crinoid holdfast adapted for rooting in soft substrate and cementing in hard substrate; spherical echinoid shape on hard or firm substrate; irregular, streamlined shapes in soft substrates	Breimer and Lane, 1978; Kier, 1972; Kier, 1974; Smith, 1984

ing above the sediment) adaptations in brachiopods. Other strategies are small and/or thin shells and changing growth direction to keep the food-gathering structures above the sediment.

Organisms living on firmgrounds have quite different problems. Burrowing forms must develop strong muscles, streamlined shells, and surface sculptures, which allow them to penetrate the firm substrate. Appendages and other features for moving along the surface are possible on firm substrates. Because substrate and water turbulence are so strongly correlated, adaptations for weak currents and soft substrates are often combined. Similarly, adaptations for strong currents and for firm substrates are often found together.

A different set of adaptations is needed for life on a hard substrate. Many organisms attach firmly to the substrate in such settings by cementation, byssus, muscle attachment, or suction. They may adapt by nestling in crevices. Some forms bore into the substrate for attachment and stability. Vagile forms develop strong locomotion and the ability to cling to the substrate in the turbulent water that often is associated with these substrates.

NONMARINE FACTORS

For the most part, the environmental variables discussed apply mainly to marine environments. Terrestrial and freshwater environments have some of the same and some different environmental parameters (Birks and Birks, 1980; Berglund, 1986). Humidity and rainfall are particularly important in terrestrial environments. Many paleoecologic studies of nonmarine fossils are concerned with climate. Climate includes a complex mixture of temperature, humidity, and rainfall, as well as the variability and seasonality of these factors. Studying ancient climates has been a major goal of plant paleoecologists, especially those studying pollen. A major subfield of paleoecology with much published research has developed around interpretation of climates based on palynology (Webb, 1985). Palynologic studies have been especially useful in making climatic interpretations based on study of Quaternary lakes and bogs

(Adam and West, 1983). Midlatitude lake deposits typically contain pollen characteristic of cold, harsh climates (such as spruce and fir) during glacial epochs and pollen characteristic of milder, more equable climates (such as oak and other deciduous pollen) during interglacial epochs. Changes from dominantly arboreal to herbaceous pollen often record a change from moist to dry climates in studies of this type.

Angiosperm leaf characteristics are extensively affected by climatic conditions (Wolfe, 1978). Botanists have long noted that large, thick leaves with smooth, entire margins (i.e., lacking lobes and teeth) are most common in warm climates (Fig. 2.26; Bailey and Sinnott, 1915). Leaves from cool climates are more often small and thin and have highly incised margins. In fact, the percentage of plant species with entire margins has been shown to vary directly with mean annual temperature (Fig. 2.27). Wolfe (1978) and others have used these relationships to interpret Tertiary climates of North America (Fig. 2.28). Birks and Birks (1980) give a more complete discussion of the use of the uniformitarian approach with plant fossils.

Morphology and distribution of vertebrate animals is strongly influenced by environmental conditions. The taxonomic uniformitarian approach thus potentially can be used in paleoenvironmental interpretations, especially for geologically relatively recent times. However, the rapid rate of evolution of vertebrates (Romer, 1961) limits their usefulness for detailed application of

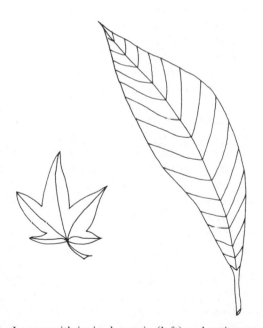

Figure 2.26   Leaves with incised margin (left) and entire margin (right).

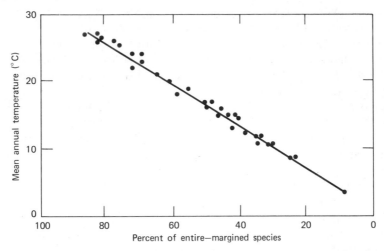

Figure 2.27 Variation with mean annual temperature in percentage of species with leaves having entire margins in forests of eastern Asia. From Wolfe (1978), reprinted by permission of American Scientist, Journal of Sigma Xi, the Scientific Research Society.

the taxonomic uniformitarian approach. Few vertebrate species are older than Pleistocene and most genera are no older than Neogene. Thus strict application of the uniformitarian principle is largely limited to Pleistocene fossils. Vertebrate fossils are also used less than invertebrates because of their lower abundance and usually proper preservation.

An example of more general application using vertebrate fossils is the determination of paleotemperatures from the distribution of large reptiles (Colbert, 1964). Mammals and birds are of less use in temperature determination because of their warm-blooded (endothermic) natures. Of the terrestrial vertebrates, reptiles and amphibians are cold blooded (ectothermic), and thus their body temperature is strongly influenced by the environment. Among modern amphibians and reptiles, smaller forms such as frogs, snakes, and turtles, can survive in cold climates by hibernation. Large reptiles, such as the crocodilians, are restricted to the tropics and subtropics, with a very few warm-temperature representatives (Fig. 2.29). Thus the presence of abundant large reptiles in the fossil record probably indicates warm temperatures. This approach leads to the conclusion that during most of the Mesozoic the tropical and subtropical zones were broader than today (Fig. 2.30). Other workers, such as Bakker (1975), have explained the wide latitudinal distribution of some large reptiles (dinosaurs) as evidence of their being endothermic. The uniformitarian approach to paleotemperature determination is certainly more safely applied to still-extant crocodilians.

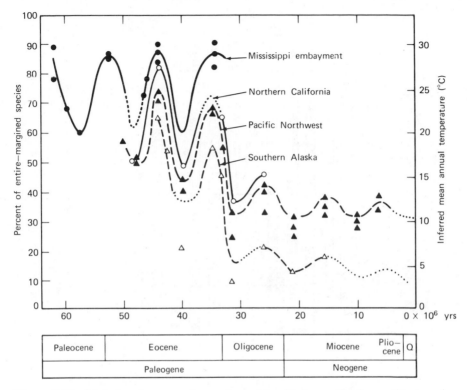

Figure 2.28 Variation in percentage of species with leaves having entire margins during the Cenozoic in four areas of North America. The curves are dotted when data are lacking or uncertain. From Wolfe (1978), reprinted by permission of American Scientist, Journal of Sigma Xi, The Scientific Research Society.

ANCIENT TAXONOMIC UNIFORMITARIANISM

The taxonomic uniformitarian approach is very useful in interpreting geologically young fossil assemblages. It becomes increasingly more difficult to apply the older the fossil assemblage. For example, the method may be very useful in dealing with Neogene assemblages in which most of the genera and even some of the species are still extant and presumably mostly living under the same environmental conditions. However, the strict application of taxonomic uniformitarianism is of little use in studying Cambrian biotas in which only some phyla are still extant. Indeed, environmental interpretations of very ancient biotas are difficult and very tenuous at best. A method that we call ancient taxonomic uniformitarianism can provide a way to interpret the environmental requirements of fossil taxa that have long been extinct.

Ancient taxonomic uniformitarianism might be paraphrased as: "The past is the key to the past." If a given fossil taxon can be shown to have lived under a

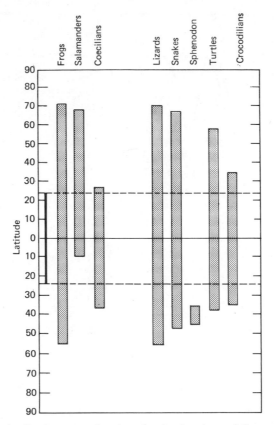

Figure 2.29 Latitudinal range of orders (and suborders of Squamata) among the modern amphibians and reptiles. The dashed horizontal lines show the tropical limits. From Colbert (1964).

particular range of environmental conditions, that group of fossils is assumed always to have lived under those conditions. For example, various lines of evidence indicate that trilobites lived under marine conditions. Therefore, whenever we find trilobite-bearing sediments, we assume that they were deposited in a marine environment. Of course, we can use the method in a more refined way by dealing with lower taxonomic categories. For example, a given trilobite genus may always be found in sediment which was apparently deposited under deep, quiet-water conditions. We then assume that that particular trilobite genus always indicates deposition under those conditions. Obviously, one must beware of circular reasoning when using this approach. The environmental range of the particular taxon must be firmly established before it can be used with confidence to interpret environments.

One might also ask why we need bother to interpret ancient environments with this approach if we already know the environmental conditions from

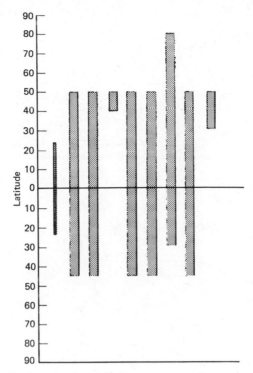

Figure 2.30 Latitudinal range of certain Cretaceous reptile groups. The narrow bar at left shows the limits of the modern tropics. From Colbert (1964).

some other relationship. We may not always have other information available on the environmental conditions. In some cases we may have only the fossils available, or we may have postmortem mixing of fossils that we need to sort. Perhaps the most basic reason for wanting to know the environmental conditions under which an ancient organism lived is for a better understanding of the ecology of ancient organisms, how they have evolved through time, and how they adapted to their environmental setting (Fig. 2.31).

The key to this method is being able to interpret the environmental conditions under which the fossil organism lived. There are many methods that can be used, most of which are based on sedimentologic features of the rocks containing the fossils, the mode of occurrence of the fossils, stratigraphic relationships, and geochemical methods. Many of these methods are discussed in later chapters of this book. The ancient taxonomic uniformitarianism approach is closely related to what Ager (1967) has called the "empirical approach." He used this phrase to describe all methods that could be used to relate fossils to the environmental conditions under which they lived. He included the following methods: (1) functional morphology, (2) mode of occurrence (in concentrations?, articulated?), (3) orientation (in life

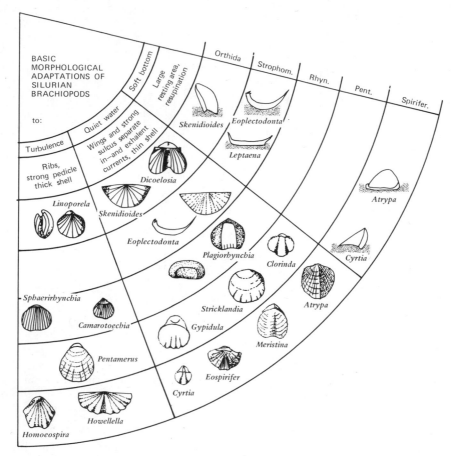

Figure 2.31 Morphological adaptations to environment in Silurian brachiopods. Strophom., Strophomenida; Rhyn., Rhynchonellida; Pent., Pentameridina; Spirifer., Spiriferida. From Fürsich and Hurst (1974).

position? current oriented?), (4) sedimentary setting, (5) organic associates (do they always occur with some other taxon?), and (6) distribution (have they been transported from their life habitat?). Ager (1967) gives several examples of the use of these factors in interpreting brachiopod paleoecology.

Other methods could be added to Ager's list. Stratigraphic relationships may be informative. For example, does the fossil always occur in a particular position in a cyclothem? Does the occurrence of the fossil in a particular part of a postgradational sequence suggest a certain depositional environment? The paleographic setting of the fossil may also suggest something about its life environment. For example, a particular fossil may occur only or predominantly in a lagoonal, back-reef setting. On the larger scale, paleobiogeographical setting may suggest some aspect of environmental conditions, especially temperature (see Chapter 10).

Geochemical properties of the fossil or associated sediment may also provide useful information about the environmental conditions under which the fossil lived. Various population and community characteristics of the fossil (e.g., diversity of the fossil assemblage) may also yield valuable information on environmental conditions. In short, all of the relationships discussed in later chapters plus various other sedimentologic and stratigraphic approaches can be used to interpret ancient environments. These interpretations can then be applied to particular fossil taxa in order to develop the method of ancient taxonomic uniformitarianism.

We might even use the method of taxonomic uniformitarianism itself to develop the ancient taxonomic uniformitarianism approach. We could call this approach "chaining." A given geologically fairly young fossil biota will include many genera and perhaps some species that are still extant. It may also include many species and some genera that are now extinct. Based on the extant genera and species, we can use taxonomic uniformitarianism with some confidence to interpret the environmental setting. We can then use this information to determine the conditions under which the extinct taxa lived. We can then use the method of ancient taxonomic uniformitarianism to extend environmental information concerning these now extinct taxa still further back in time. In concept this method could be used to extend environmental interpretations progressively further back in time. Of course, the degree of resolution is likely to become progressively poorer the older the fossil biota under consideration.

3
Biogeochemistry

In this chapter we deal with the use of chemical properties of fossils to determine paleoenvironmental conditions. The mineral composition of fossil skeletons, their trace and minor element chemistry, and their oxygen and carbon isotopic composition are discussed. Another type of geochemical technique of considerable potential that is not considered in this book is the geochemistry of preserved organic compounds in the rock record.

The chemistry of the environment in which skeletons form has a strong influence on the chemistry of the skeleton itself. In turn, the chemistry of the skeleton and the processes by which it forms may strongly influence the environmental chemistry (Lowenstam, 1974). These interrelationships form the basis for using chemical features of fossils and shells to better understand the chemistry of the environment.

Historically, the study of skeletal geochemistry has proceeded at an uneven rate. Most early references to skeletal chemistry were of an incidental nature, secondary to the main purpose of the study in which they were contained. The study of Clarke and Wheeler (1922) was a milestone and the first major American contribution to this field. Subsequently, little work was done until the early 1950s. Two major developments at this time caused the field of skeletal chemistry to blossom: the oxygen isotopic paleotemperature technique developed by H. C. Urey and co-workers (1951) at the University of Chicago and the work of H. A. Lowenstam and his students (most notable K. E. Chave) on skeletal mineralogy and trace chemistry at approximately this time (e.g., Lowenstam, 1954a,b; Chave, 1954). A concurrent milestone was the publication of a comprehensive study of the chemistry of organisms (including their skeletons) by A. P. Vinogradov (1953).

Studies of the stable isotopic composition of invertebrate skeletons have been used rather extensively and successfully over the last 35 years (especially by geochemists and oceanographers). Application of organic geochemistry has accelerated in the last decade. After initial interest in the 1950s and 1960s, only a few paleoecological studies based on skeletal trace chemistry and mineral compositions were conducted. In the last few years there has been a minor revival of interest in this subject (e.g., Brand, 1981a, 1987a,b; Graham et al., 1982; Popp et al., 1986a and b). Many paleontologists seem to have become disenchanted with geochemical techniques (if, indeed, they were ever enchanted with them in the first place!). This is evidenced from the brief treatment or nontreatment that these topics receive in most textbooks and reviews of paleontology and paleoecology. This may be in part due to the lack of understanding, training, or interest in geochemistry by many paleontologists. The complexity of geochemical relationships may also be discouraging to someone not fully conversant with the processes involved. The results may not be subject to unique interpretation. We tend to be subject to a "black box syndrome" which makes us want to be able to drop the fossil into the instrument and watch the dials point to the temperature and salinity of growth of the fossil. Finally, the negative effect of diagenesis on the chemical properties of the fossil limits the usefulness of the techniques, particularly in older fossils.

Actually, these problems are not limited to geochemical techniques. Biological relationships within a community of organisms can be as complex as geochemical relationships or more so, and that does not hinder our application of paleoenvironmental interpretations based on community distribution or structure. Often the functional significance of the morphology of fossils is not subject to unique interpretation, but we attempt to make such interpretations anyway. Diagenesis also has negative effects on the biological composition of fossil biotas, but we do the best we can with the imperfect record. We hope this chapter will help to dispel some of the mystery behind the geochemical techniques and perhaps encourage their wider use among paleoecologists. Paleontologists have traditionally been most comfortable describing and studying morphological features of fossils that they can easily see. Yet the unseen chemical attributes of the fossil are also an important aspect of that fossil and can potentially tell the paleontologist as much about certain features of the environment in which the fossil lived as can the fossil's morphologic features. The most complete possible description of the fossil and its environment would include chemical as well as morphological data.

The factors controlling the chemical properties of fossils can be divided into four categories: physical chemical, environmental, physiologic (genetic), and diagenetic (Dodd and Schopf, 1972). These four effects can be pictured as the four apicies of a tetrahedron (Fig. 3.1). The relative importance of each of the four categories of effects in determining the chemistry of any fossil can be imagined as a point somewhere within the tetrahedron. Conceptually, the paleoecologist must determine the location of that point in order to explain the chemistry of the fossil completely. In practice, that may be difficult or impossible, but the concept is useful in understanding skeletal chemistry. Actually, these categories are not completely independent; all effects should

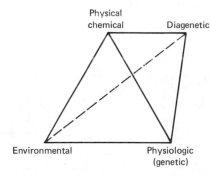

Physical
chemical   Diagenetic

Environmental   Physiologic
                           (genetic)

Figure 3.1 Schematic representation of the factors controlling the chemistry of fossil skeletons. Physical chemistry, physiology (genetics), environment, and diagenesis all affect chemistry. The relative contributions of these factors can be visualized as a point within the tetrahedron.

ultimately be explainable in physical chemical terms. For example, the environmental effects on oxygen isotopic composition are the result of environmental influence on physical chemical relationships and are not truly independent. Physiologic effects which are under genetic control presumably have a biochemical basis that could be explained in physical chemical terms if they were adequately known. Diagenetic effects are really physical chemical processes that have occurred after the formation of the skeleton. Nevertheless, this subdivision of factors is useful in facilitating discussion and is used in this chapter.

SKELETAL MINERALOGY

Strictly speaking, mineralogy is not a chemical property of the skeleton but refers to the physical arrangement of atoms in space. But this is so closely tied to the chemical nature of the atoms that for convenience we may consider mineralogy in conjunction with the chemical characteristics of fossils. The mineral composition of skeletal material has been of considerable interest to paleoecologists both as a potential tool for paleoenvironmental analysis and because of its effect on the preservation of fossils and potential fossils. Although little success can be claimed in using skeletal mineralogic composition directly to interpret paleoenvironments, information on mineralogy has been extremely useful in interpreting geologic processes and history, especially when used in combination with other information.

Mineral Components

Many different minerals are found in invertebrate skeletons (Table 3.1). By far the most important minerals among the invertebrates are the two calcium carbonate polymorphs, calcite and aragonite, although the opaline silicates and phosphates also make major contributions. In addition, several less common minerals have also been described in biological systems (Lowenstam,

TABLE 3.1 Diversity and Phylum Distribution of Biomineralization Products in Extant Organisms

Kingdom → Phylum	Monera	Dinoflagellata	Haptophyta	Bacillariophyta	Phaeophyta	Rhodophyta	Chlorophyta	Zygnematophyta	Rhizopodea	Siphonophyta	Charophyta	Heliozoata	Radiolariata	Foraminifera	Mixomycota	Ciliophora	Basidiomycota	Deuteromycota	Porifera	Coelenterata	Platyhelminthes	Ectoprocta	Brachiopoda	Annelida	Mollusca	Arthropoda	Sipuncula	Echinodermata	Chordata	Bryophyta	Trachaephyta
(Kingdom)	Monera	Protoctista															Fungi		Animalia											Plantae	
CARBONATES																															
Calcite	+	+	+			+				+	+			+	+				+	+	+	+	+	+	+	+	+	+	+	+	+
Aragonite	+		?		+	+	+			+				+					+	+		+		+	+	+	+		+		+
Vaterite						+																			+	+			+		+
Monohydrocalcite	+																								+				+		
Amorphous Hydrous Carbonate																									+	+			+		
PHOSPHATES																															
Dahllite	+														+	+		+			+								+		?
Francolite																													+		
CO$_3$Mg$_3$(PO$_4$)$_4$																							+	+	+						
Brushite																															
Amorphous Dahllite Precusor																						+				+			+		
Amorphous Brushite Precursor																									+						
Amorphous Whitlockite Precursor																					+			+	+						
Amorphous Hydrous Ferric Phosphate																								+	+			+			

HALIDESI

Fluorite

Amorphous Fluorite Precursor

OXALATES

Whewellite

Weddelite

SULFATES

Gypsum

Celestite

Barite

SILICA

Opal

FE-OXIDES

Magnetite

Maghemite

Goethite

Lepidocrocite

Ferrihydrite

Amorphous Ferrihydrates

MN-OXIDES

Todorokite

FE-SULFIDES

Pyrite

Hydrotroilite

From Lowenstam (1981), Science, 211:1126–1131, figure 1, copyright 1967 by the American Association for the Advancement of Science.

1981; Lowenstam and Weiner, 1989). Probably, additional minerals will be found. These minor minerals often serve a particular function for which they are better adapted than the primary mineral found in the skeleton of that group. For example, the chitons have magnetite cappings for their radular teeth (Lowenstam, 1962). This mineral is much harder than the aragonite that makes up the basic skeleton of the chitons. Some of the echinoids have dolomite in their teeth (Schroeder et al., 1969). Dolomite is also slightly harder than the calcite that makes up the majority of the echinoid test. The same explanation might apply to the fluorite gizzard plates of some gastropods (Lowenstam and McConnell, 1968). In many cases we do not know why the particular mineral is being used by the organism rather than one of the more common skeletal minerals (e.g., brucite in algae, gypsum in jellyfish statoliths). The explanation is probably related to the biochemistry of the organism. A certain mineral is simply easier to precipitate biochemically than another. The discussion in this chapter will concentrate on the minerals calcite and aragonite, as they are the most common of the skeletal minerals. Lowenstam (1974, 1981) has published an extensive discussion of other skeletal minerals, to which the reader is referred for further details.

In addition to skeletons that consist of only pure calcite or pure aragonite, many invertebrates have skeletons consisting of a combination of these two polymorphs. Indeed, the skeletons of sclerosponges are in part carbonate (aragonite) and in part opaline silica spicules (Hartman and Goreau, 1970). Fish have skeletons that are largely phosphatic, but they also have ear bones (otoliths) composed of aragonite. The minerals are always present in separate microarchitectural units rather than in intimate intermixtures. They may be separate units within the skeleton (e.g., shell layers) or they may form separate, isolated parts (e.g., fish bones and otoliths). Crystal sizes, shapes, orientations, and interrelationships within these microarchitectural units may be quite varied and complex (Bøggild, 1930), giving rise to another important area of study: the variation in these skeletal structure units and the factors responsible for that variation (Chapter 4). Skeletal calcites may contain considerable amounts of magnesium substituting for calcium (up to 30 mol % $MgCO_3$), greatly affecting the properties of the mineral. Many authors have considered high magnesium calcite (greater than about 4 ml % $MgCO_3$) as a separate mineral. Here we discuss chemical variation in the carbonate minerals separately.

The data shown in Table 3.1 are entirely for the skeletons of living organisms. The mineralogy of now-extinct groups cannot always be determined with certainty. Some fossil groups that were originally aragonite may since have been converted to the more stable calcite. For example, the original mineral composition of tabulate and rugose corals and the stromatoporoids cannot be determined with certainty. All are calcite now, but some workers have suggested that they were originally aragonite, as are the scleractinians. However, details of their skeletal structure suggest that no major diagenetic alteration has occurred (Sorauf, 1971; Sandberg, 1984). At the few Paleozoic localities where aragonite is preserved in some groups (nautiloids, bivalves, and gastropods), the tabulate and rugose corals are calcite (Stehli, 1956; Brand, 1981a,b). Details of the preservation of stromatoporoids suggest that they were originally aragonite (Stearn and Male, 1987).

Factors Controlling Skeletal Mineralogy

That $CaCO_3$ should be by far the most common skeletal material is not surprising on a physical chemical basis for several reasons:

1. Shallow sea water is saturated or supersaturated with respect to $CaCO_3$ almost everywhere and in some places $CaCO_3$ is precipitating inorganically (Milliman, 1974).
2. Precipitation of $CaCO_3$ is more readily controlled biologically than most minerals. This is because the concentration of the CO_3^{2-} ion (and thus the degree of saturation with respect to $CaCO_3$) is strongly dependent on pH and on the concentration of CO_2, and metabolic processes readily affect these parameters.
3. Many organisms also build skeletons of opal or of phosphate minerals, but the concentration of dissolved Si^{2+} and PO_4^{3-} is much lower than Ca^{2+} (Goldberg, 1957), requiring a much larger volume of seawater to precipitate a unit of mass of SiO_2 or $Ca_3(PO_4)_2$ than of $CaCO_3$.
4. Shallow seawater is usually not saturated with respect to SiO_2 or $Ca_3(PO_4)_2$, requiring strong biological intervention to precipitate opaline or phosphatic skeletons.

Calcium carbonate occurs in skeletons in one of the three polymorphous forms: calcite, aragonite, or vaterite. Skeletal vaterite is rare, so we concentrate our discussion on calcite and aragonite.

Calcite but not aragonite is stable under surface temperature–pressure conditions (Fig. 3.2; Fyfe and Bischoff, 1965; MacDonald, 1956). The instability of aragonite under surface conditions is indicated by its slightly larger free energy of formation than that of calcite [-269.5 versus -269.7 at 25°C (from Latimer, 1952)]. This is expressed in the greater solubility of aragonite [solubility product $= 6.0 \times 10^{-9}$ (from Krauskopf, 1967)] than calcite [solubility product $= 4.5 \times 10^{-9}$ (from Krauskopf, 1967)]. The generally decreasing abundance of aragonite in rocks of increasing geologic age is a further expression of the instability of aragonite.

On the other hand, aragonite is an extremely common mineral in modern marine sediments, probably the most common carbonate mineral in the shallow marine environment (e.g., Flügel, 1982). Much of it is of skeletal origin. This apparent paradox needs an explanation. The most probable one is that because of the high concentration of magnesium in seawater, aragonite is in fact not unstable in the marine environment. Precipitation of calcite in the

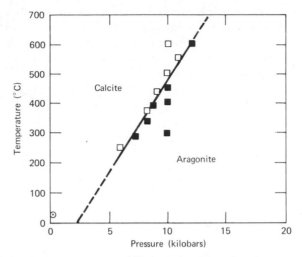

Figure 3.2 Temperature–pressure stability relationships of calcite and aragonite based on experimental studies with high-pressure–temperature bombs. The solid squares are data from experiments producing aragonite and the open squares are data from experiments producing calcite. The dashed portion of the line extrapolates the results to lower temperatures and pressures. The encircled dot shows surface temperature–pressure conditions. From MacDonald (1956), American Mineralogist, *41:* 749, copyrighted by the Mineralogical Society of America.

marine environment results in the coprecipitation of a large amount of Mg in the calcite lattice (approximately 10 to 14 mol % $MgCO_3$ or more under most surface conditions). High magnesium calcite of this approximate range of composition is a fairly common precipitate from seawater, especially as intergranular cement (Flügel, 1982). This high-Mg calcite is more soluble and thus less stable than aragonite. Berner (1975) has shown on thermodynamic grounds that high-Mg calcite with more than 8.5 mol % $MgCO_3$ in solid solution is less stable (more soluble) at 25°C than is aragonite. Calcite containing less than 8.5 mol % $MgCO_3$ in solid solution is more stable (less soluble) than aragonite, but such low-Mg calcite cannot form in full marine seawater under tropical or subtropical temperatures, due to the abundance of Mg in seawater. Consequently, aragonite and high-Mg calcite form instead.

An alternate explanation for the precipitation of aragonite in seawater is that the relative crystal nucleation and growth rates of calcite and aragonite are differentially affected by foreign ions. These ions are sometimes called crystal poisons (Simkiss, 1964). Mg^{2+} appears to be especially effective in inhibiting growth of calcite relative to aragonite. The abundance of Mg^{2+} in seawater may explain the common precipitation of aragonite in the oceans.

The mineralogic diversity of different groups of organisms (Table 3.1) indicates that the mineral composition of skeletal material is in part genetically controlled. The different minerals in different parts of a single organism is a further expression of the physiologic control. Mineral composition also changes during the life of the individual in some groups. For example, the larval shell of the bivalve *Crassostrea* is entirely composed of aragonite, whereas the adult shell is almost entirely calcite (Stenzel, 1963). The mineralogy of the shells of the bivalve *Mytilus* also changes during its life (Dodd, 1963). Possible long-term evolutionary trends in mineralogy have been described in the corals, which apparently changed from predominantly calcite in the Paleozoic to predominantly aragonite today (Lowenstam, 1974). Indeed, aragonitic groups increase in abundance after the Paleozoic (Railsback and Anderson, 1987). Also, primitive gastropods seem to have been predominantly calcite, whereas the more advanced forms are mainly aragonite (Lowenstam, 1964a).

The process by which organisms control their skeletal mineralogy is still poorly understood. Lowenstam (1981) points out that some organisms form minerals by chemically modifying inorganic solutions (biologically induced mineralization). The mineralogy in these organisms shows a minimum of physiologic control. Other organisms deposit minerals on an organic matrix, which forms a framework strongly controlling mineral composition (organic matrix-mediated mineralization). Physiologic control is much stronger in these groups.

Lowenstam (1954a,b) first noted a temperature effect on mineral composition of invertebrate skeletons. This effect is expressed in three ways (see Fig. 3.3):

1. *Group I.* Some groups of organisms having aragonitic skeletons are far more abundant in the tropics than in the higher latitudes. The reef-building scleractinian corals are perhaps the best example of this effect.

2. *Group II.* Certain groups of organisms having aragonitic skeletons are found only in the tropics and semitropics. The best examples of this effect are the calcareous green algae.

3. *Group III.* In taxa having skeletons composed of a combination of aragonite and calcite, the proportion of aragonite increases as temperature increases. Examples of the latter effect are found in the mollusks (Fig. 3.4), annelids, coelenterates, and bryozoans. In these groups the relationship between temperature and mineral composition is specific to individual genera, species, or even subspecies (Lowenstam, 1954a; Dodd, 1963).

The relation between temperature and the aragonite/calcite ratio is not always obvious because of the effects of other environmental parameters, particularly salinity (Lowenstam, 1954a; Dodd, 1963). For example, in the bivalve genus *Mytilus*, reduced salinity is accompanied by an increased propor-

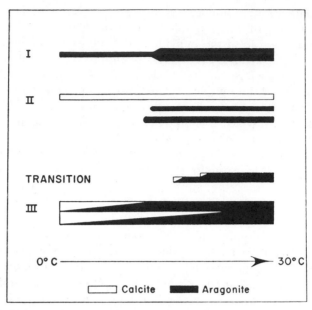

Figure 3.3 Schematic representation of three ways in which temperature affects the carbonate mineralogy of invertebrate skeletons. I, Groups of organisms having aragonitic skeletons that are now abundant in warmer waters; II, groups with aragonitic skeletons that are found only in warmer waters, whereas calcite groups occur in both cold and warm water; transition, groups with aragonite skeletons that secrete trace amounts of calcite in the colder part of their ranges; III, groups containing a combination of calcite and aragonite but with increasing amounts of aragonite in the warmer part of their ranges. From Lowenstam (1954a), copyright University of Chicago Press.

tion of aragonite (Fig. 3.5). Other more individual effects, such as growth rate, also seem to affect mineral composition.

The most common diagenetic effect on the mineral composition of fossils is solution by groundwater. Only a small portion of the skeletal material originally buried in sediments is preserved as fossils, as indicated by the great volume of normal marine sedimentary rocks that do not contain fossils. Groundwater undersaturated with respect to the mineral of which the fossil is composed will in time dissolve the fossil. Solution is especially likely to occur in a permeable rock and when the fossil is in a matrix that differs in mineralogy from the fossil. Thus a carbonate fossil buried in a quartz sand has a high probability for solution. If the matrix has the same mineralogy as the fossil, the water will quickly become saturated with respect to that mineral by dissolving matrix as well as fossil and the fossil has a greater chance of escaping complete solution. Thus a calcite fossil buried in a limestone has a good chance of preservation.

Undersaturated water in contact with both calcite and aragonite shell mate-

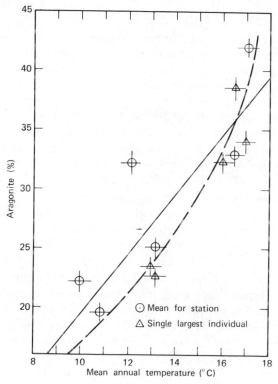

Figure 3.4 Variation in percent aragonite in *Mytilus californianus* with mean annual temperature of collecting locality. Solid line, regression line fit to data; dashed line, trend line drawn through data points. From Dodd (1963), copyright University of Chicago Press.

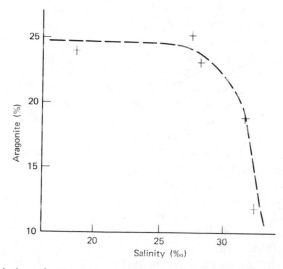

Figure 3.5 Variation of percent aragonite with salinity at collecting time in *M. edulis edulis* from Hood Canal, Washington area. From Dodd (1963), copyright University of Chicago Press.

rial will dissolve both minerals. Eventually, however, the water will become saturated with respect to calcite and the calcite will stop dissolving. Aragonite will continue to dissolve because it is more soluble than calcite, so the solution has not yet reached saturation with respect to aragonite. However, as aragonite continues to dissolve, the Ca^{2+} and the CO_3^{2+} concentrations of the solution rise, exceeding the solubility product of calcite. Calcite should then precipitate. Eventually, all the aragonite should dissolve and be reprecipitated as calcite. This process has been demonstrated in the laboratory (Kitano et al., 1976). Mixtures of calcite and aragonite fossils are thus chemically unstable and should become increasingly rare with time, as indeed they do. The most ancient examples of preserved aragonitic fossils are in situations where they are effectively removed from contact with groundwater, such as where buried in asphalt or fine shale (Stehli, 1956).

If the solution of the aragonite fossil occurs before lithification, the resulting void will collapse and all trace of the fossil may be lost, with the possible exception of an imprint. If lithification has already occurred before the fossil dissolves, a mold will be preserved, and if the mold is later filled with another mineral, a cast of the original fossil will be preserved. Fossil assemblages commonly consist of a mixture of originally calcite fossils with preservation of original structure and calcite casts of originally aragonite fossils (Horowitz and Potter, 1971). In some cases mollusks show preservation of original structure in some shell layers and casts of others, indicating a mixed mineralogy of the original shell. Also common are assemblages consisting entirely of originally calcite forms, the aragonite forms apparently having been removed by solution in an unlithified matrix (Lawrence, 1968; Chave, 1964). Obviously, care must be taken in making paleoecologic interpretations based on such greatly modified assemblages.

The diagenetic processes described above take place after burial. Solution of fossils also occurs before burial or soon after burial in the surficial sediments. Except in the polar regions, the shallow waters of the oceans are saturated or supersaturated with respect to $CaCO_3$ (Broecker, 1974), so carbonate skeletons are not dissolved there. With increasing depth, however, the oceans become undersaturated with respect to $CaCO_3$ so that solution occurs before the *lysocline* (the zone of rapid increase in $CaCO_3$ solution). The lysocline for calcite depends on oceanic condition and ranges from 3500 m depth in the Pacific to 5000 m in the Atlantic (Fig. 3.6; Broecker, 1974). Because it is slightly more soluble than calcite the lysocline for aragonite is shallower, ranging from about 500 m in the Pacific to 2500 m in the Atlantic. Berger (1970) and others have pointed out that this solution is not limited to selection between calcite and aragonite shells; but at least in the case of planktic foraminifera, can also selectively remove the more delicate shells while leaving the more robust forms. This differential preservation may have important effects on interpretation of the planktic foraminifera. The more delicate shells are usually those living in the shallowest, warmest water, so if they are selectively dissolved from an assemblage, the remaining heavier-

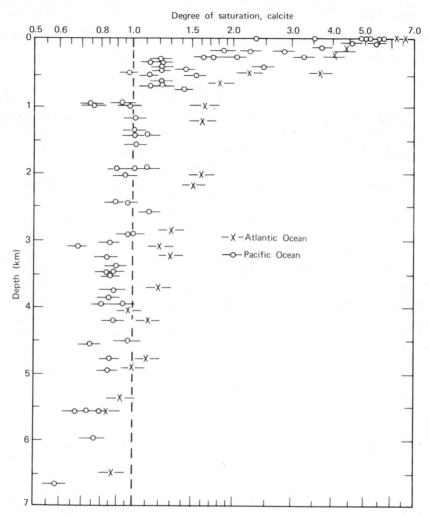

Figure 3.6   Degree of saturation with calcite of seawater samples from various depths in the Atlantic and Pacific Oceans. From Li et al. (1969), Journal of Geophysical Research, *74:* 5521, copyright the American Geophysical Union.

shelled forms give the impression of colder surface-water conditions than actually existed (Berger, 1970).

The oceans are essentially everywhere undersaturated with respect to opaline silica (Broecker, 1974). Siliceous skeletons in the oceans thus begin to dissolve as soon as they are exposed to seawater. As in the case of carbonate skeletons, the heaviest, most robust forms are most likely to be preserved. Solution of opal continues within the sediment, so that unless sedimentation is rapid, complete loss of opaline fossils is likely (Johnson, 1976).

Geologic Application of Skeletal Mineralogic Data

Paleotemperatures have never been successfully determined on the basis of skeletal mineralogy for three reasons:

1. The mineralogy–temperature effect differs from species to species and perhaps even within genetic variants of a single species; therefore, a given proportion of aragonite may correspond to a certain temperature in one species and to another temperature in another (Dodd, 1963). This limits the technique to species that still have living representatives for which the precise temperature–mineralogy relationships can be determined.
2. Several factors in addition to temperature influence skeletal mineralogy, and the interaction of these factors may be complex.
3. Because aragonite is not stable in diagenetic environments, finding well-preserved specimens is difficult.

Although use of skeletal mineralogy may not be a practical method for detailed paleotemperature determination, it can be of value in identifying temperature trends. The presence of aragonitic taxa not found outside the tropics or the abundance of aragonitic taxa not normally abundant outside the tropics would suggest a tropical climate and could be used in establishing an endpoint for such general trends. Gradients in the aragonite/calcite ratio in skeletons with mixed mineralogies could also be used. This might be done even with fossils in which the aragonite had converted to calcite if the originally aragonitic nature of the converted unit could be determined from its method of preservation (Bathurst, 1975).

Probably the most useful paleoecologic application of skeletal mineralogy data is in evaluating the degree of preservation of fossil assemblages. The examples from the New Jersey Cretaceous (Chave, 1964) and the South Carolina Oligocene (Lawrence, 1968) mentioned previously are just two examples that show how differential removal of aragonitic fossils can markedly change the composition of the original skeletal assemblage. The most extensive use of skeletal mineralogy data has been in the study of the origin and diagenesis of carbonate sediments and rocks (Flügel, 1982; Bathurst, 1975; Milliman, 1974). The mineral composition of a carbonate sediment is determined largely by the organisms that contribute their skeletons to the sediment because the major portion of carbonate sediments is of biological origin (Lowenstam, 1974). The relative proportion of calcite and aragonite as well as the amount of Mg in the calcite has a large bearing on the way the sediment behaves during lithification. The solution of aragonitic skeletal grains can result in the development of secondary porosity in carbonate rocks, a process of great practical import in the development of oil reservoirs. Solution of aragonite may be an important source of Ca^{2+} and CO_3^{2-} ions for formation of calcite cement in carbonate rocks. Bathurst (1975) has extensively reviewed the relationship of skeletal mineralogy to carbonate sediment formations and diagenesis.

Lowenstam (1974) has emphasized the importance of organisms in adding a variety of minerals to sediments. Calculations show that on a global basis enormous quantities of opaline silica, phosphate minerals, fluorite, and magnetite are contributed by organisms to the sediments. Skeletal siliceous deposits (diatomites) and phosphates (phosphate rock) are sometimes of economic importance.

TRACE CHEMISTRY

The minerals precipitated by organisms are never absolutely pure but contain trace amounts of foreign ions. They may contain a considerable concentration of the foreign ion, several mole percent in some cases. In these examples they may be more appropriately called minor or accessory elements rather than trace elements. The ions may be present in solid solution in the crystal lattice, absorbed to the crystal surface, incorporated in the organic matrix, or incorporated as separate mineral phases (perhaps trapped as impurities from the surrounding environment) (Dodd, 1967). In the discussion that follows, we concentrate on ions included in solid solution. The concentration of trace elements in a fossil is a function of the physical chemistry of the skeletal formation process, of environmental variables, of the physiology of the organism, and of the diagenetic processes. Although these factors apply to all skeletal minerals, the study of skeletal trace chemistry has dealt largely with carbonates. Most of this research has been on the concentration of the two most abundant trace metals, Mg and Sr, and we will also concentrate our discussion on these elements. The reasons for the greater abundance of these elements are threefold: (1) The ionic radii of Mg^{2+} and Sr^{2+} are close enough to that of Ca^{2+} to allow ready substitution in either the calcite or the aragonite crystal lattice; (2) the ionic charge is the same as that of Ca^{2+}, again facilitating substitution; and (3) Mg^{2+}, and to a somewhat lesser extent Sr^{2+}, are abundant in natural waters, expecially seawater.

Factors Controlling Trace Chemistry

Other factors being equal, the trace cation-to-Ca ratio in a carbonate skeleton will be proportional to that ratio in the water in which the skeleton formed. In other words, the higher the trace cation concentration relative to Ca in the water in which the shell grows, the higher will be the concentration of that element in the shell. This relationship can be expressed mathematically by

$$[Me/Ca]_{skeleton} = K[Me/Ca]_{water} \qquad (3.1)$$

in which Me is the molar concentration of the trace cation and Ca is the molar concentration of Ca. K is a proportionality constant commonly called the distribution or partition coefficient. The concept of distribution coeffi-

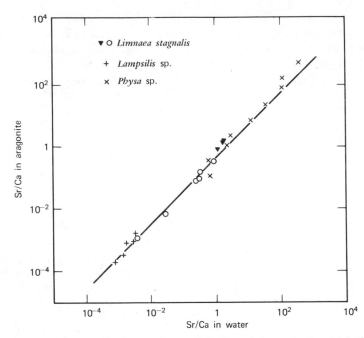

Figure 3.7   The Sr/Ca ratio in aragonite snail shells and the water in which the shells grew. From Buchardt and Fritz (1978). copyright American Association for the Advancement of Science, Science 199:292, text-figure 2.

cients is especially useful in studying the trace chemistry of inorganic precipitates (e.g., Kinsman and Holland, 1969; Füchtbauer and Hardie, 1976; Veizer, 1982). This relationship has also been demonstrated for skeletal carbonates by various researchers, including Buchardt and Fritz (1978), who determined the Sr/Ca ratio in snail shells grown in water with differing Sr/Ca ratios (Fig. 3.7).

The distribution coefficient concept is useful in paleoecology for two reasons:

1. If the trace cation-to-Ca ratio (Me/Ca) in a fossil can be determined, and if the distribution coefficient for that Me/Ca ratio is known, the Me/Ca ratio of the water in which the fossil grew can be determined. Me/Ca ratios in natural waters depend on environmental conditions and can yield useful information about depositional environments.

2. The distribution coefficient is not constant but is temperature dependent for some ions. Thus if the Me/Ca ratio is known for both fossil and water in which it grew, the distribution coefficient can be determined, and from the relationship of the coefficient to temperature, the temperature of formation can be determined. This approach has been used

extensively in igneous and metamorphic petrology to determine temperatures of mineral formation.

The foregoing applications depend on mineral and water being in chemical equilibrium. Kinetic and biochemical parameters may complicate this approach, as demonstrated for Sr/Ca ratios in the bivalve *Mytilus edulis* shells by Lorens and Bender (1980).

Distribution coefficients are also dependent on the mineral form of the $CaCO_3$. Thus calcite and aragonite have different distribution coefficients for a given Me/Ca ratio. Other factors being equal, the Mg/Ca ratio is higher in calcite than in aragonite skeletons, and the Sr/Ca ratio is higher in aragonite than in calcite skeletons. This is because the small Mg^{2+} ion (radius 0.66 Å) substitutes more readily in the calcite lattice, which has sixfold coordination and is isostructural with magnesite ($MgCO_3$), than it does in the aragonite lattice. Similarly, the larger Sr^{2+} ion (radius 1.12 Å) substitutes more readily in the aragonite lattice, which has eightfold coordination and is isostructural with strontianite ($SrCO_3$).

Biologically formed calcite and aragonite often do not have the same trace element concentrations as those of inorganically precipitated minerals. This means that the distribution coefficients for biologically formed minerals (biological distribution coefficients) may be different from those for inorganic precipitates. The reason for this is not known but must be related to the physiology of the skeletal formation process. The physiologic effect is shown by (1) differences between different groups of organisms growing under the same environmental conditions (phylogenetic effect), (2) differences that develop during the life of the individual organism (ontogenetic effect), and (3) differences between skeletal units within a single organism (microarchitectural effect). The modern genus *Nautilus* shows all three of these effects (Crick and Mann, 1987).

Distribution coefficients for Mg and Sr relative to Ca in inorganically precipitated calcite and aragonite are temperature dependent. Sr concentration decreases with increasing temperature in both calcite (Holland et al., 1964) and aragonite (Fig. 3.8; Kinsman and Holland, 1969). Mg concentration in inorganically precipitated calcite increases with increasing temperature (Mucci, 1987). Similar temperature effects on biologically modified distribution coefficients should also be reflected in the trace chemistry of skeletal carbonates. Such effects have indeed been demonstated. However, no universal relationship between trace chemistry and temperature can be established because the values of the biologically modified distribution coefficients differ from those for inorganic precipitates and even differ between taxonomic groups of organisms. Therefore, the temperature–trace element relationship must be determined for each taxonomic group. A given temperature–trace element relationship may hold for an entire class or phylum (e.g., the brachiopods) (Lowenstam, 1961), but others may only apply to a genus or species, as in the case of the mollusks (Dodd, 1965; Schifano, 1984).

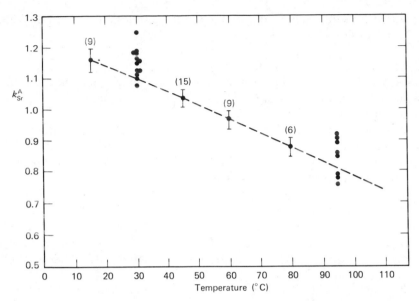

Figure 3.8   Variation with temperature of the distribution coefficient between Sr and Ca in aragonite precipitated from seawater. Numbers in parentheses indicate the number of experimental runs included in the average value. From Kinsman and Holland (1969).

The effect of temperature on Mg concentration in calcite is particularly well documented (Clarke and Wheeler, 1922; Chave, 1954; Zolotarev, 1974). Mg concentration correlates positively with temperature in many groups of organisms and is especially apparent in forms characterized by a generally high Mg content (Fig. 3.9). The effect of temperature on Sr concentration is not as

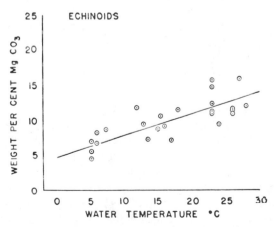

Figure 3.9   Variation with temperature of Mg concentration in echinoid skeletons. From Chave (1954), copyright University of Chicago Press.

clear but has been demonstrated in a few taxa. In some cases Sr correlates positively with temperature, as in the brachiopods (Lowenstam, 1961), and in other cases the correlation is negative, as in echinoids (Pilkey and Hower, 1960). Sr concentration and temperature are positively correlated in the calcite layer of *Mytilus* and negatively correlated in the aragonite layer (Fig. 3.10; Dodd, 1965). Scleractinian corals show a negative correlation between Sr concentration and temperature which is very close to that found in inorganic precipitates of aragonite (Smith et al., 1979). This variability in correlation between trace chemistry and temperature in different groups of organisms points up the difficulty in using these relationships as a paleotemperature determination tool. The distribution coefficient relationship suggests that skeletal carbonates might be used to monitor the chemistry of the water in which the skeletons formed (e.g., De Deckker et al., 1988).

Differences in elemental ratios are very apparent in comparing Mg/Ca and Sr/Ca ratios in marine and freshwater mollusk shells. These ratios are both usually much higher in seawater than in fresh water, and this is reflected in the shell chemistry (e.g., Dodd and Crisp, 1982). Both Mg/Ca and Sr/Ca ratios are relatively constant in seawater, so that other factors being equal, these ratios will not vary much in marine skeletons. These ratios may be quite variable in fresh water (Skougstad and Horr, 1963; Livingstone, 1963) and thus in skeletons forming in these waters. Trace chemistry of well-preserved

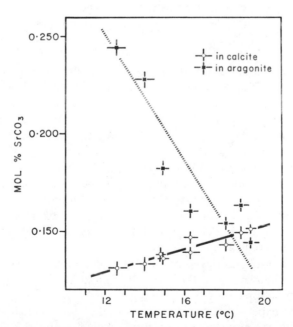

Figure 3.10 Variation with temperature of Sr in the last-formed portion of the calcite outer prismatic layer (open circles) and aragonite nacreous layer (solid squares) of *Mytilus*. From Dodd (1965).

freshwater fossil shells is thus potentially useful in studying the chemistry of ancient freshwater bodies (Chivas et al., 1985).

One might expect that Mg/Ca and Sr/Ca ratios in fossils could be used to measure paleosalinity. Eisma et al. (1976) and Dodd and Crisp (1982) have reviewed this question and show a rather poor correlation between salinity and skeletal trace chemistry. A major reason for this is that Mg/Ca and Sr/Ca ratios are not a linear function of salinity because most of the Ca, Mg, and Sr in brackish water comes from seawater. Thus seawater dominates the composition (and trace element ratios) in brackish water and the Mg/Ca and Sr/Ca ratios are nearly constant above a salinity of about 10‰.

In this discussion we have concentrated on Mg/Ca and Sr/Ca ratios. However, many other elements have also been studied in modern and fossil skeletons. Mn/Ca and Fe/Ca ratios are usually much higher in freshwater than in marine water. This difference is reflected in skeletal carbonates (Brand, 1987b).

Cd/Ca ratios have been studied in benthic foraminifera (Boyle, 1988; Hester and Boyle, 1982; Delaney and Boyle, 1987). This ratio appears to correlate with phosphate concentration (Fig. 3.11) and thus with productivity. Delaney and Boyle (1987) used this relationship to study productivity variation and deep oceanic circulation in deep-sea cores of late Miocene sediment. Variation in Cd/Ca ratios in modern corals from the Galapagos Islands appears to reflect variation in upwelling (and thus nutrient content) related to the El Niño currents (Shen et al., 1987). Krantz et al. (1988) used Cd/Ca ratios in mollusks in conjunction with oxygen and carbon isotopic analysis to determine circulation patterns (especially upwelling) on the Atlantic shelf of southeastern United States.

In some settings the Cd/Ca ratio appears to reflect release of Cd from modern industrial sources. Pb concentration and perhaps other trace elements reflect similar patterns (Shen et al., 1987; Shen and Boyle, 1987). By measuring the Pb/Ca ratio in growth bands in corals, Shen and Boyle showed the increase in Pb concentration in the oceans in the early and mid-twentieth century and its subsequent decline after about 1970, correlating with decreased use of leaded gasoline.

Fürst (1981) noted that the concentration of boron in siliceous sponge spicules correlates crudely with salinity and perhaps productivity. Wright et al. (1987a,b) have studied rare earth elements in fossil phosphates as a measure of redox potential in the ancient seas. Many other examples of studies of various trace elements in skeletons could be cited. In most cases the potential of paleoecologic application seems to be limited.

A large literature has developed on the topic of carbonate diagenetic effects on trace chemistry [see Bathurst (1975) and Veizer (1982) for reviews]. We can cover the topic only briefly here.

As might be expected, the older the fossil, the more likely it is to be altered chemically. But the nature of the matrix in which the fossil is preserved may be of equal or greater importance. Preservation is most likely in a matrix that

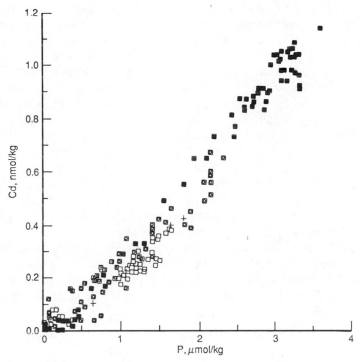

Figure 3.11 Cadmium and phosphate concentration in ocean water samples from Pacific and Arctic Oceans. Samples with the same symbol came from the same area. From Boyle, Paleooceanography, Vol. 3., p.473, 1988, copyright by American Geophysical Union.

excludes contact between fossil and diagenetic fluid. A few spectacular examples of Paleozoic fossil preservation have been described in which the fossils were embedded in an asphaltic or fine, calcareous shaley matrix (Hallam and O'Hara, 1962; Brand, 1981a,b). Mg seems to be more subject to loss than does Sr. Most of the examples of possible Mg preservation are in calcite skeletons that were orginally low in Mg (e.g., Lowenstam, 1961). In these cases preservation is difficult to demonstrate conclusively. Examples of apparently preserved Sr concentration are more common (e.g., Lowenstam, 1961; Stanton and Dodd, 1970; Brand, 1981a; Popp et al., 1986a).

During diagenesis, the carbonate mineral reequilibrates with the diagenetic fluid. This reequilibration probably usually occurs as a fine-scale solution–precipitation process. The most important controlling factors determining the trace chemistry of the resulting calcite are the ionic ratios in the diagenetic fluid as compared with those ratios in the carbonate mineral and the degree of openness of the system. If the diagenetic fluid is derived from fresh water, it is likely to have lower Mg/Ca and Sr/Ca ratios than the water from which the skeleton formed. Thus, diagenesis in such waters tends to lower the Mg/Ca and Sr/Ca ratios in the fossil. In a completely open system, the ions originally in the fossil will have no effect on its final composition, as the water composition will completely dominate the system. In a completely closed system the chemistry of the fossil will alter the composition of the diagenetic water to such an extent that the composition of the fossil will not change.

Although Mg/Ca and Sr/Ca ratios are lower in most diagenetic waters than are those ratios in seawater, other ionic ratios, especially Mn/Ca and Fe/Ca ratios, are usually higher. Thus, during diagenesis, Mn/Ca and Fe/Ca ratios usually increase in the fossil. The Mn/Ca ratio and to a lesser extent the Fe/Ca ratio has been used as a measure of the amount of open system alteration which has affected a fossil. A strong negative correlation between Sr/Ca and Mn/Ca ratios (Fig. 3.12) has been demonstrated by Brand and Veizer (1981). They interpret this as being due to lowering of the Sr/Ca ratio and raising of the Mn/Ca ratio during open system diagenesis. This has given rise to a useful method of estimating the degree of alteration of a fossil. The amount of cathodoluminescence in calcite is strongly dependent on its Mn concentration (Machel, 1983). Thus strongly cathodoluminescent fossils have a relatively high Mn concentration and have probably been altered extensively. They should thus be avoided for original trace chemical or stable isotopic study (Popp et al., 1986a).

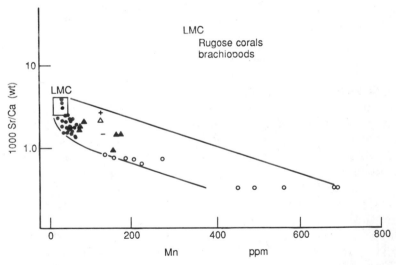

Figure 3.12 Covariation of Sr and Mn concentration in fossil rugose corals and brachiopods. Different symbols represent specimens from different locations. The LMC field delineates the theoretical possible range of low-magnesium calcite in inorganic and organic equilibrium with present-day seawater. Fossils with the lowest Mn concentrations are considered to be the least altered. From Veizer (1982), used by permission of SEPM.

Geologic Application of Trace Chemical Relationships

Data on trace chemistry of skeletal carbonates have been used in studying a number of different types of geologic problems.

1.   Interpretation of the trace chemistry of carbonate sediments depends on a knowledge of the trace chemistry of the skeletal carbonates that usually comprise the principal portion of the sediments. The major features of the trace element composition of sediments can be explained in terms of the relative proportions and taxonomic position of the major biotic contributors to the sediment [see Milliman (1974) for examples].

2.   Studies of the diagenesis of carbonate sediments and rocks are also greatly aided by knowledge of the trace chemistry of the skeletal materials that usually are their major constituent. The approximate starting chemistry of an altered carbonate rock can be predicted from the biotic composition of the rock. Knowing the original composition allows one to propose models to explain the observed chemical composition. Such an approach has been used in many studies of carbonate diagenesis [see Bathurst (1975) for examples].

3.   Skeletal minerals form an important part of the geochemical cycle of several elements (Lowenstam, 1974); hence an understanding of the chemistry of skeletal minerals is an aid in understanding the geochemical cycles of these elements.

4.   A better understanding of the chemical composition of the oceans in the past has long been a goal of geologists. The chemical composition of skeletal minerals that formed in those oceans is one of our best methods of determining ancient oceanic chemistry.

5.   Of particular interest to paleoecologists is use of skeletal chemical data to make environmental interpretations. Trace-element concentrations might be used to distinguish between freshwater and marine fossils or perhaps to determine salinity. In some cases trace-element concentrations in skeletons are correlated with temperature, potentially allowing trace chemistry of fossils to be used for paleotemperature determinations.

A number of studies have utilized trace chemistry of fossils in an effort to determine paleoenvironmental conditions (e.g., Lowenstam, 1964b; Dodd, 1966; Stanton and Dodd, 1970; Brand, 1981a, 1986, 1987a,b; DeDeckker et al., 1988). Because of the physiologic and diagenetic effects, such studies are most likely to yield reliable results when performed on geologically young fossils.

ISOTOPIC TECHNIQUES

The oxygen isotopic determination of paleotemperatures is the best known and most widely applied of the various geochemically based techniques of paleoenvironmental interpretation. This is probably because the theory behind its use is largely explainable in physical chemical terms and the effects of physiologic factors are relatively minor.

The idea of using the oxygen isotopic composition of fossils to determine paleoenvironmental conditions was first developed by H. C. Urey (1947). The original application that he had in mind was to differentiate between marine and freshwater fossils, but from a consideration of the theory of isotopic behavior the possibility of determining paleotemperatures also soon became apparent. Although theory indicated that the oxygen isotopic composition of fossils should vary with the growth temperature of the skeleton, the then-existing mass spectrometers were not capable of measuring the small isotopic differences that would result. A mass spectrometer capable of measuring differences as small as 1 part in 10,000 was eventually developed by Urey and his colleagues at the University of Chicago and the first paleotemperature determinations were published (Urey et al., 1951).

Very precise determinations of carbon isotopic composition became possible as a by-product of the oxygen isotopic project. Although the factors controlling carbon isotopic composition of fossils are more complex and less predictable than those controlling the oxygen isotopic composition, use can be made of the carbon isotopic results in studying various geologic problems such as differentiating between marine and nonmarine conditions, variation in productivity, bottom-water circulation, variation in the carbon cycle, and stratigraphic correlation.

Oxygen isotopic composition of fossils has been studied more extensively than have any of the other geochemical properties because physiologic effects on isotopic composition are small relative to other biogeochemical techniques. Thus interpretation is largely a matter of considering the interplay of the physical chemistry of the system as mediated by the environmental effects. The problem then becomes a matter of interpreting which environmental characteristics produced the physical chemical effect. Although diagenesis often causes problems in interpreting isotopic results, diagenetic effects can often be evaluated more readily than in the case of other chemical properties.

Physical Chemistry

Approximately 99.76% of atmospheric oxygen consists of the isotope ^{16}O: 0.04% is ^{17}O, and 0.20% is ^{18}O (Nier, 1950). Two stable isotopes of carbon, ^{12}C and ^{13}C, also occur in natural materials in the approximate proportions of 98.9 to 1.1. The radioactive isotope ^{14}C also occurs in nature in very small amounts but is not considered in this chapter.

Because isotopes of an element differ only in number of neutrons in the atomic nucleus, they behave very similarly, but not identically, in chemical reactions. The slight difference in weight also causes them to behave differently in physical reactions. Both chemical and physical differences result in differences in the proportion of oxygen isotopes in skeletal material. The greater the difference in atomic weight, the greater is the fractionation by physical and chemical processes. Thus in the case of oxygen, differences in relative abundance between ^{18}O and ^{16}O will be greater than those between ^{17}O and ^{16}O or ^{17}O and ^{18}O. In the subsequent discussion we consider only differences in the relative abundance of ^{16}O and ^{18}O, but the principles involved also apply to differences in the relative concentration of ^{17}O.

Both oxygen and carbon atoms are present in carbonate skeletons in the CO_3^{2-} ion. During the skeleton formation process these ions were presumably in chemical equilibrium with the oxygen and carbon in the water in which the skeleton formed through a series of chemical reactions:

$$CO_2 + H_2O \rightleftharpoons H_2CO_3 \rightleftharpoons H^+ + HCO_3^- \rightleftharpoons 2H^+ + CO_3^{2-} \qquad (3.2)$$

Due to their slight chemical differences, the different isotopes of oxygen and carbon do not behave identically in this series of reactions, and thus to minimize the free energy of the system, the isotopic ratios in the various components of the reactions differ. As we are only interested in the end members of this series of reactions, we can simplify the equation to show isotopic exchange between water and carbonate ion:

$$H_2^{18}O + \tfrac{1}{3}C^{16}O_3^{2-} \rightleftharpoons H_2^{16}O + \tfrac{1}{3}C^{18}O_3^{2-} \qquad (3.3)$$

The equilibrium constant K for this reaction can then be written as

$$K = \frac{[H_2^{16}O]\,[C^{18}O_3^{2-}]^{\frac{1}{3}}}{[H_2^{18}O]\,[C^{16}O_3^{2-}]^{\frac{1}{3}}} \qquad (3.4)$$

with the bracketed quantities indicating molecular concentrations. For simplicity, and by convention, the *fractionation factor* α is usually used in discussion of isotopic separation or fractionation. The fractionation factor for reaction (3.3) is

$$\alpha = \frac{(^{18}O/^{16}O)_{CO_3^{2-}}}{(^{18}O/^{16}O)_{H_2O}} \qquad (3.5)$$

or

$$(^{18}O/^{16}O)_{CO_3^{2-}} = \alpha(^{18}O/^{16}O)_{H_2O} \qquad (3.6)$$

with $(^{18}O/^{16}O)_{CO_3^{2-}}$ meaning the ratio of ^{18}O to ^{16}O in the carbonate ions and $(^{18}O/^{16}O)_{H_2O}$ meaning the ratio of those isotopes in the water. If the oxygen isotopes behave identically chemically, both α and K would equal 1. In fact, at 25°C, $\alpha = 1.021$ for this reaction (McCrea, 1950), indicating that ^{18}O concentrates slightly in carbonate ions relative to water. As discussed below, the temperature dependence of the fractionation factor makes possible determination of paleotemperatures. Note the similarity between the fractionation factor in isotopic studies and distribution coefficients in trace chemistry studies. The fractionation factor for oxygen isotopes varies with temperature just as does the distribution coefficient between a trace element and Ca in calcite and aragonite. These same types of relationships also apply to carbon isotopes. The carbon isotopes will fractionate between CO_2 and CO_3^{2+} in much the same way as do the oxygen isotopes.

This relationship in (3.5) shows that at a given temperature the $^{18}O/^{16}O$ ratio of the carbonate should vary directly with the $^{18}O/^{16}O$ ratio of the water with which the carbonate has equilibrated. In other words, in equation (3.6), if α has a fixed value, the $(^{18}O/^{16}O)_{CO_3^{2-}}$ value must vary directly with $(^{18}O/^{16}O)_{H_2O}$. Thus any factor causing $(^{18}O/^{16}O)_{H_2O}$ to change will be reflected in the $(^{18}O/^{16}O)_{CO_3^{2-}}$ value. Epstein and Mayeda (1953) and many subsequent workers have shown that the $(^{18}O/^{16}O)_{H_2O}$ does indeed vary and in a predictable way that can be correlated with environmental factors. The basic cause of the variation is that in the physical process of evaporation, water molecules fractionate on the basis of their oxygen isotopic content. This fractionation can be represented schematically as

$$(H_2^{16}O)_L + (H_2^{18}O)_V \rightleftharpoons (H_2^{18}O)_L + (H_2^{16}O)_V \tag{3.7}$$

in which $(H_2^{16}O)_L$ is the abundance of this molecular species in the liquid phase, $(H_2^{18}O)_V$ is the abundance of that molecular species in the vapor phase, and so on. The fractionation of the isotopes can then be represented as

$$\alpha = \frac{(^{18}O/^{16}O)_L}{(^{18}O/^{16}O)_V} \tag{3.8}$$

At 25°C, the α value for this physical reaction is 1.008. That is, the ratio of $H_2^{18}O$ to $H_2^{16}O$ in the liquid is 0.8% higher than the $H_2^{18}O/H_2^{16}O$ ratio in the vapor phase. This equilibrium relationship for the two types of water molecules holds both for the evaporation process and the precipitation process. This physical fractionation is due to the difference in vapor pressure of the $H_2^{16}O$ molecule, which has a slightly higher vapor pressure and hence evaporates into the vapor phase more readily than does the heavier $H_2^{18}O$ molecule. When water evaporates, the vapor phase will have an $H_2^{18}O/H_2^{16}O$ ratio which is 0.8% lower than that in the liquid water. Similarly, when liquid water condenses from vapor as rain, the liquid will have an $H_2^{18}O/H_2^{16}O$ ratio that is 0.8% higher than the ratio in the vapor from which it formed. This is analogous to a mixture of water and ethyl alcohol. The alcohol has a lower vapor pressure than that of water and hence evaporates more readily. The vapor above a cocktail has a slightly higher alcohol-to-water ratio than would the cocktail itself.

Measurement of Isotopic Ratios

Both oxygen and carbon isotopic ratios are measured by mass spectrometry using an instrument which is basically the design of A. O. Nier. This system requires that the sample be in the form of a gas. Carbon dioxide is used for the gas because of the ease with which it can be prepared and handled. The CO_2 is prepared from the sample by reaction with orthophosphoric acid. The method of preparation and purification is critical because chemical fractionation occurs during the acid reaction.

$$6H^+ + 2PO_4^{3-} + 3CaCO_3 \rightleftharpoons 3CO_2 + 3H_2O + 3Ca^{2+} + 2PO_4^{3-} \quad (3.9)$$

As can be seen from this reaction, only two-thirds of the oxygen in the sample goes into the CO_2. The rest reacts with the hydrogen ions to form water. Fractionation of the oxygen isotopes between the water and CO_2 occurs in the reaction, but the CO_2 can be used as a reliable sample of the original carbonate if the reaction is always conducted under the same conditions (fractionation adds a constant error to all samples and standards). Detailed discussions of sample preparation techniques are given by Bowen (1966). More recently, Shackleton (1973) has introduced a modified, somewhat simpler method of sample preparation.

The CO_2 is introduced into the mass spectrometer through a small gas leak (Fig. 3.13). It enters the source, where it is ionized by electron bombardment to CO_2^+ ions. The ions are accelerated in an electrostatic field and collimated and emerge as an ion beam that passes through a magnetic field. The ions are deflected by an amount that is dependent on their mass, the lighter ions being deflected more than the heavier. Collectors for each mass to be measured are

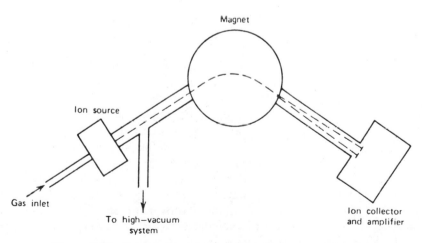

Figure 3.13 Schematic representation of mass spectrometer.

placed at the appropriate location beside the magnetic field. The ions discharge on the collectors, and the amount of charge, which can be measured by the electronic circuitry, will be proportional to the number of ions of each mass. Precision of the isotopic measurements is improved by comparison of the sample with a standard by alternately analyzing sample and standard. This comparison is facilitated by a special magnetically operated valve allowing rapid switching between sample and standard. The usual maximum precision claimed for analyses using this techinque is $\pm 0.1\%_0$ or less (standard deviation).

The ionic masses of special interest in oxygen and carbon isotopic studies are masses 44, 45, and 46. Mass 44 is the most common and consists entirely of $^{12}C^{16}O^{16}O$. Mass 45 consists largely of $^{13}C^{16}O^{16}O$, and mass 46 is largely $^{12}C^{16}O^{18}O$. The mass 46/mass 44 ratio will be a close approximation of the $^{18}O/^{16}O$ ratio, and the mass 45/mass 44 ratio will approximate the $^{13}C/^{12}C$ ratio. These mass ratios are obviously not perfect measures of the isotopic ratios because all of the isotope of interest is not in the ions of the mass measured. For example, a small portion of the ^{18}O will be in the following molecular species: $^{12}C^{18}O^{18}O$, $^{13}C^{16}O^{18}O$, and $^{13}C^{18}O^{18}O$. None of these ions will be measured, but they will contain a very small portion of the total ^{18}O. Similarly, the ^{16}O is distributed among several molecular species, all much less abundant than $^{12}C^{16}O^{16}O$. Similar distribution problems exist in connection with the $^{13}C/^{12}C$ determination. Finally, the presence of small amounts of ^{17}O further complicates the problem. Some of the ions of mass 45 will come from $^{12}C^{16}O^{17}O$, not from $^{13}C^{16}O^{16}O$. The error produced by these distribution problems is small, and Craig (1957) has derived correction formulas for them.

For most purposes the difference in the $^{18}O/^{16}O$ or $^{13}C/^{12}C$ ratio between samples is of more interest than is the absolute value of the ratio. Thus the values reported in the literature are usually δ (delta or del) values or parts per thousand or per mil ($\%_0$) deviation from a standard. The $\delta^{18}O$ value is defined mathematically as

$$\delta^{18}O(\%_0) = \frac{(^{18}O/^{16}O)_{sample} - (^{18}O/^{16}O)_{standard}}{(^{18}O/^{16}O)_{standard}} \times 1000 \qquad (3.10)$$

The $\delta^{13}C$ value is similarly defined. The δ value must be referred to a certain standard. Many different standards have been used, but for paleotemperature work, δ values are usually referred to the PDB standard. This standard gets its name from the fact that it was prepared from specimens of the belemnite *Belemnitella americana* collected from the Peedee Formation of South Carolina. Actually, the supply of the PDB standard has long been exhausted and samples are now calibrated against secondary or tertiary standards whose relationship to PDB has been determined. Commonly in the more recent literature, samples (especially water samples) are referred to the SMOW standard (Craig, 1961). SMOW is the acronym for Standard Mean Ocean Water. As the name implies, this standard is average open marine seawater. The $\delta^{13}C$ values are almost always referred to PDB.

Physiologic (Genetic) Factors

Fortunately for the use of oxygen isotopic analysis for paleoenvironmental interpretation, physiologic effects on isotopic composition are predictable and usually can readily be detected, at least in modern skeletal material. The oxygen isotopic composition of inorganic carbonates can be determined theoretically (Urey, 1947) or experimentally (McCrea, 1950). Any departure of the oxygen isotopic composition of a modern skeleton from inorganic carbonate formed under the same environmental conditions should be due to physiologic factors. These have been called *vital effects* (Urey et al., 1951) and *nonequilibrium precipitation* in the literature. At one time we had very little knowledge of factors involved in the vital effect. In recent years increased work on this subject has improved our understanding of this once "mysterious" parameter (e.g., Grossman, 1987; McConnaughey, 1989a,b).

Fortunately, several important groups of organisms in the fossil record, such as the mollusks, brachiopods, and foraminifera (especially planktic genera), seem to have little physiologic control. To avoid the problem of vital effects, paleotemperature studies have largely been confined to these three groups. Increased study has shown that physiologic effects do sometimes exist even in these groups (e.g., Tourtelot and Rye, 1969; Duplessy et al., 1970; Grossman, 1984; Romanek et al., 1987).

Large physiological effects are noted in some groups, such as the calcareous algae, corals, and echinoderms. In most cases the $^{18}O/^{16}O$ ratio is lower than the equilibrium value in groups showing a vital effect. Two processes appear to cause disequilibrium precipitation in biogenic carbonates: a *kinetic effect* and a *metabolic effect* (McConnaughey, 1989a) The kinetic effect results from lack of time for attainment of equilibrium between CO_2 and carbonate in reaction (3.2) (Erez, 1978; Turner, 1982). This effect is most pronounced in rapidly calcifying organisms and in rapidly growing parts of skeletons (Fig. 3.14; McConnaughey, 1989a). This effect increases the relative concentration of ^{16}O and ^{12}C and results in a positive correlation between $\delta^{18}O$ and $\delta^{13}C$ values. McConnaughey (1989b) has reproduced this effect experimentally. The kinetic effect has been identified in many groups of organisms, including echinoderms, corals, calcareous algae, and forams. The metabolic effect appears to influence the $^{13}C/^{12}C$ ratio only and is due to changes in dissolved inorganic carbon (DIC) caused by photosynthesis or respiration. During photosynthesis, ^{12}C is preferentially incorporated into organic compounds, leaving the DIC reservoir slightly depleted in ^{12}C. Calcium carbonate forming from this reservoir will thus be enriched in ^{13}C. During respiration, organic compounds are oxidized producing DIC, which is enriched in ^{12}C. Calcium carbonate formed from the DIC will thus be enriched in ^{12}C. The photosynthetic aspect of the metabolic effect is especially pronounced in hermatypic corals, in which it has been recognized in many studies (e.g., Weber and Woodhead, 1970; Buchardt and Hansen, 1977; Swart, 1983; McConnaughey, 1989a). The kinetic and metabolic effects are difficult to separate (Grossman,

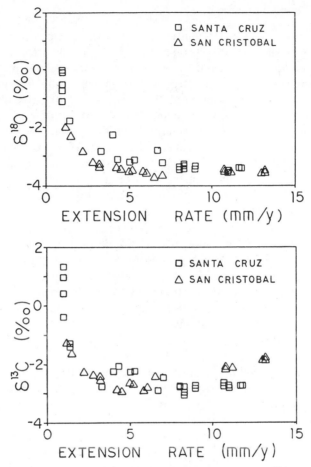

Figure 3.14   Variation with growth (extension) rate in corals of $\delta^{18}O$ and $\delta^{13}C$ showing kinetic vital effect. Values for slowest growth rates are closest to equilibrium. From McConnaughey (1989a), reprinted with permission from Geochimica et Cosmochimica Acta, *53:* 156, 1989, Pergamon Press, Inc.

1984, 1987) which has led to slow development of our understanding of the physiologic effect on isotopic compositon.

Because the amount of dissolved CO_2 in water is so much less than the oxygen in water, carbon isotopic composition is usually more variable than oxygen isotopic composition. Microenvironmental effects, which may appear to be vital effects, are likely to be more important in determining carbon isotopic composition. For example, some benthic forams live slightly below the substrate surface (Corliss, 1985). Because of oxidation of organic material in the sediment, the $^{13}C/^{12}C$ ratio in the pore water is lower than that in the ambient water above the surface. This may appear as a vital effect, whereas it is actually a microenvironmental effect. Krantz et al. (1987) suggest a similar effect in burrowing bivalves.

Environmental Factors

Temperature

The extent of fractionation of different isotopes of an element in a chemical reaction is temperature dependent. The lower the temperature, the more differently isotopes of an element behave. This can be seen from the thermodynamic relationship for the equilibrium constant of the reaction

$$\ln K = -F°/RT \tag{3.11}$$

in which K is the equilibrium constant, $F°$ the standard Gibbs free-energy change for the reaction, R the gas constant, and T the absolute temperature. (See any textbook in physical chemistry for a more detailed explanation of this relationship.) The smaller the value of T, the greater will be the value of K and thus the greater the fractionation. K goes to infinity as T approaches absolute zero. Theoretically, the isotopic species should separate completely at absolute zero. As T increases, ln K approaches zero and thus K will approach unity, meaning no fractionation. As indicated above, the fractionation factor α for the reaction involving H_2O and CO_3^{2-} at 25°C is 1.021, whereas α at 0°C is 1.025 (McCrea, 1950). The fractionation is such that the ^{18}O tends to concentrate in the carbonate ion relative to the water in which it formed. Figure 3.15 shows the pattern of variation with temperature of the $^{18}O/^{16}O$ ratio in carbonate if the $^{18}O/^{16}O$ ratio of the water is constant. The relationship

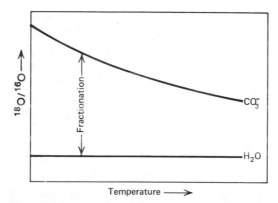

Figure 3.15 Schematic representation of the difference in the $^{18}O/^{16}O$ ratio in water and carbonate precipitated in equilibrium with that water at different temperatures. The separation of the lines is a function of temperature-sensitive fractionation.

of temperature and fractionation indicates that one should be able to determine the paleotemperature by determining the amount of fractionation (e.g., the difference in $^{18}O/^{16}O$ ratio between a fossil and the water in which it formed). But how can we determine the $^{18}O/^{16}O$ ratio in the water of an ancient sea? Obviously, we cannot, so we must assume some value for this ratio in the water. This is clearly a serious weakness of the method, but various relationships allow us to make reasonable estimates under many circumstances. The most important factor allowing us to make reasonable estimates of the isotopic composition of water is the large volume of water in the oceans. The open, well-mixed part of the modern ocean varies little in its oxygen isotopic composition; localized, small-scale processes have little effect on the oceanic ratio. Thus, if paleotemperature determinations are restricted to specimens that have grown in the open ocean, the assumption is usually made that the composition of the water was the same as that of the modern open ocean. A correction should be made, however, for the water held on the continents as glacial ice, as discussed below.

Epstein et al. (1951) empirically determined the relationship between the $^{18}O/^{16}O$ ratio in mollusk shells and growth temperature by analyzing a series of samples that had either been grown in temperature-controlled aquaria or collected at natural sites where the temperature had been carefully monitored. The $^{18}O/^{16}O$ ratio of the water in which the shells grew was also determined to take into account variation in this ratio. Over the temperature range of about 5 to 30°C, the relationship between the $^{18}O/^{16}O$ ratio and temperature is practically linear (Fig. 3.16). The best-fit equation for this relationship is

$$T(°C) = 16.9 - 4.2(\delta^{18}O_s - \delta^{18}O_w) + 0.13(\delta^{18}O_s - \delta^{18}O_w)^2 \quad (3.12)$$

with $\delta^{18}O_s$ being the deviation of the $^{18}O/^{16}O$ ratio in the sample from the PDB standard in parts per thousand and $\delta^{18}O_w$ is a measure of the oxygen isotopic composition of the water (actually the $\delta^{18}O$ value relative to PDB determined from CO_2 equilibrated at 25°C with the water). This equation has been verified for inorganically precipitated calcite by O'Neil et al. (1969). This form of the calcite temperature equation is in most common use today although other modifications have been suggested (e.g., Shackleton, 1973).

Early work on the temperature effect on oxygen isotopic composition of water did not take into account any difference in fractionation between water and calcite or aragonite. The equation of Epstein et al. (1951) is based on a mixture of calcite and aragonite samples. The equation is actually very close to that later determined by O'Neil et al. (1969) for inorganically precipitated calcite. Tarutani et al. (1969) showed that aragonite and calcite have different fractionation factors relative to water. They showed theoretically and experimentally that at 25°C, inorganic aragonite is enriched in ^{18}O relative to calcite by 0.6‰. Sommer and Rye (1978), Grossman and Ku (1981), and others have confirmed this relationship for biogenic aragonite. Grossman and Ku (1986)

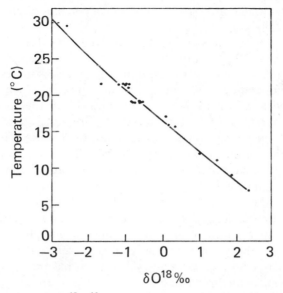

Figure 3.16 Variation in the $^{18}O/^{16}O$ ratio of skeletal carbonate (‰ deviation from the PDB standard) with growth temperature. From Epstein et al. (1953), Geological Society of America Bulletin.

have quantified the temperature–fractionation relationship for biogenically precipitated aragonite, proposing the following equation:

$$T(°C) = 20.6 - 4.34 (\delta^{18}O_s - \delta^{18}O_w) \qquad (3.13)$$

A similar temperature effect has been demonstrated for fractionation of oxygen isotopes between water and the oxygen in phosphate and silica. Longinelli and Nuti (1973) quantified the relationship for phosphate, presenting the following equation:

$$T(°C) = 111.4 - 4.3(\delta^{18}O_p - \delta^{18}O_w) \qquad (3.14)$$

This relationship has been used by Kolodny and Raab (1988) and others to determine paleotemperatures from fish teeth and bones. Luz et al. (1984) have determined paleotemperatures from conodonts using this relationship. A possible advantage of the phosphate paleotemperature scale is that skeletal phosphates are less subject to diagenesis than carbonate minerals.

Research also has been done on establishing a paleotemperature scale based on siliceous skeletons (Leclerc and Labeyrie, 1987):

$$T(°C) = 17.2 - 2.4(\delta^{18}O_s - \delta^{18}O_w - 40) - 0.2(\delta^{18}O_s - \delta^{18}O_w - 40)^2 \qquad (3.15)$$

This scale has been used to determine paleotemperatures from diatoms (Leclerc and Labeyrie, 1987).

Urey and his co-workers (1951) suggested a method for avoiding the necessity of assuming a $^{18}O/^{16}O$ ratio for the water. They suggested determining the temperature relationship for fractionation of oxygen isotopes between water and some compound other than a carbonate mineral (such as a phosphate mineral or silica) which formed in equilibrium with the same water as the carbonate mineral. In effect, this would give us two equations each with the same two unknowns, the temperature and the $^{18}O/^{16}O$ ratio of the water. Simultaneous solution of the equations should thus allow us to determine both unknowns provided that the slopes of the equations are different. This relationship is shown graphically in Fig. 3.17. The difference in the $^{18}O/^{16}O$ ratio between the carbonate and phosphate minerals that have formed in the same water is temperature dependent. Unfortunately, the slope of the temperature–fractionation curve for the phosphate–water system is quite similar to that for the carbonate–water system. Especially with experimental errors in making the $^{18}O/^{16}O$ measurements, only very low precision paleotemperature measurements can be made by this technique, based on differences in the $^{18}O/^{16}O$ ratio

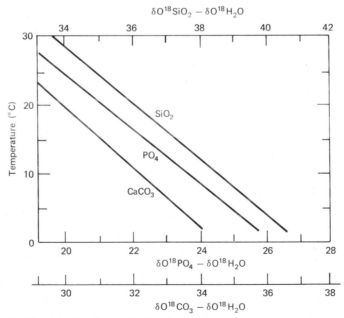

Figure 3.17 Empirically determined relationship between water temperature and $^{18}O/^{16}O$ ratios in the shells of carbonate, phosphatic, and siliceous organisms. From Hecht (1976), with permission from Foraminifera, *2:* 7; copyright Academic Press Inc. (London) Ltd.

between phosphate or silica and carbonate. In Fig. 3.17 this is shown by the fact that the slopes of the phosphate, silicate, and carbonate curves are nearly the same. Because of differences in the fractionation–temperature relationship between calcite and aragonite, temperatures potentially could be determined from calcite and aragonite fossils which formed at the same time. Unfortunately, the slopes of these curves are also very similar. This method of determining the temperature of mineral formation by analyzing fractionation of oxygen between different coexisting mineral phases has become common practice in the study of igneous and metamorphic minerals (Faure, 1977). Karhu and Epstein (1986) have recently used differential fractionation of oxygen in coexisting cherts and phosphate minerals to study ancient surface temperatures and oxygen isotopic composition of the oceans through time.

Carbon isotopes are subject to the same physical chemical principles as oxygen and other isotopes and fractionate in chemical reactions such as those shown in equation (3.2). Thus, in theory, paleotemperatures could be determined from the $^{13}C/^{12}C$ ratio in carbonate fossils. Two factors make such a paleotemperature method less practical than one based on oxygen isotopes: (1) the amount of fractionation between the isotopes of an element is dependent on the relative mass difference between the isotopes, and the mass difference between ^{13}C and ^{12}C is obviously less than that between ^{18}O and ^{16}O; (2) the $^{13}C/^{12}C$ ratio in the dissolved bicarbonate is more variable than is the $^{18}O/^{16}O$ ratio and is thus more difficult to estimate reliably to determine the amount of fractionation in the precipitation of $CaCO_3$. The lower variability of the $^{18}O/^{16}O$ ratio is due to the enormous mass of oxygen in a given volume of water compared to carbon in the dissolved bicarbonate in that water. The mass of dissolved carbon is orders of magnitude less and thus the carbon reservoir is much more likely to be affected by local chemical processes than is the oxygen. Nevertheless, temperature variation in fractionation of carbon can be detected in nature (Emrich et al., 1970). The $^{13}C/^{12}C$ ratio in a carbonate skeleton is more likely to reflect differences in CO_2 or HCO_3^- with which the CO_3^{2-} is equilibrated. Many factors affect the carbon isotopic composition of the CO_2 and HCO_3^-, making precise interpretation of the carbon isotopic composition difficult. $CaCO_3$ precipitation rate may also be important in determining carbon isotopic composition (Turner, 1982). Grossman and Ku (1986) have shown a temperature effect on the difference in carbon isotopic composition between aragonitic and calcitic foraminifera which potentially may be useful in paleotemperature determination.

Salinity

The oxygen isotopic composition of the hydrosphere varies considerably because of the different vapor pressure of $H_2^{16}O$ and $H_2^{18}O$. This variation is due largely to fractionation that occurs during evaporation and as water vapor condenses back to the liquid phase, as discussed in the section on physical chemical factors. Imagine a mass of water vapor that has originated over the ocean. That vapor should initially have an $^{18}O/^{16}O$ ratio which is 8‰ lower than the ratio in the seawater from which it formed. When the vapor first starts to condense, the rain should be preferentially enriched in $H_2^{18}O$ because its lower vapor pressure favors its condensation into the liquid phase. Indeed, the rain should have an $^{18}O/^{16}O$ ratio which is 8‰ higher than that in the vapor and thus be the same as the ratio in the seawater from which the vapor formed. The removal of $H_2^{18}O$-enriched water as rain from the vapor will result in the remaining vapor being somewhat more depleted in $H_2^{18}O$. The next rain to fall will still have an $^{18}O/^{16}O$ ratio that is 8‰ higher than the vapor, but it will have a lower ratio than the first rain because it formed from vapor that had been somewhat depleted in $H_2^{18}O$. As this process continues the vapor and thus the rain falling from it becomes increasingly depleted in $H_2^{18}O$. The last rain (or snow) to come from our imaginary vapor mass should be very light indeed.

This process of progressive lowering of the $^{18}O/^{16}O$ ratio during condensation with immediate removal of the condensate can be described theoretically by the Raleigh equation (Epstein, 1959), which is shown graphically in Fig. 3.18. The x-axis shows the percentage of the original vapor mass that has condensed to rain or other precipitation, and the y-axis shows the $^{18}O/^{16}O$ ratio (expressed as $\delta^{18}O$). The solid line shows variation in the composition of the liquid water as the precipitation process proceeds, and the dashed line shows the composition of the vapor phase. The two lines are always vertically separated by 8‰, the fractionation in the evaporation–condensation process. In nature, the process is probably not this simple. Reequilibration of rain as it falls through the atmosphere complicates the process (Gat, 1980).

This precipitation model has many implications for the pattern of $^{18}O/^{16}O$ ratios in precipitation and freshwater bodies resulting from it (Dansgaard, 1964). The condensation of water vapor is, of course, largely dependent on cooling, the atmosphere being able to retain less water vapor as it cools. Secondary factors are the time required for condensation to occur and the presence of nucleation sites. Rain condensing from oceanic vapor should be heavier (have a higher $^{18}O/^{16}O$ ratio) near the shore and become lighter inland as the vapor progressively condenses. As the vapor rises and cools in going over mountains, the rain should become lighter. As the vapor gradually cools in going from low to higher latitudes, the condensation process gradually goes further to completion and the rain (or snow) becomes progressively lighter. The lowest $^{18}O/^{16}O$ ratios described are those from polar snow (Epstein, 1959). Thus to a first approximation the $^{18}O/^{16}O$ ratio of fresh water varies directly with the temperature at which condensation occurs. This relationship has been used in studies of glacial snow and ice. Winter snow can be readily differentiated from summer snow on the basis of its $^{18}O/^{16}O$ ratio. The approximate minimum and maximum temperature of the year can be determined from the $^{18}O/^{16}O$ ratio in the ice, allowing determination of the relative temperature history on a glacier for the last several thousand years (Dansgaard et al., 1969).

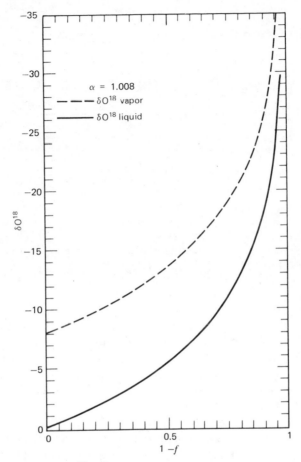

Figure 3.18 Variation in the $^{18}O/^{16}O$ ratio of water and vapor with degree of condensation of the water-vapor system. The condensed phase is continuously removed from the system. f is the fraction remaining in vapor phase. See text for detailed explanation. From Epstein (1959), Researches in Geochemistry, p.226.

The very low $^{18}O/^{16}O$ ratio of glacier ice leads to some complications in paleotemperature determinations. As shown schematically in Fig. 3.19, water is withdrawn from the oceans to produce the glaciers. The glaciers in effect store an excess of $H_2^{16}O$ on the continents, causing a rise in the $^{18}O/^{16}O$ ratio in the oceans. Craig (1965) estimates that if all glacial ice were to melt, the $^{18}O/^{16}O$ ratio of the oceans would be lowered by 0.5‰. If the volume of glacial ice were to increase, the $^{18}O/^{16}O$ ratio in the oceans would rise due to preferential removal of $H_2^{16}O$. Craig (1965) estimates the maximum difference in the oceans in $^{18}O/^{16}O$ ratio between glacial and interglacial ages to be about 1.5‰. This is equivalent to the fractionation difference produced by a 6°C change in

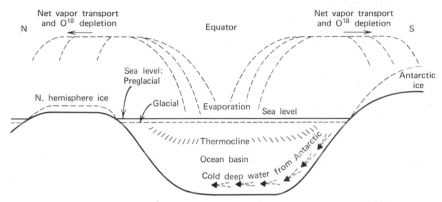

Figure 3.19 Schematic diagram illustrating global changes in the $^{18}O/^{16}O$ ratio of seawater between glacial and nonglacial times. During glacial times ice, which is depleted in ^{18}O, is retained on land. This causes the $^{18}O/^{16}O$ ratio of the water remaining in the oceans to rise. From Hudson (1977), Scottish Journal of Geology, Scottish Academic Press Limited.

temperature. Clearly, a correction is necessary before a comparison can be made between paleotemperatures for different times in the Pleistocene and for any time when glaciers of significant size have existed. Variation in the $^{18}O/^{16}O$ ratios in fossil foraminifera from deep-sea cores is more a reflection of variation in isotopic composition of the oceans than of temperature variation (Shackleton, 1967).

Evaporation of seawater continually preferentially removes $H_2{}^{16}O$. The $^{18}O/^{16}O$ ratio might thus be expected to increase constantly in the seas. But of course isotopically lighter water is continually being returned to the seas by rivers so that a steady state has been reached. Circulation is vigorous enough in the open ocean that the $^{18}O/^{16}O$ ratio varies relatively little. In restricted, nearshore areas the composition may vary markedly. The $^{18}O/^{16}O$ ratio may be high in restricted lagoons in which evaporation exceeds the inflow of fresh water and will be low in brackish water bays and estuaries. The $^{18}O/^{16}O$ ratio in seawater thus correlates crudely with salinity; in an individual estuary the ratio may correlate rather precisely with salinity (Mook, 1971). If seawater of a constant oxygen isotopic composition mixes with fresh water which also has a constant oxygen isotopic composition, the $^{18}O/^{16}O$ ratio in the resultant mixture should vary linearly with salinity (Fig. 3.20).

As discussed above, the $^{18}O/^{16}O$ ratio in a carbonate skeleton forming at a given temperature has a constant relationship to the ratio in the water from which it formed. Thus if the oxygen isotopic composition of the water varies, this should be reflected directly in the composition of the skeleton (Fig. 3.21). If the temperature is constant across the salinity gradient, paleosalinities should be determinable from oxygen isotopic analysis of fossils. The use of

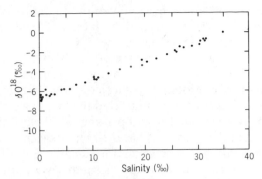

Figure 3.20 Variation of $\delta^{18}O$ values with salinity for water samples from the western Scheldt estuary, the Netherlands (data from Mook, 1971). From Dodd and Stanton (1975), Geological Society of America Bulletin.

this technique for paleosalinity determination requires knowing the $^{18}O/^{16}O$ ratio of the seawater and fresh water that are being mixed to form the brackish intermediates. Seawater composition might be assumed to be that of normal, open marine water, but estimation of the freshwater composition will be more difficult. One solution to this problem is to determine the $^{18}O/^{16}O$ ratio in a fossil that is known to have grown in contact with the fresh water that was responsible for the dilution.

An ingenious but more complex system of determining paleosalinities which utilizes both oxygen and carbon isotopic composition has been developed by Mook (1971). He found that if he made a plot of the $\delta^{18}O$ versus $\delta^{13}C$

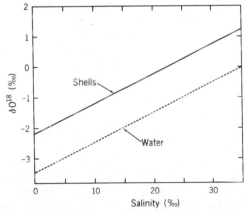

Figure 3.21 Relation between $\delta^{18}O$ and salinity for shells and water at 12°C assuming simple mixing of full marine and fresh water. From Dodd and Stanton (1975), Geological Society of America Bulletin.

values for modern shells collected at various salinities, they lay along a straight line (Fig. 3.22). This is because both the $\delta^{18}O$ and $\delta^{13}C$ values for the brackish water are linear functions of the $\delta^{18}O$ and $\delta^{13}C$ values of the sea water and fresh water mixed to yield the brackish intermediates. Shells from another bay or estuary in the same region (the coastal area of the Netherlands in the case of the work of Mook) lay along a different straight line on the $\delta^{18}O$ versus $\delta^{13}C$ plot. The second line intersected the first at a $\delta^{18}O$ and $\delta^{13}C$ value that corresponds to the value for an open marine shell. This is because the seawater involved in both mixtures has the same oxygen and carbon isotopic composition, whereas the fresh water for the two estuaries had different compositions. If one assumes that the seawater has a normal, open marine isotopic composition, the $\delta^{18}O$ value at the intersection of the lines can be

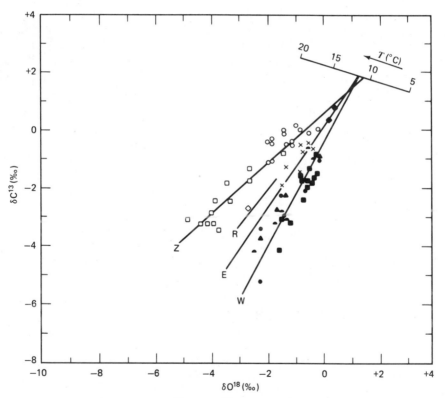

Figure 3.22 Relation between $\delta^{13}C$ and $\delta^{18}O$ of skeletal carbonate for four Netherland estuaries: western Scheldt (W), eastern Scheldt (E), mouth of the Rhine (R), Zuiderzee–Waddenzee (Z). The line with the temperature scale indicates the isotopic composition of full marine carbonates formed at those temperatures. From Mook (1971), Paleogeog., Paleoclimat., Paleoecol. *9*: 257, Copyright Elsevier Scientific Publishing Co.

used to calculate a paleotemperature for the area. This value for the seawater can be used along with the value from a freshwater shell from each of the estuaries to calculate paleosalinities using the method described above.

Carbon isotopic composition also varies with salinity (Mook, 1971; Eisma et al., 1976). The $^{13}C/^{12}C$ ratio is usually higher in marine shells than in freshwater forms; the skeletons of brackish water forms are likely to have intermediate values. This is because marine bicarbonate is derived from the decay of marine organic matter and equilibration with atmospheric CO_2, both of which are characterized by relatively high $^{13}C/^{12}C$ ratios. Bicarbonate in fresh water, on the other hand, is largely derived from the decay of terrestrial and freshwater organic material, which is characterized by low $^{13}C/^{12}C$ ratios. Also, freshwater bodies (except perhaps for large lakes, especially saline lakes) have not had sufficient time to reach equilibrium with atmospheric CO_2. The carbon isotopic composition is also altered by solution of previously deposited carbonate sediment or ancient carbonate rock with which the water body is in contact. This usually has the effect of raising the $^{13}C/^{12}C$ ratio in the bicarbonates.

Carbon isotopes fractionate markedly in various organic processes, especially during photosynthesis. These processes differ between marine and nonmarine and aquatic and terrestrial plants (Degens, 1969). A number of studies of the carbon isotopic composition of the organic compounds in sediments and sedimentary rocks have indicated that this can be a useful method for distinguishing marine from nonmarine rocks and even distance from shore (Sackett and Thompson, 1963).

Diagenesis

Unfortunately for the paleoecologist, the oxygen and carbon isotopic composition of fossils has often been altered diagenetically. In a very general sense, the older the fossil, the more likely it is that it has been altered. Unaltered fossils as old as Paleozoic are probably least common, although recent work suggests that little-altered Paleozoic fossils are more common than once thought (e.g., Popp et al., 1986a; Viezer et al., 1986). Preservation in Mesozoic fossils is more common and is quite common in Cenozoic fossils. The lithology of the matrix in which the fossil is buried is as important in determining the isotopic preservation of a fossil as it is in determining trace-element preservation. In general, the less permeable the sediment, the greater is the probability that the original isotopic composition of the fossil will be preserved. Some exceptional cases of preservation, such as in the Pennsylvanian Buckhorn Formation of Oklahoma, result from the early impregnation of the matrix with asphalt, forming an extremely effective seal against water movement through the formation (Crick and Ottensman, 1983). Alteration of the oxygen and carbon isotopic composition occurs as a result of reequilibration of the oxygen and carbon in the carbonate of the fossil with that in the water and dissolved carbonate in the diagenetic environment. The same physical

chemical rules apply during the alteration process as during the original formation of the fossil. Both the isotopic composition of the water and the temperature of alteration are likely to be very different from those that prevailed
during formation of the fossil; hence the reequilibrated composition will usually be different from the original. Because fresh water usually has a lower
$^{18}O/^{16}O$ ratio than seawater, marine fossils that are altered by contact with
fresh water will have their $^{18}O/^{16}O$ ratio lowered and the apparent paleotemperature raised.

An important aspect in the degree of alteration of the fossil is the openness
of the diagenetic alteration system. In a completely open system (or water-
dominated system), alteration occurs in contact with an infinite reservoir of
water. The isotopic composition of the $CaCO_3$ will be determined completely
by the composition of the water, not by the composition of the original
$CaCO_3$. In a closed system (or rock-dominated system) alteration occurs in
contact with a limited amount of water. Equilibration of this water with an
infinite volume of $CaCO_3$ results in water that reflects the composition of the
rock and that will not change the isotopic composition of the rock. Most
diagenesis occurs in systems between these two extremes, although near-
surface diagenesis in permeable rocks probably approaches the open system.
Deep, burial diagenesis will probably be close to the closed system.

Oxygen isotopic composition is likely to be more extensively altered than
carbon isotopic composition because the amount of oxygen in the carbonate
of a fossil is likely to be much less than that in the diagenetic pore water with
which it is in contact. The oxygen diagenetic system will thus be open or
dominated by the composition of the water. The amount of carbon in the
bicarbonate of the altering water will be much less than the oxygen, and the
carbon isotopic contribution of the fossil to the carbonate–water system will
be greater (i.e., the system will be more nearly closed). More extensive alteration will be necessary before the original carbon isotopic composition is
completely obliterated.

Alteration of the isotopic composition does not necessarily affect the physical appearance of the fossil, so if the extent of alteration is slight, it may be
difficult to detect. A number of criteria (similar to those discussed in connection with trace chemical alteration) have been used or suggested for detecting
alteration (Dodd and Stanton, 1976):

1. Is the skeleton morphologically preserved? Does it still show its original
microstructural relationships, or is it replaced by secondary calcite? This is
usually a necessary but not sufficient test; that is, a fossil that does show a
recrystallization texture will not have its original oxygen and carbon isotope
composition preserved, but a fossil that is apparently well preserved morphologically may in some cases be altered. An exception to this rule would occur
in cases of closed system alteration, which might affect texture but would
preserve the original isotopic composition.

2. Is the original mineral composition of the fossil preserved? This test is especially used for aragonitic fossils, which should retain their original isotopic composition.

3. Is the original trace chemical composition of the fossil preserved? Several recent studies (e.g., Popp et al., 1986a) have utilized cathodoluminescence as a measure of alteration. Preserved marine skeletal material should be low in Mn and Fe and thus have no cathodoluminescence. Alteration, particularly in contact with reducing fresh water, results in addition of Mn to the calcite and thus causes cathodoluminescence (Machel, 1983). This is a relatively easy test to apply in searching for fossils or parts of fossils that are preserved. Other chemical analytical techniques can of course also be used. For example, relatively high Sr or Mg concentrations characteristic of marine calcite may suggest isotopic preservation. Brand and Veizer (1981) suggest using a combination of Mn, Sr, and Mg concentration to measure preservation of isotopic composition.

4. Does the isotopic composition of the fossil contrast with that of the surrounding matrix? Extensive alteration in contact with diagenetic water should affect carbonate in the matrix to the same extent as carbonate in the fossil if the process has gone to completion. Popp et al. (1986a) and others have described examples of apparent partial alteration even within an individual fossil. This test is, of course, limited to fossils preserved in a carbonate-containing matrix.

5. Is the paleotemperature based on the $^{18}O/^{16}O$ ratio "reasonable," that is, between O and 30°C for normal marine organisms? This test is useful in detecting gross alteration but would not detect minor alteration, nor would it be adequate for fossils of organisms that lived in water that differed from modern normal marine in its isotopic composition.

6. Does the isotopic composition vary within the fossil, especially in such a way that can be related to seasonal variation in environmental conditions? This is an excellent way to test for alteration and was first used by Urey et al. (1951) to test for isotopic preservation in a belemnite from Jurassic rocks of Scotland. Variation in the oxygen isotopic composition within the fossil was interpreted as reflecting temperature variation during the life of the organism. Diagenesis should remove the original, systematic variations and either make the composition of the fossil more uniform or variable in a manner reflecting the diagenetic process and not original variation during growth.

7. Is the interpretation based on the isotopic results geologically reasonable in comparison with interpretations based on independent evidence? Many of the studies of oxygen and carbon isotopic composition of fossils have used this criterion. The work by Emiliani (1955, and other papers) and several other workers on the oxygen isotopic paleotemperatures from foraminifera in deep-sea cores makes extensive use of this test. The temperature variation pattern shown by isotopic analysis is in agreement with the pattern of varia-

tion in foraminiferal assemblages, coiling directions, and sedimentological properties.

Geologic Applications of the Isotopic Technique

Paleotemperature and Glacier Ice Volume

The earliest and best known applications of isotopic data have been in determination of paleotemperatures. Hundreds of studies have been made reporting paleotemperatures since the first such study was published by Urey et al. (1951). We review only a few of these studies here.

The first fossil group to be studied with the isotopic technique was the belemnites. This group was selected because of its heavy, chemically stable skeleton, composed of large, low-Mg calcite crystals. These features maximize the chance of preservation of the original isotopic composition of this group. [Spaeth et al. (1971) have argued that belemnites may not be adequately preserved.] Another advantage of the belemnites is that they were apparently normal marine, relatively open water forms. This minimizes problems associated with the isotopic composition of calcite forming in brackish and/or restricted environments. The first results published were from a belemnite from Jurassic rocks of the Isle of Skye (Urey et al., 1951). A sequence of samples was analyzed from a single specimen that appeared to show seasonal variation in the growth temperature, confirming the preservation of the isotopic composition in the specimen.

Later work by Lowenstam and Epstein (1954) gave results from Cretaceous belemnites from many locations in North America and Europe. These specimens showed both a stratigraphic and a geographic variation in temperature. The general stratigraphic variation pattern for upper Cretaceous time was a gradual increase in temperature from Albian through Coniacian–Santonian time followed by decreasing temperatures through the end of Cretaceous time (Fig. 3.23). In general, the Cretaceous temperatures were about the same as, or warmer than, temperatures at the same latitude today. A comparison of paleotemperatures for Maastrichtian time from many North American and European localities showed a poorly defined temperature gradient with a general decline in a northward direction. Lowenstam and Epstein (1954) also determined paleotemperatures from various other fossil groups, especially bivalves and brachiopods, as well as from bulk chalk samples. These temperatures were usually higher than belemnite temperatures and were considered possibly to be partially altered diagenetically. Another interpretation is that the mobile belemnites spent part of their lives in deeper, colder water than the other fossil groups.

Bowen (1966) published results of analyses of belemnites, especially Jurassic specimens, from many localities around the world. One of the main objectives of this study was to document the effect of continental drift on the planetary temperature gradient. Bowen noted some such effects, but in gen-

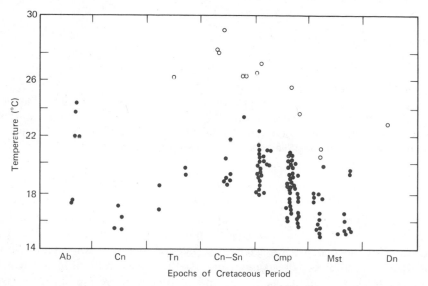

Figure 3.23 Paleotemperatures determined from the $^{18}O/^{16}O$ ratio of brachiopods (open circles) and belemnites (closed circles) from the Cretaceous of western Europe. From Lowenstam and Epstein (1954), copyright University of Chicago.

eral the data for any given geologic time are too sparse to delineate adequately the planetary temperature gradient.

Closely following the belemnite studies have been a long series of studies of foraminifera, especially of specimens from deep-sea cores. The pioneer in these studies was Cesare Emiliani. Foraminifera from deep-sea cores are nearly an ideal subject for study by the isotopic method. The planktic types as well as many benthic forms are composed of chemically stable low-Mg calcite. The foraminifera are also preserved in a very stable environment in contact with seawater rather than highly $CaCO_3$ undersaturated fresh water. The deep-sea specimens grew in the open ocean in contact with well-mixed, average seawater. When conditions are right for their preservation, specimens are likely to be abundant through a considerable section of core. Most of the earlier studies of foraminifera concentrated on Pleistocene and Holocene specimens in an effort to clarify the temperature history of the oceans during that time span. One of the big surprises of the early work in this field was the discovery by Emiliani (1955) of many temperature cycles during the Pleistocene rather than the then-traditionally accepted four or five based on continental Pleistocene stratigraphy. Emiliani continued to refine his work on the Pleistocene temperature history and developed a standard temperature curve for Pleistocene and Holocene time (Fig. 3.24; Emiliani, 1966). Although some have argued with the details of this curve and the extent to which it reflects ice volume rather than temperature, the usefulness of the oxygen isotopic technique for studying Pleistocene climatic history is now firmly established.

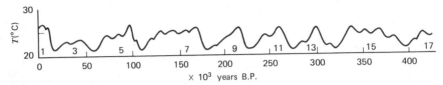

Figure 3.24   Generalized variation pattern in temperature and $^{18}O/^{16}O$ ratio of surface water in the central Caribbean during the last 425×10^3 years. The numbers below the curve refer to time divisions recognized on the basis of the oxygen isotopic temperature data. From Emiliani (1966), copyright University of Chicago.

Determination of paleotemperatures of the Pleistocene is difficult because of the uncertainty of the $^{18}O/^{16}O$ ratio of the oceans. Variation in the mass and average $^{18}O/^{16}O$ ratio of the glacier ice can have a significant effect on the $^{18}O/^{16}O$ ratio in the oceans. To calculate a paleotemperature for a Pleistocene fossil, a correction for the composition of the water must be made which requires an estimation of the volume and isotopic composition of the ice. No general agreement has been reached on what these correction factors should be; however, most present workers (e.g., Dansgaard and Tauber, 1969; Shackleton and Kennett, 1975; Prentice and Matthews, 1988) believe that most of the variation in the $^{18}O/^{16}O$ ratio in Pleistocene foraminifera is due to variation in ice volume, not temperature. The foraminifera are thus largely monitoring ice volume and not temperature. In terms of studying the geologic history of glaciation, this is equally valuable, if not more so (Broecker and van Donk, 1970).

During times when there is a significant volume of polar ice, polar temperatures can be assumed to be near O°C, at least in the modern and late Cenozoic oceans. Warmer bottom water from nonpolar sources may have prevailed before this time (Prentice and Matthews, 1988). Oceanic bottom waters derived from these polar surface waters should also have temperatures near O°C. Thus variation in oxygen isotopic composition of deep-benthic foraminifera should be due to the composition of the water, not due to temperature. Planktic foraminifera are subject to this same water effect but also an additional temperature effect. The temperature and ice volume effects act in the same direction; thus the magnitude of variation in oxygen isotopic composition of planktic foraminifera is usually greater than that for benthic types. The difference in $\delta^{18}O$ variation between benthic and planktic foraminifera is one basis for our estimation of the relative importance of temperature versus water composition in determining the isotopic composition of planktic foraminifera and variation in ice volume through time.

The relative importance of temperature versus ice volume in explaining variation in the oxygen isotopic record of pre-Middle Miocene deep-sea sediments has been a continuing source of debate (Matthews and Poore, 1980). Recent work tended to emphasize the importance of the ice volume factor, at least as early as Oligocene time (Miller et al., 1987). Major development of North American continental ice sheets seems to have begun at the end of Early Miocene time, some 15 m.y. ago (Woodruff et al., 1981). Apparent times of increased ice buildup on the continents have been correlated with times of sea-level fall as deduced from seismic stratigraphy and erosion on the continental margins (Miller et al., 1985). Cyclic variation in the oxygen isotopic composition of both benthic and planktic foraminifera is clearly developed in upper Neogene sediments. A change from relatively low to higher amplitude variation occurs at about 900,000 years B.P. Prell (1982), and others have interpreted this as due to an increase in the amount of glacial ice (and greater lowering of sea level) during the more recent glacial stages.

Another important contribution by Emiliani in his early work on the foraminifera has been in measuring the oceanic bottom water temperatures through the Cenozoic from deep-sea benthic forams. Because the bottom water is the most dense water in the oceans, a likely place of origin is in the polar regions. Thus bottom-water temperatures may give us a measure of the minimum seawater temperatures present on earth at a given time. Emiliani (1954) assumed a polar source for bottom water and suggested that polar temperatures declined from a high of about 10°C in late Cretaceous time to near zero by Pliocene time. Others (e.g., Savin et al., 1975; Shackleton and Kennett, 1975) in more recent years have refined our knowledge of this trend.

Recently, Brass et al. (1982) and Prentice and Matthews (1988), among others, have challenged the idea that bottom waters have always formed in polar regions. The other likely source is high-salinity, lower-latitude sources as seawater density increases with increasing salinity as well as decreasing temperature. Bottom water with this origin would obviously not be the coldest water in the oceans, and fossils that formed in contact with it would not be recording polar temperatures or variation in ice volume. Prentice and Matthews (1988) suggest that variation in $\delta^{18}O$ in shallow water and tropical planktic forams reflect ice volume variation. They assume that the temperature of the shallow tropics away from upwelling areas is constant through time at about 28°C. Thus variation in the composition of forams living in this water reflects variation in composition of the water, which in turn is controlled by ice volume. Using this relationship, they have estimated ice volume and sea level throughout Cenozoic time. By combining data from both planktic and deep benthic forams, they have estimated the temperature contrast between the deep and shallow ocean (Fig. 3.25).

With the advent of the Deep Sea Drilling Project (DSDP) the time scope of isotopic paleotemperature studies of the deep sea expanded greatly. Cores can be obtained from the deep sea that extend back into the Mesozoic. Oxygen isotope paleotemperatures have been obtained from many of these cores, giving the general pattern of temperature change for approximately the last 60 million years (Fig 3.26; Savin et al., 1975; Shackleton and Kennett, 1975). As was also indicated by the belemnite data, the temperature dropped in late Cretaceous time before rebounding slightly in Paleocene–Eocene times. The

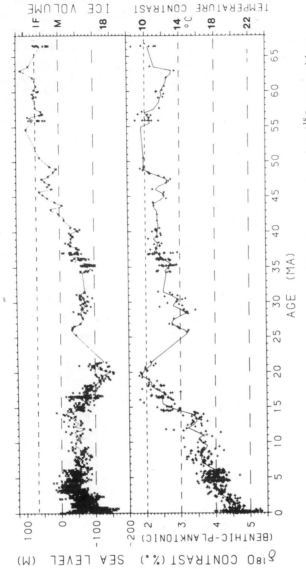

Figure 3.25 Glacioeustatic sea level for the last 65 million years based on $\delta^{18}O$ composition of planktic foraminifera (upper). IF, ice-free; M, modern glacial volume; 18, ice volume 18,000 years B.P. Temperature difference between surface and bottom for tropical oceans based on the difference in $\delta^{18}O$ between benthic and planktic foraminifera (lower). Temperature contrast scale assumes no $\delta^{18}O$ difference between deep and surface water. From Prentice and Matthews (1988), Geology *16:* 965.

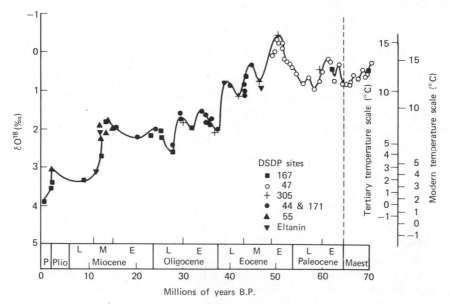

Figure 3.26 Oxygen isotopic composition of multispecies assemblages of benthic foraminifera. The Tertiary temperature scale applies to samples older than middle Miocene. This scale is based on the assumption that there were no extensive glaciers at this time. The modern temperature scale applies to the present oceans. This scale assumes that the glaciers were comparable in volume to those existing today. Temperatures for Middle and Late Miocene and Pliocene lie between these two scales. Abbreviated time intervals are P, Pleistocene; Plio, Pliocene; Maest, Maastrichtian. Illustration courtesy of S. Savin and R. G. Douglas.

temperature again dropped sharply in Oligocene time before rising again in the Miocene and then dropping rather abruptly into the Pleistocene.

Pleistocene deep-sea paleotemperatures have also been the subject of detailed study on the basis of oxygen isotopic analysis of foraminifera. Rather detailed maps of temperatures at various times in the Pleistocene have been prepared as part of the CLIMAP program (CLIMAP Project Members, 1984). These temperatures are in part determined on the basis of isotopic analysis.

A number of studies have been conducted on fossil bivalves, gastropods, and brachiopods, but these groups have not received the work that the foraminifera have. These groups are often shallow-water, inshore forms occurring in relatively restricted habitats. The estimated $^{18}O/^{16}O$ ratio for the water is thus often less certain. On the other hand, many species of these groups do have diagenetically stable skeletons of low-Mg calcite and have proved useful for oxygen isotopic studies, particularly in the Cenozoic. Many studies have interpreted Pleistocene temperatures on the basis of oxygen isotopic analysis. For example, in a recent study of mollusks from Pleistocene terrace deposits from the channel islands off southern California, Muhs and Kyser (1987) attempted to separate the effect of temperature from ice volume on the oxygen isotopic composition of the fossils. The magnitude of the ice volume effect was estimated from the age of the fossil and independently determined sea level for that time. Any residual effect on the isotopic composition was considered to be due to temperature. From this approach, they concluded that the temperature during the last interglacial was somewhat lower than the present temperature at this locality.

Dorman and Gill (1959) studied the variation in temperatures through Cenozoic time in Australia based on analysis of fossil bivalves. The pattern they observed is very similar to that found later in forams from deep-sea cores. Buchardt (1977, 1978) has more recently reported paleotemperatures for the upper Cretaceous and Cenozoic for the shelf area of the North Sea on the basis of analysis of mollusk shells. His results show the same general climatic trends as do the planktic foraminifera in deep-sea cores. A sharp cooling in early Oligocene time is especially prominent (Fig. 3.27).

An especially noteworthy example of a study in part involving older fossils is that of Lowenstam (1961) in which he demonstrated that except for the glacier ice effect, the oxygen isotopic composition of the oceans has apparently not changed markedly since Pennsylvanian time. Most researchers once felt that fossils older than late Paleozoic rarely retained an unaltered isotopic composition. This was based largely on the fact that even Paleozoic fossils with preserved shell structure gave paleotemperatures which usually seemed unreasonably high (Lowenstam, 1961). More recent work, such as that by Popp et al. (1986a) and Veizer et al. (1986) on brachiopods, suggests that the fossils are isotopically preserved. This has raised the question as to whether or not the temperature was indeed higher during early to middle Paleozoic time or if the oxygen isotopic composition of the oceans was different at that time.

Another approach to measuring paleotemperatures besides analysis of fossils would be to analyze ancient water. At least in a general way the oxygen isotopic composition of fresh water is a function of temperature. In general, the colder the temperature of condensation of water, the lower will be the $^{18}O/^{16}O$ ratio. This is because the colder the air mass, the less water vapor it can hold. As the air cools, it gradually loses its water, and the more water that is lost, the lighter the remaining vapor will be (Fig. 3.18). Thus, as indicated above, the isotopically lightest water is found at the poles. Dansgaard et al. (1969) have analyzed ice from cores in glaciers in Greenland to determine the temperature variation pattern for the last 100,000 years or so. The glacier ice faithfully records the lower temperature during the last glacial age and the rising temperature into the present interglacial age. Under appropriate conditions the ancient record is contained in ice exposed at the surface in the ice-sheet margin (Reeh et al., 1987). This is because subsurface ice returns to the surface at the ice margin. Deuterium/hydrogen ($^2H/^1H$) ratios in water and ice behave in a similar manner, the $^2H/^1H$ ratio decreasing with temperature of

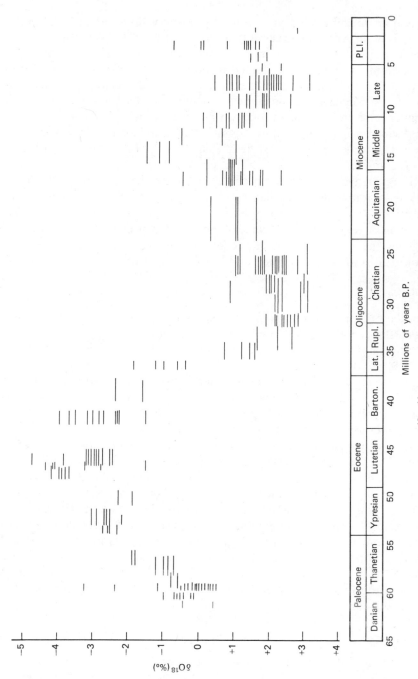

Figure 3.27 Variation in the $^{18}O/^{16}O$ ratio of Cenozoic molluskan shells from northwestern Europe. The length of the horizontal bars indicates the uncertainty in age determinations. From Buchardt (1978), by permission of Macmillan (Journals) Ltd., London.

precipitation. Jouzel et al. (1987) used this relationship to measure temperature trends in an Antarctic ice core that records the last 160,000 years.

A related approach is to determine the oxygen isotopic composition of freshwater calcium carbonate that formed in equilibrium with the temperature-
dependent rain and snow. One of the best sources of such calcium carbonate is
that of deposits in caves or speleothems. A number of studies of speleothem
carbonates have traced temperature variation patterns (e.g., Harmon et al.,
1978). An increasing number of studies have been conducted on lake marls and
fossils (e.g., Turner et al., 1983; Fritz et al., 1987). Interpretation of these
results is complicated by factors such as evaporation of the lake water, especially the residence time of water in the lake. Long residence time allows more
time for evaporation and thus an increase in the $\delta^{18}O$ value. It also allows more
time for equilibration of the water with CO_2 in the atmosphere. This usually
results in an increase in the $\delta^{13}C$ value of the water. However, determination of
the age of the samples as well as climatic interpretation is simplified by
palynological study (e.g., Gennett and Grossman, 1986).

Correlation

One of the principal uses of oxygen and carbon isotopic data in recent years
has been the correlation of deep-sea cores and even shallow-water deposits
(Shackleton and Matthews, 1977; Mangerud et al., 1979; Heusser and Shackleton, 1979). As data on the isotopic composition of foraminifera in deep-sea
sediments have accumulated over the years, a distinctive pattern of variation
of composition with time has developed. Emiliani (1966) subdivided this pattern of variation into oxygen isotopic time divisions (Figs. 3.24 and 3.28). This
allows correlation between cores on the basis of well logs of isotopic composition. Because most of the variation in oxygen isotopic composition of planktic
foraminifera during Pleistocene and Holocene time is due to variation in
oxygen isotopic composition of the ocean water, these changes in isotopic
composition should be worldwide and thus should make an excellent basis for
correlation (Shackleton and Opdyke, 1973, 1976). Recent studies (Prell et al.,
1986) have used graphic and numerical techniques for highly detailed correlation based on minor but consistent patterns in oxygen isotopic variation.
Distinctive patterns in carbon isotopic composition have also been used for
correlation. For example, a distinctive shift in $^{13}C/^{12}C$ ratio in the Late
Miocene of the Indo-Pacific region has been used for correlation in this area
(Loutit and Kennett, 1979; Keigwin, 1979).

Paleosalinity

Isotopic studies of paleosalinity have not been as extensive as those of
paleotemperature. In general, the studies that have reported paleosalinity
results have done so only in general terms and not in absolute values. Kennett
and Shackleton (1975) have documented an abrupt decrease in salinity in the
Gulf of Mexico between about 17,000 and 13,500 years B.P. They interpret

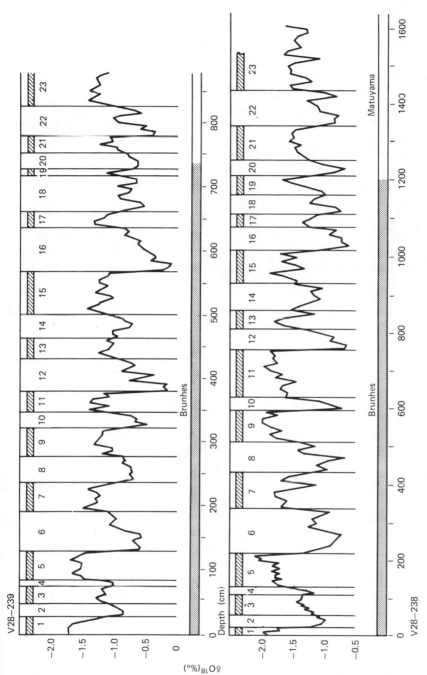

Figure 3.28 Variation in the $^{18}O/^{16}O$ ratio of two cores from the Pacific Ocean showing the use of oxygen isotopic composition in correlation. The dark and light bars below the curves show the paleomagnetic record. Dark, normal magnetization: light, reversed magnetization. The numbers above the curves refer to the oxygen isotopic temperature time divisions of Emiliani (see Fig. 3.24). From Shackleton and Opdyke (1976), Geological Society of America Bulletin, by permission of the Geological Society of America.

this as being due to a large volume of glacial meltwater coming down the Mississippi River. Dodd et al. (1984) noted apparent salinity effects on the oxygen and carbon isotopic composition of fossil mollusks from the Neogene Eel River Basin of northern California. Kammer (1979) also postulates low-salinity conditions on the basis of oxygen and carbon isotopic analysis of forams. Tan and Hudson (1974) have worked extensively with fossil bivalves from the Jurassic Great Estuarine Series of Scotland. They have used the oxygen and carbon isotopic composition for paleosalinity as well as paleotemperature interpretations.

We have used $^{18}O/^{16}O$ ratios in mollusk shells from two zones in the Pliocene San Joaquin Formation of central California to determine paleosalinities (Dodd and Stanton, 1975). Samples from open marine assemblages were assumed to have lived at normal marine salinity. Samples of freshwater bivalves were used to establish the oxygen isotopic ratio of fresh water. Using the relationship shown in Fig. 3.21, we determined the salinity associated with intermediate values from specimens that apparently lived in brackish water. This approach allowed us to determine the salinity gradient within the Pliocene embayment in which the fossils lived.

Other Applications

Several applications have been made of oxygen isotopic data in addition to the applications discussed above. In a pioneering study, Eichler and Ristedt (1966) used both oxygen and carbon isotopic data to study the life history of modern *Nautilus* (Fig. 3.29). Later analyses by other workers have shown similar patterns for modern (e.g., Taylor and Ward, 1983) and fossil (Landman et al., 1983) cephalopods. Eichler and Ristedt interpreted the abrupt change in oxygen isotopic composition at shell increment 12–13 to reflect a migration to deeper water at this time in their life history. The change in carbon isotopic composition suggested a dietary change when the animal hatched from the egg. Recent work by Crocker et al. (1985) showed that the $^{18}O/^{16}O$ composition of eggwaters of *Nautilus* eggs is lower than that for seawater and thus explains the abrupt increase in the ratio between embryonic and older specimens.

Several studies showing variation in isotopic composition within bivalve shells have been published in recent years (e.g., Jones et al., 1984; Jones et al., 1986; Romanek et al., 1987). The isotopic data yield details of the temperature history during the life of the individual but also give details of life history, such as the age of the specimen, season of maximum growth, time of commencement of sexual maturity, and spawning season. As these studies are usually combined with study of skeletal structure, they are discussed more extensively in Chapter 4.

Carbon isotopic composition of skeletal calcium carbonate appears to reflect the productivity of the waters in which the organism lived (Kroopnick et al., 1977; Berger and Killingley, 1977). Large amounts of fixation of organic

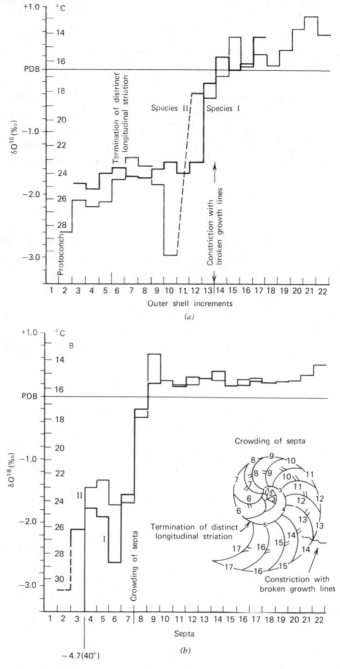

Figure 3.29 Oxygen isotopic composition and temperature equivalents in outer-shell increments (a) and septa (b) from two *Nautilus* specimens. The graphs show variation in oxygen isotopic composition during growth of the shell. From Eichler and Ristedt (1966), *Science* *153:* 734–736, copyright American Association for the Advancement of Science.

carbon occurs when productivity is high. The $^{13}C/^{12}C$ ratio in organic carbon is lower than in dissolved bicarbonate in the water. Thus, during photosynthesis, ^{12}C is preferentially removed from the bicarbonate reservoir, increasing its $^{13}C/^{12}C$ ratio. The $^{13}C/^{12}C$ ratio in calcium carbonate precipitated in equilibrium with that bicarbonate will thus increase. High productivity results in high $\delta^{13}C$ values in the calcium carbonate; lower productivity gives lower $\delta^{13}C$ values. This pattern can be seen in skeletal carbonates formed over a range of depth, $\delta^{13}C$ values being higher in skeletons from more productive shallow water than from unproductive deep water. Apparent worldwide variation in carbon isotopic composition with time may reflect global variation in productivity. Aharon (1985) used this approach to measure variation in sea-surface paleoproductivity in coral reefs on the basis of bivalve shells from tectonically elevated reefs in the New Guinea coast.

This review of the literature of the use of oxygen and carbon isotopic analysis of fossils for paleoecological interpretations is by no means complete, but should serve to indicate the usefulness of this technique. Table 3.2 lists references to other applications of oxygen and carbon isotopes in carbonate skeletons.

TABLE 3.2 Examples of Other Uses of Oxygen and Carbon Isotopic Studies of Carbonate Skeletons

Application and References	
1. *Terrestrial Climates:*	Goodfriend and Margaritz, 1987
2. *Oceanic Circulation:*	Juillet-Leclerc and Schrader, 1987; Mix and Pisias, 1988; Carriquiry et al., 1988
3. *Life History of Organisms:*	Crocker et al., 1985; Killingley and Rex, 1985; Jones et al., 1986
4. *Symbiotic Algae in Animals:*	Jones et al., 1986; McConnaughey, 1989a
5. *Early History of Life:*	Schidlowski, 1988
6. *Archaeology (Food Source):*	DeNiro, 1987
7. *Archaeology (Artifact Source);*	Herz and Dean, 1986

4

Skeletal Structure

Paleontologists and paleoecologists have traditionally been concerned primarily with the external morphology of fossilized skeletons. However, study of the internal properties of the skeletons can also be useful in better understanding the organism that formed the skeleton and the conditions of its environment. By internal properties we mean the chemical, mineralogical, and physical characteristics of the skeleton. The chemical and mineralogical properties of skeletal materials were discussed in Chapter 3. In this chapter we discuss the physical arrangement of mineral crystals within skeletal materials and the influence of the growth environment on this arrangement.

This field of study has not yet received the attention that it merits. Promising indications have been found of correlation between the growth environment and skeletal structure, but extensive applications of these to the fossil record have not been made. Many references to skeletal structure of fossils can be found scattered through the paleontologic literature, but most are more or less incidental to the study of external morphology. An early exception is the pioneering work of Carpenter (1844). Shell structure studies were really placed on a systematic, modern footing by Schmidt (1924) and Bøggild (1930). Skeletal structure studies received considerable impetus with increasing interest in the petrology of carbonate sediments and rocks in the late 1950s. Because fossil fragments are a major constituent of carbonate rocks, the identification of these fragments on the basis of their skeletal structure becomes critical. Several important publications that survey skeletal structure of major taxonomic groups appeared at least in part as a result of this interest (e.g., MacClintock, 1967; Majewske, 1969; Horowitz and Potter, 1971; Flügel, 1982). However, these studies were largely concerned with describing the structure and to a certain extent the origin, but did not deal extensively with environmental implications of shell structure. The greatest interest in shell structure versus environment has been the result of studies of banding in molluskan shells, work that was pioneered by Barker (1964) and Pannella and MacClintock (1968). The extent of interest and the potential for interpretation of skeletal banding is reflected in the papers on this subject published in a symposium volume (Rosenberg and Runcorn, 1975). Most recently, study of skeletal banding in bivalve shells has been correlated with variation in oxygen and carbon isotopic composition (e.g., Williams et al., 1982). This union of two techniques has proven to be useful for studying details of the life history of the organism as well as detailed interpretation of environmental conditions. Application of skeletal structure to paleoecology is still in its infancy and should be an area of considerable progress in the future.

Skeletal structure can be examined on at least two levels of detail. At the finer level the structure is studied in terms of the detailed shape and arrangement of the individual crystallites making up the skeleton. This has sometimes been called skeletal ultrastructure. At a coarser level the interrelationships of larger units of relatively uniform structures within the skeleton are studied. This has sometimes been called the microarchitecture of the shell. As an example of this difference in scale, one might study the details of the size, shape, and orientation of aragonite crystals within the mother-of-pearl or nacreous layer of a bivalve shell. On the other hand, one might also study the shape and distribution of the nacreous layer itself within the shell. Both levels of study may potentially provide useful information about the growth habitat of the the bivalve. Both levels of study are considered in this chapter.

The detailed structure of skeletal materials differs considerably between different taxonomic groups. However, structures may be quite uniform within some groups, such as the echinoderms, so that even small fragments of skeletal material can be identified as belonging to that phylum. The factors determining the skeletal structure are very poorly known, but clearly they are largely genetically controlled physiologic and biochemical factors. To a certain extent the structure is influenced by the physical conditions of the growth site, such as through direct physical interference between growing crystals. Environmental factors are obviously also involved, as they affect the physiology of the organism.

GROWTH MECHANISM

Skeletal structure is a function of the mechanism of growth of the skeleton. Skeletal growth can be viewed at two scales just as can the skeletal microstructure that results from that growth. On the finer scale, skeletal growth can be studied in terms of how an individual crystallite grows. Considerable progress has been made in better understanding this growth or mineralization process, but much remains to be learned (Wilbur, 1976; Lowenstam and Weiner, 1989). Although agreement is not universal, the mineralization model shown in Fig. 4.1 is widely supported.

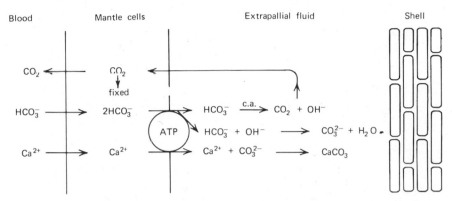

Figure 4.1 Model showing possible method of $CaCO_3$ mineralization in the mollusks. The carbonate is secreted on an organic matrix that forms from the extrapallial fluid. The composition of this fluid is controlled by the biochemical reactions occurring in the mantle. The Ca^{2+} and CO_3^{2-} are brought to the mantle by the body fluids and/or they come from the external medium (seawater in the case of marine mollusks). From K. M. Wilbur, 1964, in Physiology of Mollusca, by permission of Academic Press.

Three factors in the mineralization process are involved in determining skeletal structure.

1. The chemical composition of the solution from which the mineral precipitate (and factors controlling that chemistry) is obviously important. The mineralization site may be within cells in some of the simpler organisms or between the living tissue and previously formed shell in others (Lowenstam, 1981). It is always to some extent isolated from the surrounding water (or air for terrestrial organisms) allowing the chemistry to differ from that of the surroundings.

2. The organic matrix is important in determining the structure of the skeleton (Towe, 1972). The model shown in Fig. 4.1 proposes that crystal growth starts on the matrix, which acts as a template controlling both the mineral composition (see Chapter 3) and at least in part the physical configuration of the crystals. The ions from which the crystals grow are attracted to specific sites on the matrix. Thus the arrangement of these sites determines both growth location and crystal structure.

3. To some extent the physical constraints of the growth environment determine the size and scale of crystallites. The crystallites can only grow a certain distance before they begin to interfere with the growth of other crystallites. The polygonal shapes of the crystals in some of the structures probably result from this interference and growth to fill the available space. Raup (1968) has likened the growth of crystals in some invertebrate skeletons to the growth of soap bubbles that interfere with one another, producing polygonal shapes (Fig. 4.2).

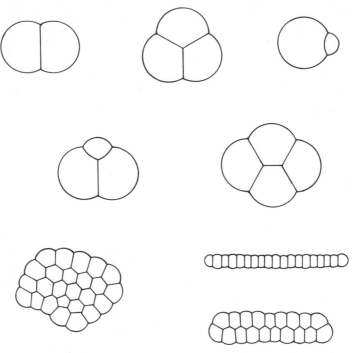

Figure 4.2 Outlines of soap bubbles that are similar in shape to crystals and plates formed during skeletal mineralization. The polygonal shapes result from interaction between the bubbles (or crystals) as they grow to fill the limited space. From Raup (1968), Journal of Paleontology, *42*:55, by permission of Society of Economic Paleontologists and Mineralogists.

On a larger scale the microarchitecture of a skeleton is a function of the overall growth pattern. Organisms show four basic methods of skeletal growth (Raup and Stanley, 1978). The most widespread method is accretion, in which growth is by addition of mineral material to the preexisting skeleton. All of the previously formed skeleton is retained throughout life and is constantly being enlarged. Growth by accretion allows the organism to be protected continuously throughout life. It also permits continuous use of the entire skeleton; none of it ever has to be discarded. The accretion method of growth is ideal for the paleoecologist because the skeleton contains a continuous record of growth throughout the life of the organism.

A skeleton can grow by accretion only if it is in contact with living tissue. This and the necessity of building from the foundation of the old skeleton place definite limits on the shape and structure of the skeleton. Organisms having this mode of growth all live in a tube or a modified form of a tube (Fig. 4.3). The tube may be coiled (gastropods, cephalopods, etc.), partially filled with structures such as septa and tabulae, allowing the organism to live only near the

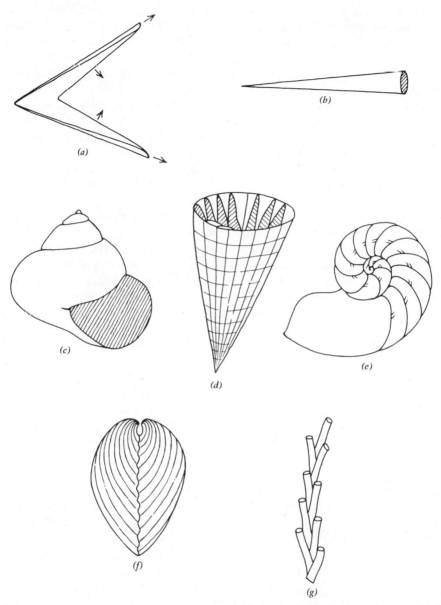

Figure 4.3 Shells shaped as cones and modified cones formed by accretion: (*a*) section of basic cone showing growth at margins which lengthens the cone and growth on the inner surface which thickens the cone, producing a basic two-layered microarchitecture; (*b*) gently tapering cone (annelid); (*c*) coiled cone (gastropod); (*d*) cone filled with structures (coral); (*e*) cone divided into chambers (cephalopod); (*f*) double, rapidly expanding cone (bivalve); (*g*) multiple colonial cones (bryozoan).

distal end of the tube (bryozoans and corals), partitioned into chambers with the animal occupying only the last chamber (cephalopods), doubled into two rapidly expanding tubes placed open end to open end and hinged to shut out the outside world (bivalves and brachiopods), grown together in colonies (corals and bryozoans), or equipped with a trapdoor (gastropods and cephalopods). The tube can grow in length and expand at its distal margins and it can increase the thickness of its walls by addition to the tube interior (Fig 4.3). No other directions of growth are possible. All shells growing by accretion have a basic two-layered structure. As the skeleton accretes, one layer forms at the growing margin of the tube and the other forms on the inner surface of the tube. The inner layer may grow by addition directly to the crystals of the outer layer so that the two are indistinguishable. On the other hand, the inner layer may develop different structures in different places within the tube, giving one or more extra layers or units. The geometry of the skeletal microarchitecture can be analyzed in terms of skeletal growth patterns.

The second method of skeletal growth is by molting. This growth form is restricted to the arthropods and probably is a major reason for their evolutionary success. In molting, the skeleton is periodically discarded and a new, larger, and perhaps differently shaped skeleton develops. This method allows much more variety of shape because the skeleton can be essentially molded to the body. It has a disadvantage, however, in that the animal must periodically rebuild an entire new skeleton. While doing so it may have to abandon normal functioning and is without skeletal protection. The basic framework of the arthropod skeleton is organic (chitinous). In groups with a mineralized skeleton this framework is strengthened by the precipitation of calcite or apatite.

The third method of skeletal growth is by addition of skeletal elements. The simplest example of this is the addition of spicules to a spicular skeleton (sponges, alcyonarians, holothurians). A more complex example is the echinoderm skeleton, which is enlarged by the addition of new plates. The new plates are formed in the apical region and are in effect pushed down the side of the skeleton as newer plates are formed above them. Echinoderm plates are surrounded by and permeated with living tissues. This also allows each plate to grow by accretion throughout the life of the specimen. Thus in effect the echinoderms grow by a combination of addition of plates and accretion.

The fourth type of growth has been called modification. This method is used in the vertebrates. It involves growth on all sides and even within the skeleton and resorption of material so that both the shape and size of the skeleton can change as needed. In terms of variability of shape this is the ultimate skeleton. It does, however, require an internal position in order to develop. Thus it provides little protection to the organism. As a result, the vertebrates have developed skin tissue and its various modifications such as scales, feathers, and horny plates for protection.

Little is known about the relationship between shell structure and environment in most invertebrate groups. The three groups that have received the most attention in this regard are the foraminifera, corals, and mollusks. Thus

TABLE 4.1   Selected Examples of Studies of Skeletal Structure

Fossil Group	Reference
Algae	Johnson, 1961; Wray, 1977
Foraminifera	Hay et al., 1963; Towe and Cifelli, 1967, Hansen, 1979
Anthozoa	Kato, 1963; Sorauf, 1971, 1972
Bryozoa	Tavener-Smith and Williams, 1972; Ross, 1976; Sandberg, 1977
Brachiopods	Williams, 1968; Williams and Wright, 1970
Mollusks	Bøggild, 1930; MacClintock, 1967; Kennedy et al., 1969; Majewske, 1969
Arthropods	Travis, 1960; Levinson, 1961; Bourget, 1977
Echinoderms	Raup, 1966a; Donnay and Pawson, 1969

we concentrate our discussion on these three groups in this chapter. Table 4.1 lists some recent and/or important studies or reviews of skeletal micro- structure in the important fossil groups.

FORAMINIFERA

Three fundamental wall structures have been described in the foraminifera: agglutinated (arenaceous), hyaline, and porcellaneous (Fig. 4.4). The aggluti- nated structure consists of sediment grains acquired from the surrounding environment and embedded in a predominantly organic matrix. The porcella- neous structure consists of randomly oriented, elongate calcite crystals with a thin surface veneer of tangentially oriented crystals (Fig. 4.5; Towe and Cifelli, 1967). It is nonporous and dense as compared to the hyaline structure. The hyaline structure consists of somewhat larger elongate calcite crystals that are oriented perpendicular to the test surface (Fig. 4.5; Towe and Cifelli, 1967). Both a granular structure and a radial structure have been described in the hyaline group, but Towe and Cifelli minimize the importance of this distinction. The hyaline wall is penetrated by numerous pores that parallel the crystals. Detailed discussions of the microstructure of foraminiferal walls are included in the work of Hay et al. (1963) and Towe and Cifelli (1967).

These three basic wall types are crudely correlated with growth environment (Greiner, 1969, 1974). In the Gulf of Mexico, foraminifera with agglutinated structure are especially common in low-salinity and relatively low temperature bays and estuaries such as Mobile Bay and Mississippi Sound (Fig. 4.6). They are also relatively more abundant in deep water than shallow and in high latitudes than low. Foraminifera with porcellaneous walls are especially abun- dant in shallow, warm environments with normal to elevated salinity, such as Laguna Madre in the south Texas coast and parts of Florida Bay in the Florida

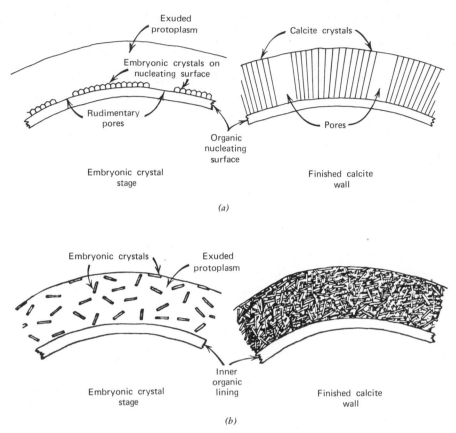

Figure 4.4 Schematic drawing of hyaline (*a*) and porcellaneous (*b*) wall type in foraminifera. From Greiner (1974), copyright President and Fellows, Harvard College.

Keys area. Foraminifera with hyaline structure are most common in areas with intermediate temperature and salinity conditions, such as San Antonio and Matagorda Bays. Greiner explains this distribution as being the result of differing solubility of $CaCO_3$. Because the solubility of $CaCO_3$ varies with both temperature and salinity, cold and low-salinity waters are generally unsaturated with respect to $CaCO_3$, and thus presumably, $CaCO_3$ tests would be more difficult for the foraminifera to secrete, even with physiologic intervention. Seawater in areas with high temperature and salinity is usually considerably supersaturated with respect to $CaCO_3$. The relatively massive, randomly nucleated porcellaneous structure can perhaps be more readily secreted in such supersaturated waters. In seawater saturated or only slightly supersaturated with respect to $CaCO_3$ the porous but more highly organized hyaline structure is secreted. This is a reasonable but simplistic explanation of the observed distribution pattern, for numerous exceptions can be found with porcellaneous foraminifera occurring in unsaturated waters and agglutinated taxa occurring in

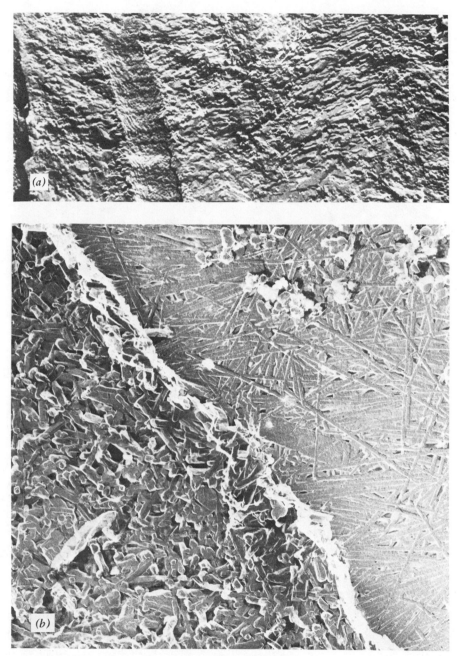

Figure 4.5 Electron photomicrographs of basic foraminifera wall types. (*a*) Hyaline wall (*Cibicides refulgens*). Note pore in left-hand side of photo. × 5000. (*b*) Porcellaneous wall (*Quinqueloculina seminulum*). Upper right is shell surface, lower left is interior of wall. × 14,000. (*c*) Surface of agglutinated test (*Karreriella bradji*). Note the circular coccoliths incorporated into the test. × 5600. Photos supplied by K. M. Towe.

Figure 4.5   (Continued)

saturated and highly supersaturated conditions. Nevertheless, the generalized distribution can be useful in paleoecologic interpretations.

The relative proportion of porcellaneous miliolid foraminifera in carbonate sediments deposited in shallow seas varies with temperature (Murray, 1987). Sediments in tropical and subtropical areas usually contain more than 40% miliolid forams, whereas sediments in warm-temperate environments usually

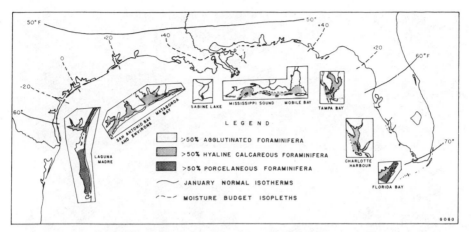

Figure 4.6 Variation in proportion of foraminifera with agglutinated, hyaline, and porcellaneous walls in selected bays and lagoons in the northern Gulf of Mexico. From Greiner (1974), copyright President and Fellows, Harvard College.

contain between 20 and 40%, and sediments in cool-temperate seas, usually less than 20% miliolid foraminifera. Based on this relationship, he concluded that the Pliocene Coralline Crag Formation in England was deposited under cool-temperate conditions and the Eocene Calcaire Grossier formation in the Paris Basin was deposited under warm-temperate conditions. Haynes (1965) suggested that the glassy, radial-hyaline wall may have evolved as an adaptation to facilitate light penetration into the test of foraminifera containing symbiotic algae. Indeed, many radial-hyaline genera contain symbiotic algae.

ANTHOZOA

The skeletons of modern scleractinian corals are composed of spherulitic clusters of aragonite needles (Fig. 4.7; Sorauf, 1972). Recent work by Gladfelter (1983) suggests that trace amounts of calcite may actually form the initial framework on which the sperulitic clusters grow. The clusters in turn are arranged in vertical series forming elongate, fan-shaped trabeculae. Among the extinct subclasses of corals, the Rugosa have a skeletal structure very similar to the scleractinians (Sorauf, 1971). Octocorals, with a spicular skeleton, have a structure of elongate calcite crystals subparallel to the long axis of the spicule (Sorauf, 1974). To date, most work on skeletal structure in the corals has dealt with detailed description, relations to the calcification process, and variation between taxonomic groups. This provides the necessary background for detailed studies of relationships between structure and environment.

Temperature and light appear to be the major environmental controls on

Figure 4.7   Skeletal structure in modern scleractinian corals. (*a*) Aragonite needles in spherulitic arrangement (*Cladocora caespitosa*). Note banding. × 3000. (*b*) Spherulitic aragonite needles and growth banding (*Balanophyllia malounensis*). × 750. Photographs courtesy of J. E. Sorauf.

the structure of coral skeletons. The skeletal characteristics that have been most studied in relation to the environment are the external features of the epitheca, which are probably related to skeletal structure, and to the spacing of internal features in colonial coral skeletons.

Growth banding is very prominent in corals and can be closely related to environmental parameters [see Dodge and Vaismya (1980) for a review of banding in corals]. Banding occurs on two scales in the epitheca of Rugosa and some genera of the Scleractinia (Fig. 4.8; Wells, 1963). On the larger scale, constrictions and expansions with a spacing of about 1 cm appear to be

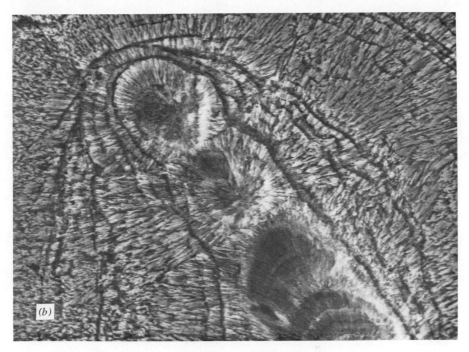

Figure 4.7   (Continued)

the result of annual variation in growth. The expansions presumably form during summer growth when conditions are especially favorable, allowing expansion of the coral polyp. Constrictions probably form during winter growth or during times of less favorable conditions. On a much smaller scale, fine ridges with a spacing of a few micrometers are the result of daily growth cycles, probably related to variation in light intensity. Calcification in corals is much more rapid during the daylight hours than at night, presumably because of the effect of the symbiotic algae within the coral tissue. During the daylight hours these algae take up CO_2 in their photosynthesis and thus raise the pH, aiding precipitation of $CaCO_3$ (Goreau, 1959). The daily bands can be grouped into fortnightly, tidal, and lunar cycles (Scrutton, 1964).

Each of these cycles has an environmental cause and thus is potentially useful in environmental reconstruction. One major application has been to determine the length of the day in the geologic past. Because of tidal friction, the speed of rotation of the earth should be gradually diminishing (Wells, 1963). The effect of this has been to decrease gradually the number of days per year. Astrophysicists (Runcorn, 1975) have in fact estimated the rate of this deceleration and have estimated the length of the day and thus the number of days per year for the past. Banding in coral skeletons offers an independent means of making this calculation (Fig. 4.9). Conversely, once the number

Figure 4.8 Growth banding on the surface of the rugose coral *Zaphrentoides pellaensis* from the Lower Chester (Mississippian) Pella Formation, Mahoska County, Iowa. × 5. Specimen supplied by A. S. Horowitz.

of days per year during geologic time has been determined, the geologic age of the fossil corals should be determinable. This can be done by counting the number of daily growth lines between the annual expansions or constrictions. By counting bands on the epithecae of specimens of the rugose coral, *Helliophyllum,* Wells (1963) determined that there were 400 days in the year during Devonian time. He checked this method by counting growth bands on modern specimens of several scleractinian genera; the modern specimens did indeed have about 365 bands per year. An alternative approach is to determine the number of days in the synodic (lunar) month. This number also has decreased with time as predicted from the decrease in the earth's rotation (Scrutton, 1964).

Determining the number of days in the year from fossil corals is not a very practical method of age determination for several reasons: (1) Growth bands can be counted only on the epitheca of very well-preserved specimens; (2) the counting procedure is somewhat subjective because some bands may be indistinct and the precise location of the maxima of the annual expansion or the minima of the constrictions is hard to determine objectively; and (3) the precision of this method is not as good as more conventional biostratigraphic

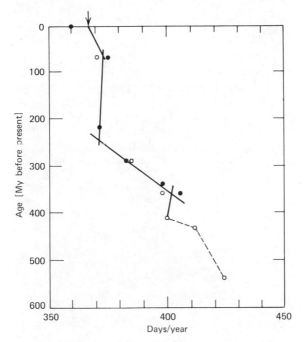

Figure 4.9 Number of days per year as estimated from growth rings in fossils. Data from various published sources. From Creer (1975).

methods. Nevertheless, the technique has confirmed the predicted slowing of the earth's rotation and has been an especially notable example of cooperation between geophysical and paleontological researchers.

Annual cycles have also been described by T. Y. H. Ma (1958) and others in the skeletons of tabulate corals. These cycles consist of a regular pattern of closely spaced tabulae alternating with more widely spaced tabulae (Fig. 4.10). The wide spacing may be the result of rapid growth under optimal conditions, probably in the summer; and the more closely spaced tabulae develop under more marginal conditions, probably in the winter. Ma used this model in an attempt to locate the position of the equator as related to the drifting continents during various times in the Paleozoic. He reasoned that specimens living near the equator should have rather uniform growth conditions and thus should not develop banding in their skeletons, but that specimens living far from the equator, where seasonality is pronounced, should have a well-developed banding pattern. Thus the banding pattern for a given time in the geologic past should be correlated with latitude and should help to locate the position of the equator. This approach has not been widely used because of the scarcity of reliable data on banding in the corals, but Ma's data have been reinterpreted in light of plate tectonic theory by Fischer (1964a).

Figure 4.10 Banding in a tabulated coral caused by cyclic spacing of tabulae. × 1.
Photo courtesy of Allen Archer.

The model appears to be basically sound and potentially useful but needs to
be more extensively tested. One problem that has developed from more
recent work (Weber et al., 1975) is that banding occurs in modern corals even
in the equatorial zone.

The growth rate of corals can be determined from the growth banding.
Growth rate is in part under environmental control and can thus be used to
make paleoenvironmental interpretations. These data can also be used for
analysis of population dynamics. Wells (1963) and Johnson and Nudds (1975)
determined growth rates for Devonian and Carboniferous corals. The Carboniferous corals from England grew more rapidly than the Devonian ones
from New York. This difference has been explained as due to England being
nearer the paleoequator during the Carboniferous than was New York during
the Devonian. Other factors must also be considered, however, in the interpretation of growth rates. For example, Johnson and Nudds also note that
monthly banding is less well developed in specimens from shaley than from

limy beds. Reduced light intensity in muddier and/or deeper water may be the cause for this difference.

Huston (1985) also used banding in modern corals to study variation in growth rate with depth. He found that in most cases growth rate decreases regularly with depth. Some genera showed little depth effect or a maximum growth rate a few meters below the surface.

Ali (1984) studied banding in Jurassic scleractinian corals from England. The growth rate indicated was approximately the same as that for modern corals of similar morphology which are restricted to the tropics. To the extent that growth rate is controlled by temperature, this suggests that Jurassic temperatures in England were similar to those in the modern tropics.

Because banding in modern corals is often difficult to recognize, the use of x-radiographs has proved valuable. Specimens of scleractinian colonies have been studied by making thin slices of the coral colonies and transmitting x-rays through them onto x-ray sensitive film (Fig. 4.11). The photographs thus produced reveal alternating light and dark bands that appear to be related to variation in the density or compactness of the skeleton (Knutson et al., 1972). The work with living corals of Weber et al. (1975), Weber and White (1977), Huston (1985), and others indicates that the dense bands, which probably result from slower growth, are produced in late summer or early autumn. In addition to annual bands, Hudson et al. (1976), in studying corals from the Florida Keys, described disturbance bands that apparently usually resulted from extremely low winter temperatures.

Growth rate, and thus banding in corals, is controlled by a number of factors, including solar radiation, suspended sediment, temperature, nutrient supply salinity, and wind and wave activity (Dodge and Vaismya, 1980). Seasonal temperature changes are probably one cause of banding, but Knutson et al. (1972) and Buddemeier et al. (1974) have noted annual growth bands in x-radiographs of corals from the tropical Pacific where the temperature varies little seasonally. These bands are probably caused by variation in light intensity between the rainy (and cloudy) summer season and the remainder of the year. Spacing between growth bands, and thus growth rate, has also been ascribed to water depth, perhaps as it is correlated with light intensity (Huston, 1985). Weber and White (1977) used the seasonal pattern of density variation in coral skeletons to determine growth rates in two species of the genus *Montastrea*. They found that the growth rate in specimens of *M. annularis* from shallow water correlated directly with the average temperature for the growth locality (Fig. 4.12). Growth-rate determinations from fossil corals can thus potentially be used for determining paleotemperatures. In fact, by counting annual bands, Hudson and Shinn (1977) determined the growth rate for a large Pleistocene fossil specimen of coral and found it to be considerably slower than conspecific modern specimens. This suggests that the fossil coral grew in either deeper and/or colder water than the modern specimens.

Study of density bands in corals is in many ways similar to the study of tree rings in the science of dendrochronology. Hudson et al. (1976) have in fact

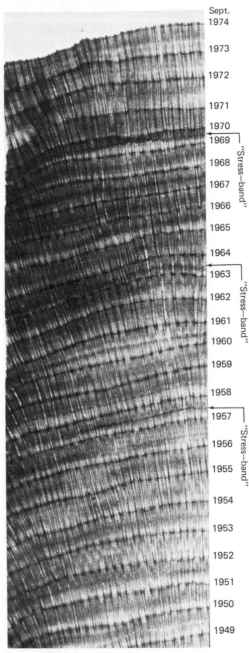

Figure 4.11 X-radiograph of a core taken from a specimen of *Montastrea annularis* living on the Hen and Chickens Reef, Florida Keys. The dates indicate the year of formation of the bands as determined by counting downward from the growing surface. "Stress bands" are formed during temperature extremes. × 1. Photo courtesy of J. H. Hudson.

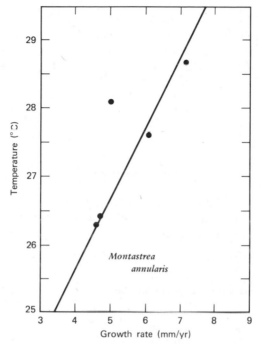

Figure 4.12 Variation of growth rate with mean annual temperature for shallow colonies of *Montastrea annularis*. The line is a least-squares regression line fit to the data. Growth rates were determined from x-radiographs showing annual banding. From Weber and White (1977).

suggested the term *sclerochronology* for the study of coral banding. This approach holds considerable potential as a paleoecologic tool for determining water depth, temperature, and perhaps other factors. An added advantage to studying banding in corals is that events occurring during the life of the coral can be put into a time framework that is longer than for other invertebrates. By correlating the banding pattern between coral specimens (as is routinely done between trees in dendrochronology), the record can potentially be extended to cover even longer time periods.

Banding in modern corals has also been used to study shorter-term events. Druffel (1982) studied banding and ^{14}C concentration in modern corals. She noted higher than expected ^{14}C concentrations in the early eighteenth century, a time characterized by slightly lower than normal temperatures. This correlation may be due to a slight decrease in solar activity at this time. The concentration of ^{14}C in corals decreased in the first half of the twentieth century, due to introduction of CO_2 into the atmosphere from fossil fuel. This CO_2 contains no ^{14}C, resulting in dilution of ^{14}C in total atmospheric CO_2. The ^{14}C concentration increased abruptly in the early 1950s as a result of production of ^{14}C in atomic bomb testing (Druffel and Suess, 1983). Toggweiler and Trumbore (1985) determined the ^{90}Sr concentration in corals, which could be dated by their annual banding. They documented the buildup of ^{90}Sr as also related to nuclear testing in the South Pacific area. By comparing the timing of the ^{90}Sr increase in various areas, they determined the current pattern allowing transfer of water between the Pacific and Indian Oceans.

MOLLUSCA

Shell Structure Types

Skeletal structure is more varied in the mollusks than in any other group of organisms. Perhaps in part as a consequence, skeletal structure has been studied more extensively in the mollusks than in any other group. Several aspects of skeletal structure of the mollusks, especially the bivalves, seem to be influenced by environmental conditions; the influence of environment on

TABLE 4.2   Shell Structural Types in the Mollusca

Type	Example of Genera with This Structure	Description
Nacreous	*Mytilus, Pinctata, Nautilus*	Tabular aragonite crystals arranged in layers with organic matrix
Prismatic	*Mytilus, Inoceramus, Pinna*	Polygonal columns of calcite or aragonite oriented normal to growth surface
Foliated	*Ostrea, Pecten*	Laths of calcite arranged in sheets subparallel to growth surface
Crossed-lamellar	*Mercenaria, Arca*	Aragonite laths arranged in alternating bundles so that crystals cross at a high angle
Complex crossed-lamellar	*Chama*	Bundles of aragonite crystals in many orientations arranged with crystals crossing in several orientations
Homogeneous	*Mya, Arctica*	Equidimentional, apparently randomly oriented crystals
Myostracal	Under muscle of most mollusks	Irregular prisms or blocks of aragonite
Periostracum	Outer layer in all mollusks	Uncalcified complex proteinaceous

banding patterns especially has been the subject of extensive research in recent years.

Seven common basic shell structural types have been described in the mollusks (Table 4.2) and several other minor types and variants have also been recognized. Comprehensive descriptions of these basic types are given in Bøggild (1930), MacClintock (1967), Majewske (1969), Taylor et al. (1969), Carter (1980), and Carter and Clark (1985).

These shell structural types are widely but not randomly distributed among the mollusks (Carter, 1980). Shell structure has evolved through time. For example, calcitic prismatic and foliated structures appear to have evolved from the more primitive nacreous structure. Similarly, the crossed-lamellar and complex crossed-lamellar structures are most common among the more advanced groups, having replaced the nacreous structure. The nacreous structure itself has evolved through time from a more porous, flexible structure to a more compact, mechanically strong structure which characterizes modern nacre (Mutvei, 1983). Thus the distribution of structural types among the bivalves is probably determined primarily by phylogeny and biological factors (Carter, 1980).

Shell structure also is related to the mode of life, and hence is sensitive to environmental influence (Table 4.3). Taylor and Layman (1972), Carter (1980), and Currey (1980) determined various mechanical properties of the bivalve shell structures and attempted to relate these to function in the organ-

TABLE 4.3 Mode of Life and Shell Structure Combinations for Bivalve Families[a,b]

	Aragonite Prisms and Nacre	Calcite Prisms and Nacre	Foli-ated	Composite Prisms c.l. and c.c.l.	Crossed-Lamellar and Complex c.l.	Homoge-neous
Free-living epifaunal			••			
Byssate		•••••	•••		•••••	
Cemented	•••		•••		•	
Boring		•			•••••	•
Shallow bur-rowing	•••••				•••••	•••••
rowing	•••••			•	•••••	•
					•••••	
					•	
Deeper bur-rowing				••••		
rowing	••••			••••	•••	••••
				•		

[a]After Taylor and Layman (1972).
[b]In cases in which a family has two modes of life, the family is entered twice. Each dot represents one family with a combination of life habit and shell structure. c.l., crossed lamellar; c.c.l., complex crossed lamellar.

ism. The nacreous structure is the most resistant to breakage in tension, compression, impact, and bending tests, whereas the composite prismatic, crossed-lamellar, and complex crossed-lamellar structures are the hardest. Crossed-lamellar structure is least elastic. These properties may in part explain the observed distribution of structural types among the different modes of life, but other factors, such as the rate and efficiency of biochemical formation of the shell, must also be involved. Because of its strength, the nacreous structure has obvious advantages in the shelled cephalopods in resisting hydrostatic pressure. The nacreous structure, perhaps also because of its strength, is common in bysally attached forms (Mytilacea) that are often exposed to strong current and wave action as well as mechanical crushing by predators. Nacreous structure is also common in shallow burrowing freshwater bivalves (Unioacea), where the structure may be advantageous in being less soluble in water that is likely to be unsaturated with respect to $CaCO_3$. The reduced solubility of the nacreous shell results from the high organic content, which shields the $CaCO_3$ from contact with the water. The nacreous structure is also found in some very thin shelled bivalves, where its high strength would be an advantage.

The foliated structure is common in cemented bivalves (Ostreacea) and some bysally attached forms (Anomiacea). The advantage of the foliated structure may be that it can be deposited quickly to form thick shells that are resistant to predators and mechanical destruction. The lower density of the foliated structure may also be an advantage to prevent sinking into the substrate by unattached oysters such as Gryphea and Exogyra, which often lived in soft sediment. Low density as well as strength and flexibility are advantages of the foliated structure in free-swimming Pectinacea. Crossed-lamellar and complex crossed-lamellar structure is found in many different bivalves but is especially common in burrowing taxa, where its hardness may be an advantage in resisting abrasion by the sediment. Crossed-lamellar and complex crossed-lamellar structures are also found in boring bivalves, in which shell hardness should be especially valuable. The homogeneous structure, also relatively hard, is found largely in burrowing bivalves.

Although it is seldom preserved, the uncalcified periostracum is functionally important. Its primary function is to form the substrate on which calcification occurs, but it has other functions as well. Bottjer and Carter (1980) suggest that it may help to stabilize the animal in the substrate, serve a camouflaging and tissue protection function, aid tactile perception, and protect the shell from solution in water undersaturated with $CaCO_3$. Corbulids have a well-developed conchiolin (organic) layer within the valve. This crops out at the commissure, allowing a hermetic seal (like a gasket). This feature is especially valuable to a group that usually lives in harsh, brackish environments. It also protects against chemical boring by gastropods, which rely on solution of $CaCO_3$ (Lewy and Saintleben, 1979).

Little is known about variations within these basic shell structure types in response to environment. Extra myostracal layers in the nacreous structure of the bivalve *Brachidontes recurvus* seem to result from life under stress situations (usually periods of reduced salinity) (Davies, 1972). A similar variation in the nacreous shell structure of *Geukensia demissa,* a modern shallow-water bivalve, may result from alternations of well-developed, typical nacreous structure in the summer months, smaller pitted crystals in autumn, and cessation of growth and solution of the nacreous surface in winter. The solution, caused by acids formed by anaerobic respiration in the animal, results in a fine-grained structure in the shell that is really a degenerated nacreous structure. When growth resumes in the spring, a thin layer of aragonite prisms (much like myostracum) results from the vertical stacking of nacreous aragonite crystals. Finally, normal nacreous structure begins to form later in the spring or summer. Thus the nature of the structure is a function of temperature as it affects the behavior and respiration pattern of the animal (Lutz and Rhoads, 1977). Factors other than temperature (such as reduced salinity) may also produce extra myostracal layers similar to those observed by Davies (1972). Zones of slightly finer crystal size that occur in the prismatic structure of Mytilus appear to result from summer growth (Dodd, 1964).

The shell structural types are arranged in discrete microarchitectural units within the shell. The shape of these units is a result of the method of growth of the shell, as can be illustrated by describing the growth of the bivalve shell. Shell growth in the other molluskan classes is a modification of this basic pattern. The bivalve shell grows by accretion in two general areas (Fig. 4.13): the mantle margin and the general mantle surface, the two areas being separated by the pallial line. Growth at the mantle margin results in growth at the margin of the shell and is thus largely responsible for increasing the length and width of the shell. Growth on the mantle surface results in thickening of the shell. Mantle margin growth in a sense produces the framework, and growth

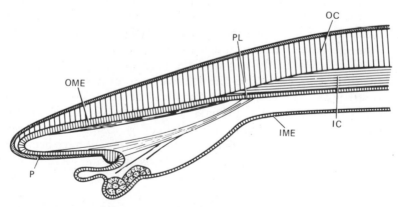

Figure 4.13 Diagrammatic cross section of bivalve shell and mantle. P, periostracum; OC outer calcified shell layer; IC, inner calcified shell layer; OME, outer mantle epithelium; IME, inner mantle epithelium; PL, pallial line. From Wilbur (1964), in Physiology of Mollusca, by permission of Academic Press.

at the mantle surface plasters and thickens the basic frame. Commonly, the shell structural types in these two areas are different and result in distinctive microarchitectural units. For example, growth at the mantle margin may produce the prismatic structure, which thus forms an outer layer of the shell, and growth at the mantle surface may produce the nacreous structure, which lines and thickens the shell. The basic two-layered calcareous structure may be complicated by subdivision of the inner layer into two or more units that may have different structural types.

In addition to the outer calcified layer produced at the mantle margin, there is a thin organic layer, the periostracum, forming the outermost portion of the shell and the surface on which the outer calcareous layer is deposited. The periostracum is produced by the outermost of three folds of the mantle margin (Fig. 4.13). It wraps around the mantle margin and onto the exterior surface of the calcareous shell itself. As growth proceeds, the periostracum in effect unrolls from the mantle margin as the shell grows outward underneath it. The outer calcareous layer grows by nucleation on the inner surface of the periostracum, and as it thickens it pushes the adjacent mantle margin away from the periostracum. Thus the area that was originally at the shell margin is left behind and eventually the pallial attachment migrates past it and inner shell layer is secreted. The distinctive myostracal shell structure forms under the attachment of the muscle to the shell, so that the location of the muscle attachment throughout the life of the animal can be traced by this structure. This method of growth produces an inner layer that thickens from a feather edge near the shell margin to a maximum in the beak area, where the layer has been forming for a longer period of time. The outer layer is thickest at the pallial line and thins both toward the shell margin and toward the beak area.

The relative proportion of these microarchitectural units may change with environmental conditions. In species with shells composed of a combination of calcite and aragonite, the two minerals are present in different shell layers or different subdivisions within a shell layer. Changing proportions of these minerals as influenced by environment are thus reflected in changing proportions in the shell microarchitectural units. In modern specimens of *Mytilus californianus* (Fig. 4.14), for example, the aragonite/calcite ratio, as determined by x-ray diffraction, and the relative proportions of the aragonitic nacreous structure and the calcitic prismatic structure, as determined from sections, are strongly correlated (Dodd, 1964). Both parameters are in part controlled by growth temperature (see Chapter 3). Using this relationship, the structures of Pleistocene specimens of *M. californianus* have been used to interpret paleotemperatures for various locations along the southern California coast (Dodd, 1966).

Skeletal Banding

The feature of the structure of mollusk shells that has attracted most attention in recent years is the banding or growth increment pattern observed in several

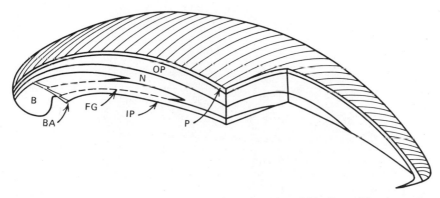

Figure 4.14 Three-dimensional cutaway view of a valve of *Mytilus californianus* showing shell structural features. P, periostracum; OP, outer prismatic layer; N, nacreous layer; IP, inner prismatic layer; BA, blocky aragonite layer; FG, fine-grained zone; B, beak area. After Dodd (1964), Journal of Paleontology, by permission of Society of Economic Paleontologists and Mineralogists.

species of bivalves. This banding is similar to the annual and daily produced growth features in coral skeletons discussed above. Prominent growth lines or shelving on the surface of bivalve shells have been recognized for many years as probably annual features caused by the slowing or stoppage of growth during winter months, and have in fact been used in studies of age and population dynamics. Weymouth (1923), for example, recognized not only growth ridges but also variation in the internal structure of the shell of the pismo clam, *Tivela stultorum,* which was reflected in variation in the opacity to transmitted light.

Barker (1964) published the first detailed description of internal banding within bivalve shells. He described banding cycles of five orders or scales, ranging from a few micrometers to several millimeters. He interpreted these banding cycles as probably due to daily tidal cycles, day-to-night variations, lunar monthly tidal cycles, seasonal storm patterns, and annual temperature variation. The larger-scale banding patterns are produced by regular variation in the smaller-scale banding; the basic, smallest-scale banding is produced by variation in the amount of organic material or conchiolin in the shell. Pannella and MacClintock (1968) recognized a similar pattern of banding in specimens of *Mercenaria mercenaria* as well as other bivalve species (Fig. 4.15). They conducted a series of growth experiments in which they notched shells of living specimens, returned them to their natural environment, and recollected them at a later date. The count of the daily cycles in the shell banding beyond the notch was the same as the number of days between notching and killing of the specimens, confirming their interpretation of the origin of these bands. They were also able to recognize the effects of storms known to have occurred

Figure 4.15 Longitudinal section of the outer shell layer of *Mercenaria mercenaria* showing daily growth bands. The growing edge of the shell is to the right. The outer surface of the shell is toward the top of the photo. × 200. Photograph courtesy of Copeland MacClintock.

during the time period as well as effects of the seasons, spawning cycles, and lunar monthly cycles in the banding pattern.

Ohno (1985) performed similar experiments with seven species of bivalves. Species from the intertidal zone produced approximately two bands per day, one for each tide. Variation in band thickness faithfully recorded monthly tidal variation. Subtidal species with symbiotic algae (*Tridacna*) produce daily bands. In some cases number of bands in subtidal species did not correspond to either tides or days.

The physiologic mechanism by which the bivalve produces bands is not known but apparently is related to variation in the relative rate of formation of organic matrix or conchiolin and calcium carbonate. The organic matrix may be forming at a relatively more constant rate than the $CaCO_3$, so that at times when little or no $CaCO_3$ forms, an organic-rich band occurs in the shell. On the other hand, the banding may result from periodic solution at the shell surface. When the bivalve shell is closed, the animal respires anaerobically and produces acids that are neutralized by solution at the shell surface. The organic matrix that was associated with the dissolved carbonate is left as a residue on the shell surface. When the valves are reopened and the animal resumes aerobic respiration, calcification is also renewed, and the organic matrix residue is incorporated within the new shell material, giving it a high organic content and a different color. The alternating pattern of shell opening and closing (and of respiration) varies regularly with environmental conditions (especially the tidal cycle and day and night), so that the banding that results from this behavioral pattern correlates with environmental conditions (Lutz and Rhoads, 1977). Jones (1983) has reviewed patterns and causes of banding in molluskan shells.

Growth lines on the outer surface of bivalve shells have also been studied in detail. Davenport (1938) first suggested that the fine growth lines on the surface of *Pecten* shells recorded daily growth intervals, and noted annual lines on *Pecten* shells resulting from winter cessation of growth. He proposed that the number of daily lines between the annual lines should increase with increasing temperature, reflecting a longer growing season. This potential method for determining paleotemperature trends has never been adequately tested on fossil *Pecten*. Clark (1968) later confirmed the daily origin of the fine growth lines by observation of *Pecten* specimens growing in aquaria as well as in nature. The same cycles and events that are recorded in the internal shell structure are preserved in these growth lines. One difference is that occasionally the *Pecten* specimens do not produce a growth line during the day so that the number of observed growth lines for any time interval is commonly a few less than the number of days.

Growth lines on the shell exterior in some cases can be correlated with internal banding. Working with the bivalve *Prothothaca*, Peterson and Ambrose (1985) noted that in specimens living in a muddy substrate, exterior lines and internal banding show annual cycles. Specimens living in a sand substrate show prominent growth lines that are more frequent than annual. Others have also suggested that some prominent growth lines, both internal and external, are produced by storms or other stress events and do not record annual cycles.

One obvious use for the banding pattern in fossil bivalve shells is to determine the number of days in the year, in much the same manner as has been done with the corals (see above). This has been done and the results are in general agreement with those from the corals (Pannella et al., 1968), although they show a somewhat more complex pattern of variation in the length of the day through time. Growth lines in nautiloids and ammonites have also been interpreted as showing daily and monthly growth patterns. Kahn and Pompea (1978) suggested that the number of days per lunar month could be determined by counting the number of growth bands between septa in nautiloids. However, Saunders and Ward (1979), Landman et al. (1988), and others give evidence that contradicts this conclusion.

External growth rings in ammonite shells show a cyclic spacing, suggesting a fortnightly (lunar monthly?) growth pattern. If this conclusion is correct, the growth rings can be used to determine the age of the shells. Doguzhaeva (1982) used this method to determine ages in the range of $2\frac{1}{2}$ to $4\frac{1}{2}$ years for mature Paleozoic and Mesozoic specimens which he examined. However, Saunders (1984) demonstrated that modern *Nautilus* may live for many years after growth ceases at maturity. Thus the age recorded by the shell may not be the true maximum age of the animal.

The shell of a bivalve or any organism having a shell that grows by accretion can be considered as a record of the life history of the organism, because shell material that was produced throughout the life of the individual is preserved in the shell. The banding pattern contains the record of the growth of the animal, and to the extent that environmental and biological factors control growth, variations in these factors will be recorded in the banding pattern.

Rhoads and Pannella (1970) discuss the factors controlling the banding pattern and indicate some of the paleoecological and paleobiological applications that can potentially be made from a study of the banding pattern. They also demonstrated a number of environmental effects on the banding pattern of modern specimens of bivalves. They showed that growth rate in *Mercenaria mercenaria* is dependent on substrate type, being higher on sandy substrate than in mud. The time of death of *Gemma gemma* was studied and mortality was found to be highest in autumn and early winter. In fossils such information could tell us something about the biology of the species and would also imply seasonal climates. Species of the genus *Nucula* in depths of 3 to 4970 m were compared. The banding in the shallow-water forms was found to be much more pronounced and irregular than that in the deep-ocean specimens. This suggests that shell banding in fossils can potentially provide information on paleobathymetry.

Jones (1980, 1981) and various co-workers have extensively studied banding patterns in the western Atlantic species *Spisula solidissima* (Fig. 4.16). This slow-growing bivalve produces two prominent bands each year, one in the summer and the other in the winter. Large specimens may have more than 30 pairs of bands. Band width varies in this species in response to temperature. Off the New Jersey coast, growth is actually more rapid during colder than warmer years. Williams et al. (1982) have expanded this work by study-

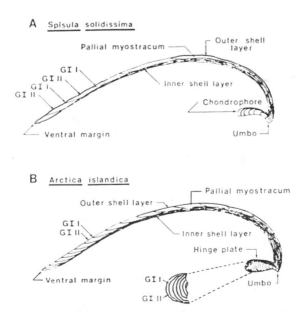

Figure 4.16 Radial cross sections through the shells of *Spisula solidissima* (*A*) and *Arctica islandica* (*B*) showing the alternation of the annual shell growth increments, GI I and GI II. After Jones (1980).

ing oxygen and carbon isotopic composition of bands and subdivisions of bands. This has confirmed the seasonal nature of the bands and allowed quantitative determination of the actual temperature. Study of variation in the isotopic composition of specimens from different depths allowed them to determine the location of the thermocline and its effect on productivity. Work with fossil specimens of *S. solidissima* suggests that these techniques will be useful in studying ancient environments as well.

Studies of banding have several potential applications in paleoecology, paleobiology, and archaeology. Variation in growth rate as determined from banding (e.g., Tanabe, 1988) may be used to determine relative paleotempera-tures, as growth rate usually varies directly with temperature. Paleolatitude might also be estimated by variation in development of seasonality in the banding pattern in much the same way as attempted by Ma in his work with fossil corals. The occurrence of storms or other high-stress situations in the geologic past might be suggested by the presence of irregularly spaced inter-ruptions or slowing of growth. Banding can reveal the season of death of the specimen, allowing identification of such features as mass mortality. Banding studies can potentially be useful in conjunction with studies of population dynamics (see Chapter 8) in determining age of specimens and season of death.

Determining the season of death of clams used as food by ancient man can aid studies of early human dietary patterns. Coutts (1970, 1975) made such a study of banding in *Chione stutchburyi* shells collected from archaeological sites in New Zealand. By studying the banding pattern in the shells he was able to determine the time of year when the bivalves were collected for food and thus the season of occupation of the midden sites. Many of the sites were occupied only during winter.

Periods of exceptionally high temperature may cause slowing of growth rate or disturbance zones in the shell banding pattern. Kennish and Olsson (1975) note such effects in the shells of *Mercenaria mercenaria* that were growing in a bay near an area of cooling-water discharge from a power sta-tion. They suggested that the shell structure of this species could be used as a natural recorder of thermal conditions near such installations. Fritz and Lutz (1986) propose a similar use of banding patterns in the freshwater bivalve *Corbicula* to monitor environmental perturbations.

SKELETAL BANDING IN OTHER FOSSIL GROUPS

To date, most work on shell banding has been confined to the bivalves and corals because banding is better developed than in other groups. However, the methods can potentially be used on any organism that forms its skeleton by accretion. Perhaps different techniques of sample preparation will aid in the study of these groups. Some effort has been made in studying banding in stromatolites (Pannella, 1972), although the reliability of data from stromatolites has been questioned (Park, 1976). Banding in stromatolites is, of course, not skeletal banding and is perhaps better treated in a discussion of biogenic sedimentary structures and trace fossils (Chapter 6). Seasonally produced growth bands have been identified in fossil coralline algae (Wright, 1985).

Rosenberg (1982) identified bands in Ordovician brachiopods. These bands appear to be produced by growth variation correlated to tidal cycles. The maximum number of cycles identified may correspond to the number of days in the lunar month. Hiller (1988) noted three types of growth lines in articulate brachiopod shells: (1) diurnal, (2) seasonal, and (3) disturbance.

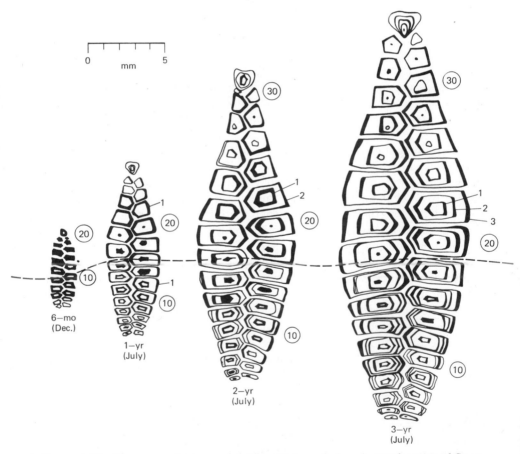

Figure 4.17   Diagrammatic representation of interambulacral growth zones of *Strongylocentrotus purpuratus* at 6 months, 1 year, 2 years, and 3 years of age. Dark areas represent translucent zones. The end of the 3 years of life is marked on representative plates. Plate position numbers, indicated in the circles, show how plates appear to shift toward oral side of test as new plates are added aborally. The dashed line shows the position of the ambitus (greatest test diameter). From Pearse and Pearse (1975), *Amer. Zool. 15:*748, by permission of the American Society of Zoologists.

Very prominent annual growth banding has been described in echinoderm plates (Fig. 4.17; Pearse and Pearse, 1975). This group might also offer promise in yielding paleoecologically useful information. Otolith (fish ear bones), and probably other vertebrate structures as well, also show an annual and tidal or daily banding pattern (Campana, 1984; Radtke, 1987).

A pattern of annual color banding in some corals was noted by Risk et al. (1987). They ascribed the bands to seasonal variation in boring by endolithic algae in the coral skeleton. Boring activity of these algae increases during the dry, sunny season and decreases during the cloudy season. They use similar banding in Devonian stromatoporoids and Ordovician rugose corals to determine annual growth rates.

Shell structure may yield much paleoecologically useful information, but its potential has just begun to be realized. The shell of an organism that grows by accretion is analogous to a tape recorder. It preserves the record of the events during the entire life of the organism as the shell material forms at the growing edge much as the tape rolls past the recording head. The task of the paleoecologist is to learn how to play back the "tape" and retrieve the information recorded in the shell structure. The shell is also somewhat analogous to a crustal plate that is formed at the midoceanic ridge or rise and then is displaced away from the rise. In the case of the accreting shell the growing edge moves or progressively recedes from the shell margin once it forms. The growth bands in the shell are somewhat like the magnetic bands in the oceanic crust formed by magnetic polarity changes.

5

Adaptive Functional Morphology

The fitness of organisms to their environment has long been recognized. The morphology, physiology, and life history of organisms are so well tuned to the environment in which they live that for many people it has been a significant demonstration of nature's providence. With the development of evolutionary theory the fitness became less a demonstration of original creation than of a continuing dynamic and progressive process of change that maintains the species in an optimal well-adapted relationship with its environment, as both the physical and biological aspects of the environment may change through time.

THEORY

The observation of the fitness of living organisms to their environment provides the initial basis for interpreting the morphology of fossils in order to determine ancient environments. The adaptation of organisms is so impressed on us from our own observations or from the natural history literature and television programs that we take it for granted, albeit sometimes with amazement. For example, plants of many different lineages are adapted to life in dry habitats; that is, they are xerophytes. They possess a suite of adaptations for obtaining and conserving moisture, such as shallow but extensive root systems, thick fleshy stems to store water, waxy cuticle to reduce transpiration, spines to inhibit browsing animals, reduced leaves to keep surface-to-volume ratio low, and chlorophyll in the stems to make up for loss of leaf surfaces.

The stresses characteristic of life in a desert habitat have presumably always been more or less the same, and by analogy with modern desert biota, the range of responses in life habit and morphology are limited and reasonably well known and predictable. Thus xerophytic plants were presumably similarly adapted in the past because the physiology of plants provides only a number of ways to cope with aridity. From this observational basis, and because plants lacking the appropriate physiologic potential will not be able to adapt to aridity and thus would not be part of the desert flora, we infer that adaptive functional morphology possess great potential for paleoenvironmental analysis. Going back in geologic time, lower-level taxa become extinct, so our level of comparison with modern taxa is at an increasingly higher taxonomic level. However, we assume that even at these higher taxonomic levels morphologic responses to the biotic and physical aspects of the environment have not changed so greatly. Is this assumption reasonable?

The theoretical basis for adaptive functional morphology is provided by the modern theory of evolution through natural selection. We know that organisms are adapted to their environment; therefore, in the analysis of a fossil we can ask ourselves what the function and possible environmental significance of the morphology was. If we can answer that question correctly we can predict something of the environmental conditions under which the fossil lived.

In any population, the full genetic content of the population—the genotype—will be transmitted unchanged to the next generation unless outside forces or chance intervene. However, all individuals in the population are not equally likely to survive because genetic and corresponding physiologic and morphologic differences will cause them to be differently adapted to the environment. Through the selection process, and in the absence of chance phenomena, only the best adapted individuals form the breeding population that gives rise to the next generation and that provides the genetic material that will determine the genotype and, therefore, phenotype of that generation. Selection is the "outside force" operating to change genetic composition, which according to the Hardy–Weinberg law should otherwise remain constant.

The phenotype of the organism, that is, its morphology, physiology, behavior, or any other expression of its genetic composition or genotype, will vary because natural selection directs evolution of the organism toward optimality or maximum efficency. It is important to remember that this optimality is in terms of the entire developmental history of the organism, not of a specific morphologic trait that might be analyzed as isolated features (Mayr, 1983). Organisms with the same genetic composition may also have variable morphology, physiology, or behavior because of direct and immediate influence of the environment during their life span. This is ecophenotypic variation, and is commonly difficult to distinguish from morphology determined by natural selection (Johnson, 1981).

For example, the morphology of fossils within a stratigraphic sequence might be genetically determined, developed through long-term selection, or it might represent ecophenotypic response, generation after generation, to the same environmental stress. An example of the difficulty of distinguishing ecophenotypic and evolutionary effects is exemplified by the micromorph fauna described in the the middle Cretaceous Grayson shale of east central Texas by Mancini (1978a,b). Mancini described the possible environmental conditions and mechanisms by which the members of an assemblage would be generally smaller than the conspecific representatives in adjacent strata: (1) evolutionary selection in response to specific and persistent environmental conditions; (2) ecophenotypic dwarfing or stunting, but not necessarily strong selection pressure and evolution; (3) transportation and sorting by size; and (4) a seasonal pattern of recruitment and mortality resulting in specimens that are juveniles of normal larger taxa, and small but adult specimens of taxa that attain normal size within the short time interval from recruitment to seasonal kill-off. Mancini described the criteria and the evidence by which one could arrive at a choice of the most probable mechanism (Table 5.1). He concluded that in the particular case of the Cretaceous fauna he studied, the cause was most probably evolutionary, with the small size representing an adaptation for survival on a soft substrate. To distinguish these possible causes is not easy, however, as indicated by the earlier interpretation of stunting by Kummel (1948) and of juvenile mortality by Britton and Stanton (1973). Because it is difficult to distinguish evolutionary from ecophenotypic effects, we can normally assume only that the organism is adapted through selection for its environment, but it is important to make this distinction because the ecophenotypic morphology is not inheritable, and thus the significance of the particular morphology within an evolutionary progression is very different in the two cases.

Under conditions of long-term stability, the genotype and phenotype should remain constant, although the amount of intrapopulation variability is unspecified. Under changing environmental conditions, however, the genotype in successive populations will change, so that an evolutionary dynamic equilibrium is maintained by maximizing at any time the average efficiency of the species within the ecosystem.

This model of adaptiveness has proven to be very useful in explaining the morphology of fossils because, as noted above, it should be relatively time independent, and because a specific anatomical-physiological system should evoke a limited range of morphologic adaptations in response to the external environment.

This evolutionary basis for adaptive functional morphology must be examined in more detail, however, before it can be used to interpret the fossil record: (1) How strongly deterministic and finely tuned is the action–interaction couplet of adaptive morphology and natural selection? (2) What is the role of other factors in addition to natural selection in determining morphology? (3) What is the the extent to which one can expect unique morphologic responses to specific environmental conditions?

TABLE 5.1 Micromorph models and diagnostic criteria[a]

Diagnostic Criteria	Stunting[b]	Transportation	Juvenility	Pedomorphosis[c]
Macrofaunal				
Population dynamics				
Age distribution	Normal		Immature animals	Normal adult and gerontic animals rare
Growth curve	Slow decline		Truncated	Rapid decline
Maturity indicators	On smaller-than-normal animals		Absent	On younger than normal animals
Size-frequency distribution	Moderately positively skewed	Narrow and symmetric	Highly positively skewed	Highly positively skewed (S); moderately positively skewed (I)
Survivorship curve	Sigmoidal		Rapidly decreasing	Decreasing then linear (S); linear (I)
Diversity	Lowest		Low	Moderate (S); low (I)
Trophic structure	High-deposit-feeding organisms, (LF); high-filter-feeding organisms (LO)			High filter feeding organisms (S)
Condition and orientation of fossils		Altered and broken, may be oriented		

Characteristic		Stunting factors[b]	Pedomorphic factors[c]
Other characteristics		Stenohaline taxa rare (LS); high nektonic/benthonic ratio (LO)	Soft bottom morphologies and behaviors (S)
Trace fossils		Deep vertical burrows (LS, HS); systematic traces (LF); rare (LO)	Wispy indistinct burrows (S); deep vertical burrows (I)
Microfaunal diversity and composition	Mixed fauna with altered tests	Low diversity; high agglutinated/calcareous benthonic ratio, miliolids rare (LS); low agglutinated/calcareous benthonic ratio, miliolids abundant (HS); high planktonic/benthonic ratio, smaller than normal individuals with smooth and thin tests (LO)	High smooth-shelled cytherellid/ornate ostracode ratio (S); low diversity (I)
Sedimentary structures, texture, and geochemistry	Stratification	Low trace metals, high Fe and organic C (LS); low P and organic C (LF); lamination, high S, Fe and organic C (LO)	Clay-size particle texture (S); sediment record of fluctuating factor (I)
Paleogeographic isolation		Absent	Present (I)

[a]From Mancini (1978a).
[b]Stunting factors: LS, low salinity; HS, high salinity; LF, low food; LO, low oxygen.
[c]Pedomorphic factors: S, substrate fluidity; I, instability.

Natural selection can evidently result in very fine morphologic detail, as indicated by the close similarities in morphology and geographical distribution for examples of mimicry. On the other hand, it can be argued that natural selection should generally not be so finely tuned to environmental conditions for a number of reasons:

1. Environmental variability on a scale of duration from much less than a generation to many generations in length means that the adaptational objective of natural selection is a moving target, and thus only a generalized result is possible within the genotype.
2. Random processes such as occasional density independent catastrophic mortality events get in the way of a steady and invariant natural selection pressure on the genotype.
3. The genotype, even if finely tuned, may not be expressed in the phenotype.
4. The morphology may be vestigial, determined by some previous selection event but now under neutral selection pressure.
5. The morphology may be an expression of selection for another, but gene-linked phenotypic characteristic.
6. A morphologic characteristic may be highly adaptive at one stage of an organism's life history, but then be of neutral or perhaps even negative value at others. Thus it may be difficult to relate the morphologic characteristic to the ontogenetically specific environmental parameter. It is important to remember in analyzing morphology that selection does not optimize each particular trait or stage of life, but the total life history of the animal—the epigenetic system—which ensures that individuals of a population will survive to the point of reproducing to generate the next population.

The phenotypic results of natural selection are not unlimited, but are constrained by historical-phylogenetic and constructional factors (Fig. 5.1). The historical-phylogenetic factor is expressed in the physiology and developmental system of an organism. Individually, organisms resemble their ancestors. Within the population, distinctive phenotypic characteristics reflect genetic characteristics that are the result of past environmentally determined selection and the nonselective factors mentioned above. The historically established unique aspects of the genotype limit the evolutionary changes that are possible in the future. This rationale applies equally well to the established and discrete populations of a species throughout its history.

The constructional constraint, also referred to in the literature as the structural or architectural constraint (Raup, 1972), is determined by the nature of the materials available and by nonselective aspects of the resulting skeleton. In building a bridge, engineers may choose from a range of materials. The form of the resulting structure will be very different, depending on whether

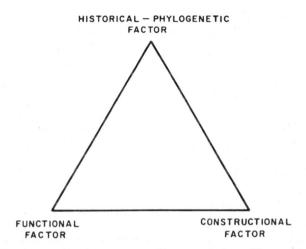

HISTORICAL — PHYLOGENETIC
FACTOR

FUNCTIONAL
FACTOR

CONSTRUCTIONAL
FACTOR

Figure 5.1 The three major factors controlling morphology. The position of the morphologic feature in the triangle represents the relative influence of the factors. After Seilacher (1970), Lethaia, *3:* 394.

they are using wood, stone, prestressed concrete, or a combination of cable and steel to build a suspension bridge. Similarly, the calcitic and/or aragonitic exoskeleton of a pelecypod limits the form and size as well as life habit of the animal.

Gould and Lewontin (1979) have used the spandrels that form between arches joining at right angles as a launching point for a broad discussion of adaptive functional morphology and the methodology of analyzing it. Spandrels are an inevitable result of the arched design, just as a pediment inevitably forms under the end of a gabled roof. Because these are not the immediate result of designing for arches or a gable, but are constructional consequences, they cannot be interpreted in terms of primary selection. (They may be utilized in a variety of ways, however, that may be of considerable adaptive significance). This emphasizes again that the entire organism must be considered as an epigenetic system, although the usual approach in morphologic analysis is to identify and analyze the individual parts or traits of an organism. The organism must integrate all of its traits into a functional unity, which necessarily implies compromises, trade-offs, and different levels of adaptiveness for different traits.

The importance of phylogenetic-historical and constructional factors is well illustrated by the morphology of an echinoderm, which is limited by its fundamental anatomy and method of building a skeleton—an internal skeleton of plates composed of single crystals of calcite with an open meshwork construction, a water vascular system, a stenohaline physiology, and a pentaradial symmetry. Even more fundamentally, Dafni (1986) and others preceding him have argued that mechanically the overall shape of an echinoid can be analyzed in terms of a balloon in which the gravitational forces on the body and the outer surface tension are balanced. These constaints leave natural selection a limited range of options, and in regular echinoids much of the adaptive morphologic consequences are expressed in size and in spine characteristics. Because of these constraints, how an echinoderm adapts to an environmental situation will perhaps be very different from an organism in another phylum with different phylogenetic and constructional constraints on the body plan. Seilacher (1970) also points out that some morphologic features have no particular adaptive function but result from the method of growth of the organism. One example is the polygonal shape of coralites in a coral colony, which may simply be the result of physical interference of the growing cylinder-shaped coralites as they impinge on one another.

Our discussion of adaptive functional morphology began with the example of xerophytic plants, which although in widely scattered deserts, possess characteristic attributes reflecting common solutions to the problem of surviving in this rigorous habitat. We implied that these morphologic features were determined by natural selection working toward a solution, with only a limited number of avenues open to it. How valid is this generality, and to what extent should it be expected in diverse taxa? One might expect, for example, that xerophytes evolving in a single lineage would be similar because of the common historical-phylogenetic and constructional bases, but what are the responses of plants belonging to different genera or families? Similarly, should we expect similar adaptive functional morphologic solutions by nautiloid and ammonoid cephalopods or by terebratuloid and rhynchonelloid brachiopods to the particular and similar environmental stresses of each of these pairs of taxa? Are the open, lacelike tests of diatoms and radiolarians a common solution by these two very different organisms to the same problems of remaining in suspension and of building an SiO_2 skeleton when this material is present in low concentrations in the ocean? If the light and open skeleton is a common adaptation, detailed differences in the test morphology presumably reflect the historical-phylogenetic constraints of their differing physiology and test-building procedures. These questions are not easily answered. As a first approximation, the degree of similarity in morphologic response should be directly correlated with the similarity in historical-physiological and architectural characteristics. The resulting limited range of possible avenues for adaptive response channels evolution into common, homeomorphic solutions. At the same time, if we survey organisms living in a particular environment, we commonly find that they are able to survive by means of diverse adaptations. In evolutionary terms, progressive change along different pathways has led to morphologically expressed multiple stable points. The consequence of the fact that most problems can be solved in more than one way is that it will generally be difficult to arrive at simple morphology–function or morphology–environmental parameter pairs.

Morphologic analysis has taken two paths: (1) The natural history approach observes living organisms, determines the geographical and habitat distributions of their morphologic characteristics, seeks to correlate these morphologic patterns with environmental parameters, and infers causation from these correlations; and (2) the engineering approach focuses on the fossil itself to try to understand how it lives and what the functional explanations of its morphology are. The engineering analysis may be at the natural history level but may also involve physical modeling or mathematical simulation. This is basically a deterministic, engineering procedure that was most clearly formulated in the paradigmatic method by Rudwick (1964a). Although seldom carried out formally, it is described here in some detail because it contains the basic assumptions and procedures inherent in all morphologic analysis.

1. The structure to be explained is identified and described. It may be the total organism, but is more commonly only a limited part of it. Morphological analysis tends to focus on abnormal structures. Examples are the aberrant morphology of rudist bivalves as compared to other bivalves, or the extreme neck length of the giraffe as compared to related mammals. Detailed description of the structure is important in the analysis and has resulted in numerous efforts to characterize form in simplified and mathematical terms—the subject of theoretical morphology.

2. All possible functions for the structure are imagined. This is perhaps the most difficult step in the analysis, for how can we hope to imagine the functional aspects of an extinct organism whose physiology is largely unknown and whose life history can only be guessed at? This step may be aided by comparing the fossil with homologous structures in living relatives or with analogous structures in unrelated taxa. The more remote the modern analog is in age or relationship, the more speculative will be the hypothetical functions. At this stage of analysis, the paradigmatic method views an organism in a very mechanical way, as an integrated assemblage of well-adapted individual parts. Because the analysis necessarily assumes that the structure has a function, difficulties arise if the structure in fact is vestigial or is genetically linked to other highly adapted structures, but is of neutral or even negative value itself.

3. The ideal structure—the paradigm—to fulfill each of the possible functions conceptualized in the preceding step is formulated. The historical-phylogenetic and constructional constraints must be taken into account at this step because they limit the range of structural paradigms. An additional consideration, as mentioned previously, is that a function may be solved in a number of quite different and independent ways. The conventional paradigm for speed in a quadruped is quite different for the placental cheetah or gazelle (long forelegs and hindlegs) than it is for the marsupial kangaroo (powerful hind legs with a large tail for balance). Also, a functional need may be solved by the combination of several structures and by physiologic and behavioral adaptations not evident in the fossil record. For example, one solution by plants to the problem of lack of water in the desert is for otherwise rather ordinary appearing plants to grow rapidly after an infrequent rainfall, complete the life cycle, and produce seeds that do not germinate until the next

rain. Such a plant would appear much like a plant from a more humid region, and its adaptations for life in the desert might not be evident. The initial formulation of the paradigm is a mental construct based on analogous structures in other organisms or on engineering considerations. Use of physical models or actual specimens in experimental analysis of structures has become an important source of information at this step, and calculations of strength and performance in engineering terms may also be useful.

4. The actual structure is compared with the paradigms, and the function of the most similar paradigm is assumed to be the function of the actual structure. To avoid being simplistic, this interpretation must consider again the complicating factors noted in step 3. In addition, because the total organism must be a well-integrated and harmoniously organized entity, individual structures may be less than the paradigm, and they may function at less than their independent optimal levels.

5. The inferred function is explained by the causative environmental parameter.

The paradigmatic method is obviously highly speculative and involves interpretations that cannot be proven. Yet the essence of the method is inherent in all adaptive functional morphologic studies. The results are most tenuous when the organism being analyzed is extinct, with no closely related living relatives for comparison, and morphologically analogous but unrelated organisms are also lacking. The resulting conclusions are at best reasonable and possible but cannot be demonstrated to be the only possible conclusions.

The analysis of the skull morphology of the Miocene North American

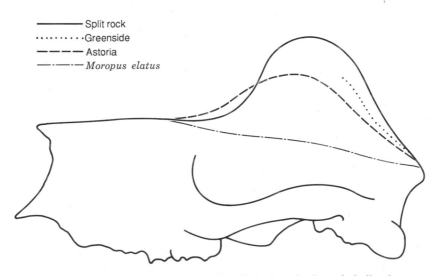

Figure 5.2 Comparison of posterodorsal skull outlines in domed chalicothere specimens and in *Moropus elatus*. From Munthe and Coombs (1979).

chalicotheres by Munthe and Coombs (1979) is an example of the formal application of the paradigmatic method. In three specimens of this large terrestrial perissodactyl vertebrate, the upper surface of the skull is distinctly domed (Fig. 5.2). The analysis of this anomalous structure is outlined in Table 5.2. In the first column, the possible functions have been listed; in the second and third columns, the similarities and dissimilarities of the actual structure to

TABLE 5.2 Possible Functional Interpretations of Chalicothere Domes, Arranged in Approximate Inverse Order of Probability[a]

Interpretation	Corroborating Information	Objections
I. Aquatic specialization		Other aquatic specializations absent in chalicotheres.
A. For buoyancy	Dome has enclosed space that could have contained air.	Enclosed volume (about 800 cm^3) too small relative to body size to increase buoyancy; position of dome too high on body; most mammals neutrally or positively buoyant and none known to develop skeletal structure to increase buoyancy.
B. For increased feeding time under water	These functions long used to explain hadrosaur crests (see Hopson, 1975, for historical review).	
1. As a snorkel		Dome has no opening to serve as snorkel entrance.
2. U-tube traps water and prevents its entering lungs		No evidence of elongated tubes in dome; Ostrom (1962) discredited U-tube hypothesis for hadrosaurs.
3. Storage area for air to be drawn into lungs		Volume of dome miniscule compared to lungs (see Ostrom, 1962); hypothesis would require special valves to regulate water and air.
II. High-impact combat and defense	Analogy with fossil and living dome-skulled forms; some supporting strutwork in dome.	No skull or vertebral modifications for holding opponent or absorbing large impact forces.

TABLE 5.2   (Continued)

Interpretation	Corroborating Information	Objections
III. Myological specialization		
A. For support of proboscis	Wilfarth (1938) suggested a proboscis associated with hardrosaur crests.	Ostrom (1962) and others have countered Wilfarth's idea; living animals with proboscis have no skull doming (minor doming present in *Tapirus bairdii;* nasal bones of domed chalicothere skulls not reduced or retracted; snout area well ossified; no scars for proboscis muscles on or near dome.
B. For mastication	Temporalis muscle originated on domed part of skull; Kurtén (1976) hypothesized skull doming of some *Ursus spelaeus* associated with variable positioning of temporalis to emphasize different parts of dentition.	Upper limit of temporal fossa well below top of dome (i.e., temporalis origin unexpanded, unlike arctoid carnivores with domed skulls); chalicothere dome too far posterior on skull to alter significantly direction of force to coronoid process; little concomitant modification of dentition and/or coronoid process.
C. For holding head erect	Chalicotheres may have browsed on high tree leaves.	
IV. Desert specialization		No evidence of other desert adaptations in chalicotheres (despite Matthew's suggestion in 1929 that claws used to dig for subsurface water).
A. For water retention by countercurrent exchange (as outlined by Schmidt-Nielsen et al., 1970)	Elaborated sinuses in dome may have communicated with nasal passages; dome allows additional space in which long narrow tubes for exchange could be elaborated.	No evidence of long tubular structures within dome; nasal passages unmodified from condition in *Moropus elatus.*

TABLE 5.2   (Continued)

B. For better humidifying or cleansing of air	Possible extra surface area for these functions (if dome bore nasal epithelium and communicated with nasal passages).	No evidence of turbinal bones (usually support nasal epithelium) in dome; no evidence of special connection between dome and nasal passages; nasal passages unmodified.
V. Enchanced olfactory acuity	Dome could have provided extra surface area for olfactory nerve endings.	No evidence of turbinal bones in dome; olfactory bulbs and cribriform plate (preserved in Greenside skull) similar to those of *Equus;* olfactory acuity in living mammals not generally associated with skull doming.
VI. Acoustic signaling (functioning in intraspecific recognition, attraction, or challenge)	Hadrosaur crests have been suggested as sound producers or resonators (see Hopson, 1975, and others); frontal sinuses may enhance vocalization in living mammals (e.g., in humans).	Partitioned (strutted) dome could not have functioned as *single* resonating chamber.
VII. Visual recognition of conspecifics	Dome is high on skull and easily visible; sexual dimorphism (Coombs, 1975) and concentrations of chalicothere remains at Agate and elsewhere suggest some groupings of conspecifics and possible group structure.	No negative evidence known.
VIII. Visual display (males attracting females, males threatening competing males)		Sex of animals unknown (males expected to have larger domes).
IX. Low-impact butting (low velocity ritualized butting)		Chalicothere dome unassociated with bony protuberances (horns, ossicones, antlers), unlike most living low-impact combatants with skull domes; sex of animals unknown; orientation of internal partitions of dome unknown.

[a]From Munthe and Coombs (1979).

the paradigm for each hypothesized function are presented. The conclusions from the analysis are presented, appropriately, in probabilistic terms: The most probable function for the hollow dome was behavioral—for low-impact butting and for visual display; its use for visual and acoustic signaling was also possible; other functions are considered to be unlikely in varying degrees. The authors also emphasize that this analysis does not preclude the possibility of multiple functions for the domed skull. Finally, of course, could this strongly deterministic analytical technique overlook the possibility that the doming is pathogenic and that an adaptive explanation is being sought where none exists? This analysis of the chalicothere gives the impression of being overly structured and consequently artificial. Its strength, however, is that its logic is clearly laid out, and a multiple-hypothesis method is established by listing all the imaginable functions. From this, the analysis leads to clear preferences among the possible choices. More commonly in studies of adaptive functional morphology, the possible functions are not spelled out; instead, the favorite function is evaluated against the structure and if the function and the structure match satisfactorily, the analysis is concluded. The resulting sufficient but not necessary conclusion is unsatisfactory for just this reason, and has lead to the perception of adaptive functional morphology as generating "just so stories" with no rigorously established validity and little in the way of conclusions that contribute to general rules with predictive value.

TABLE 5.3 **Selected examples of adaptive functional morphology, arranged by taxonomic group**

Fossil Group	Reference
Calcareous algae	Bosence, 1976
Foraminifera	Malmgren and Kennett, 1978; Wetmore, 1987
Sponges	Bergquist, 1978; Stearn, 1984
Coelenterata	Grauss et al., 1977; Haggerty et al., 1980
(Corals)	Stearn, 1982; Kershaw, 1984
Bryozoa	Rider and Cowan, 1977; McKinney, 1984
Brachiopods	Rudwick, 1964b; Grant, 1972; Alexander, 1984
Mollusks	Hickman, 1985; Ward and Westermann, 1985; Seilacher, 1985; Hewitt and Westermann, 1986; Hayami and Oka-moto, 1986; Savazzi, 1989
Arthropods	Stitt, 1976; Schmallfuss, 1981; Jaanusson, 1984
Echinoderms	Seilacher, 1979; Ausich, 1983; Smith, 1984
Vertebrates	Hopson, 1975; Stein, 1975; Thomson, 1976
Theoretical	Beatty, 1980; Gould and Vrba, 1982; Mayr, 1982; Reif, 1984; Fisher, 1985

Table 5.3 lists examples of studies of adaptive functional morphology, organized by the specific taxa that are the subject of each study. The subsequent discussion, however, presents examples arranged by methodology rather than by the kind of organism analyzed.

ADAPTIVE FUNCTIONAL MORPHOLOGY STUDIES

Morphologic Distribution Patterns and Gradient

The simplest approach to adaptive functional morphology is to recognize gradients or discrete discontinuous distribution patterns of morphologic features among modern organisms. If these patterns can be related to environmental parameters, they are potentially valuable in paleoenvironmental analysis, even if the reason for them is poorly understood. Some of these relationships, such as the general latitudinal increase in size for animals, presumably reflecting slower growth, longer life span, and later attainment of sexual maturity in colder regimes, have been recognized for many years.

A similarly well established example is the correlation between climate and the shape of angiosperm leaves. Angiosperm leaves can be separated into those with entire, smooth margins and incised, lobed, and toothed margins (Fig. 2.26). Botanists have long noted that large, thick leaves with entire margins are more common in warm climates, whereas smaller and thinner leaves with highly incised margins are more common in cool climates. (Bailey and Sinnott, 1915). The percentage of plant species with entire margins can be correlated directly with mean annual temperature (Fig. 2.27) and have been used to determine Tertiary climates of North America (Fig. 2.28; Wolfe, 1978).

Correlation of bathymetry with the morphology of benthic foraminifers was proposed by Bandy (1960a,b;1964). This approach has been used commonly in the petroleum industry to arrive at bathymetric interpretations of Tertiary foraminifers because it eliminates much of the tedious task of identifying the microfossils and provides results that have proven to be useful. The cause of morphologic changes with depth, however, is very poorly understood. Depth, in itself, probably plays only a minor role in determining morphology; a more satisfactory explanation would be in terms of the environmental parameters correlated with depth.

In a study of benthic foraminifers on the Texas shelf, Severin (1983) recognized six morphologic groups that had distinctive depth distributions (Fig. 5.3). The overall morphologic gradient is from rounded symmetrical forms predominating in the shallower part of the depth range to angular and asymmetric forms and forms with greater amounts of ornamentation and spinosity predominating in deeper water. Severin attributes this morphologic gradient with increasing depth to decreasing sediment disturbance by bioturbation and wave action.

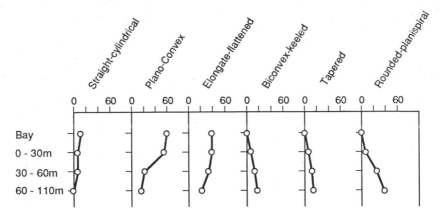

Fig. 5.3   Foraminiferal test morphology: water-depth relation, Texas Gulf coast. Data from Severin (1983).

In a similar study of deep-water benthic foraminifers in the Norwegian Sea, Corliss and Chen (1988) have also demonstrated depth partitioning of foraminiferal morphology (Fig. 5.4). They separate the foraminifers into epifaunal, living in the uppermost 1 cm of the sediment, and infaunal, living at a depth greater than 1 cm in the sediment. By means of this distinction, they show that forminiferal morphology is linked primarily to the depth at which the organism lives within the sediment. Infaunal forms are largely restricted to water depths of less than about 1500 m; epifaunal forms are predominantly at greater depths. A potential interpretation is that morphology is correlated primarily with life position relative to the sediment surface, and that some depth-dependent parameters determine the depths at which life within the sediment is possible. Corliss and Chen relate the distribution pattern to carbon flux and carbon content in the sediments, as these two parameters reflect food availability for the foraminifers. They suggest that the infaunal position would be less advantageous at greater water depth because of the decreased rate of food influx.

These two studies are interesting because they provide morphologic trends that are correlated with water depth. The value of the results would be strengthened if analysis of the data were as comprehensive as that by Munthe and Coombs (1979) for the chalicothere. Only with improved understanding of the causes of the morphologic trends can morphology be tied quantitatively to depth and thus be generally applicable.

Theoretical Morphology

A major requisite of morphologic analysis is a clear and thorough description of the overall form of the organism or the particular structure that is being investigated. Two desirable aspects of this description are (1) to separate

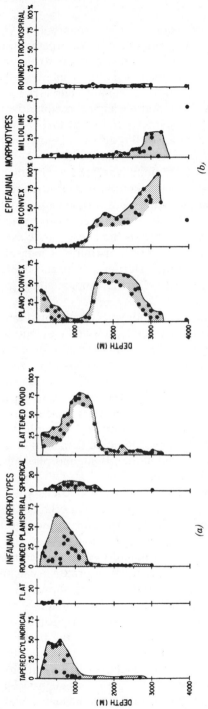

Figure 5.4 Depth distribution of deep-sea benthic foraminiferal morphotypes based on 45 Norwegian Sea samples. Morphotypes are defined by shape of foraminiferal test and nature of coiling. Total percentage of each morphotype is plotted for each sample. (*a*) Five morphotypes associated with infaunal microhabitat; (*b*) four morphotypes associated with epifaunal microhabitat. From Corliss and Chen (1988), Geology, *16*: 718; by permission of the Geological Society of America.

levels of form—that is, for example, the primary form from secondary ontogenetic changes, successively finer degrees of ornamentation, ecophenotypic effects, and pathologic irregularities; and (2) to do this as simply as possible. An example of this distinction that was mentioned previously was the test of diatoms or radiolaria. The primary characteristic is the minimal amount of skeletal material used, and perhaps consequently, the open meshwork structure. This has a general interpretation probably valid for both these very different groups. The detailed differences in the test, on the other hand, require more specific and lower-order explanations.

Bayer (1977a,b, 1978) has also emphasized the importance of recognizing the different levels of morphologic adaptation. He describes form as determined jointly by an internal growth program than can be defined by simple mathematical expressions and by the external ecophenotypic and selectional processes. The internal growth pattern is the conservative aspect of form and incorporates the historical-phylogenetic constraints. Changes in morphology represent "weak points" where the external factors can override the growth program and can be best understood if the fundamental growth program is first established.

Early efforts in theoretical morphology concentrated on describing the growth form of organisms such as cephalopods and gastropods in terms of the logarithmic spiral. A major stimulus in theoretical morphology was Raup's (1966b) description of gastropod growth by means of three parameters: the expansion in size of the generating curve or cross section of the whorl, the translation of the generating curve along the axis around which it spirals, and the change in distance of the generating curve from the axis (Fig. 5.5). This morphologic characterization has the advantage of simplicity, of providing a basis for examining second-order morphologic details as departures from or variations on the general form, and of identifying morphologic characteristics of potential environmental significance.

This last point is illustrated by the fact that shells with lower rates of expansion are more efficient in the use of shell material because, by reducing the rate of whorl expansion, the amount of shell required to build the inner whorl wall is minimized. Consequently, Raup suggested that this should be the preferred form in areas of low availability of calcium carbonate, such as at higher latitudes where the seawater is not as saturated in calcium carbonate as in more tropical latitudes, and thus that the degree of whorl expansion might constitute an indicator of paleolatitude or paleoclimate.

The other advantage of the mathematical description of the shell is that it helps to identify the limits of the morphology of the particular group being analyzed, which leads to the logical question of why other potential morphologies are apparently nonadaptive. For example, the limitations of the realized morphology for gastropods and other organisms that basically have a spiral form helps in the search for the historical-phylogenetic and constructional constraints that are imposed on each of these taxa (Fig. 5.5).

The theoretical morphology approach can be applied to a wide range of

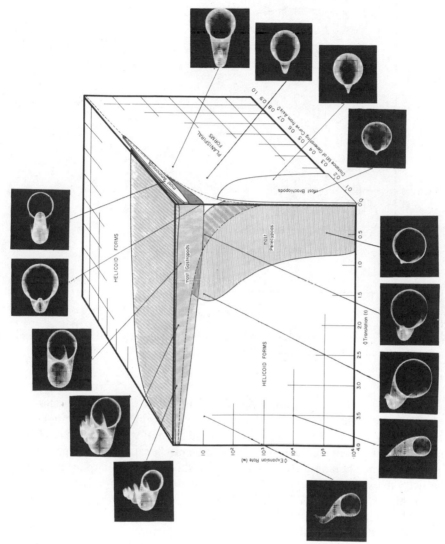

Figure 5.5 Three-dimensional block illustrating the spectrum of possible shell forms. The shape of the generating curve is assumed to be constant. The regions occupied by the majority of species in four taxonomic groups are outlined in the block. Species of these groups are not commonly found in the blank regions of the block. From Raup (1966b).

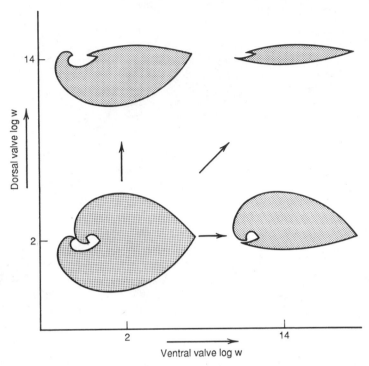

Figure 5.6 Hypothetical (computer-produced) combinations of potential biconvex brachiopod morphologies. From McGhee (1980).

taxa. McGhee (1980), for example, analyzed in detail the morphologies of four brachiopod orders. He used valve convexity as his basic form parameter, expressed by W, the whorl expansion rate, a measure of valve curvature (Fig. 5.6). This single shell characteristic is significant because it also provides insights about the shell area-to-volume relationship, the degree of asymmetry of the two valves, and the degree to which the shell growth extended out away from the hinge line (Fig. 5.7). The differences in morphology of the four orders of brachiopods are summarized in Fig. 5.8. Comparison of Figs. 5.6 and 5.8 shows that the analysis more rigorously clarifies the differences between the orders than had been done in the past, and provides a basis for explaining the selection pressures and environmental parameters responsible for these diverse morphologies. The forms might be explained by differences in ontogenetic development, in life habit, or in adaptive advantages. McGhee interprets the morphology by differences in growth program (growth of some shapes, for example, requires resorption of previously deposited shell) and in life habits. The ratio of body volume or mass to shell surface area is comparable to the ratio of gastropod volume to shell area, and thus also may be correlative with paleotemperature. However, the relationship of form and life

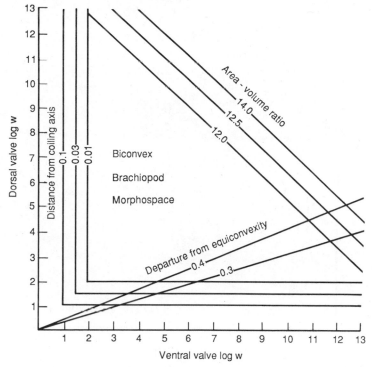

Figure 5.7   Biconvex articulate brachiopod morphospace. From McGhee (1980).

habit to substrate characteristics seems to be the most promising way to explain the functional significance of these morphologic differences.

Analysis of Specific Structures

The role of theoretical morphology in the study of adaptive functional morphology of brachiopods is illustrated by comparing McGhee's results with those of Fürsich and Hurst (1974) based on sedimentologic distribution and inferred life habits and those of Alexander (1975) based on flume studies.

Modern brachiopods occur from the lower intertidal to abyssal depths but are most abundant on the continental shelf. A number of morphologic features have been related to depth, especially in fossils, but apparently most of these features are basically controlled by depth-dependent decrease in grain size, water turbulence, and food supply (Fig. 2.31; Fürsich and Hurst, 1974). The efficiency of the food-gathering system in the brachiopods, determined by the area of the lophophore relative to the volume of the organism, should be reflected in shell shape because the volume of the organism increases as the third power of linear dimension ($v \approx l^3$) whereas area only increases as the

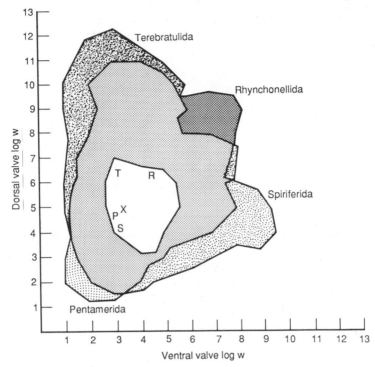

Figure 5.8　Composite of the outermost contour boundaries for the four orders of brachiopods. Regions occupied exclusively by a single order are indicated, regions of overlap between two and three orders are shaded, and the region shared by all four orders is left blank. The modal morphology for each order, and for the entire sample, is indicated by symbols. P, pentamerides; R, rhynchonellides; S, spiriferides; T, terebratulides; X, brachiopod composite peak. From McGhee (1980).

square of linear dimension ($a \approx I^2$). Thus, with increasing size, the surface-to-volume ratio of the animal, the relative proportion of lophophore surface area for feeding, and consequently feeding efficiency all decrease. Support for these inferences is provided by deeper-water present-day taxa, which do, indeed, tend to be smaller and thinner than shallow-water types. Ager (1965) noted this same trend in Mesozoic deep-water brachiopods.

In addition to small size, another strategy to improve feeding capability would be to increase the lophophore area in order to increase the number of current-producing cilia and food-trapping filaments. Unfortunately, the lophophore is never preserved in fossils, so its size can only be estimated indirectly. Two methods of estimating lophophore size are from the size and complexity of the lophophore supports (brachidia) and from the volume of the cavity in modern brachiopods. On these bases, Fürsich and Hurst suggested the following order of increasing lophophore area for the Silurian brachiopods they studied: orthids, strophomenids, rhynchonellids, pentamerids, and spiriferids. The environmental distribution of these groups generally supports this correlation between lophophore area with depth or water turbulence and correspondingly, food supply.

Most modern and fossil brachiopods are attached to a hard substrate either with the pedicle or by cementation. However, many fossil taxa lived on unlithified sediment using a variety of adaptations of morphology and life habit. If the sediment is relatively firm, present-day brachiopods may simply lie free on the surface (perhaps after an early stage of pedicle attachment to a small object) or they may have a pedicle that splits into rootlets that penetrate the sediment and attach to many small grains. Brachiopods living on a soft, soupy substrate have a problem of keeping the shell from sinking into the substrate. Thayer (1975a) indicated four strategies that a brachiopod (or other organism) might use to keep from sinking into a soft substrate: (1) reduce the density by having a thin, smooth shell; (2) develop a broad flat shape to distribute the body weight over a maximum area (the snowshoe strategy); (3) retain a small size into adulthood to reduce total mass (the micromorph strategy); and (4) keep the commissure high so that feeding can continue even if the shell is largely buried (the iceberg strategy). Many brachiopods have thin flat valves (Fig. 5.9 strategy 1). Some brachiopods have spines to help distribute their weight on the sediment (Fig. 5.9b, strategy 2). Others have developed a large interarea to accomplish this (Fig. 5.9c). The extended hinge

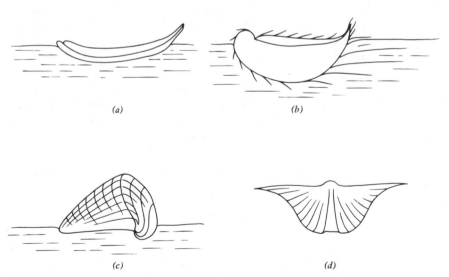

(a)   (b)

(c)   (d)

Figure 5.9 Morphological adaptations in brachiopods to prevent sinking into soft substrate: (a) thin, flat, concavoconvex shape; (b) spines to increase surface area: also deep, cup-shaped lower valve; (c) large flat interarea; (d) extended hinge (mucronate).

area or wings in brachiopods such as *Mucrospirifer* also provide support on a soft substrate (Fig. 5.9d). Small size (strategy 3) is a common feature in brachiopods living on soft substrates. A concavoconvex shape, with the lower valve convex downward to form a cupped shell and the upper valve concave upward), maintains the commissure above the level of the sediment surface in some brachiopods (Fig. 5.9b, strategy 4).

Alexander (1975) noted many of these features in Ordovician specimens of *Rafinesquina alternata*. Specimens that lived in quiet water on soft substrates are small, broad, and flat, and have a concavoconvex shape; some also have a long hinge line. Specimens of the same species that lived in more turbulent water on a firmer substrate are larger and more globose, with a biconvex shape. These shapes are so different that some earlier workers had called them different species; however, they intergrade in some environments, suggesting that the morphologic variation is an environmental adaptation.

Alexander used a combination of three approaches in analyzing the functional significance of shape and size of *Rafinesquina*. First he considered, on a theoretical basis, the adaptive reason for a particular morphology. Next he used the sedimentary features and stratigraphic setting to determine the environmental conditions under which the brachiopods lived. In this way morphology could be indirectly related to the environment. Finally, he conducted studies of models in a flume to determine the stability of shells of differing morphologies in currents.

The maximum size of *Rafinesquina* differs considerably from population to population. Several environmental factors theoretically might affect the specimen size, as has been described previously in connection with dwarfed or micromorph faunas (Mancini, 1978a). Three shape parameters for *Rafinesquina* were determined: perimeter-to-volume ratio (P/V), length-to-height ratio (L/H), and alation. P/V, the ratio of the distance along the commissure to the volume, is a measure of the thinness of the brachiopod: A high value for P/V indicates a thin specimen with large surface area but little body volume. L/H, the ratio of the distance between the anterior and posterior margins to the maximum distance from the commissure to the pedicle valve exterior, is a measure of the convexity of the pedicle valve: Low L/H values correspond to convex valves. These parameters are different from those used by McGhee but lead to the same description of the brachiopod form. Very convex valves that drastically change curvature near the anterior margin are termed geniculate. Alation is measured by the ratio of the length of the hinge to the width of the valve at midlength. It is a measure of the elongation of the hinge or "wingedness."

Alexander reasoned that a high P/V ratio would be an adaption for low oxygen and/or low food concentrations because a thin, broad body provides a larger surface area for respiration and food gathering. A high L/H ratio corresponds to a flat shell that presents a maximum area in contact with the sediment surface and should be an adaptation for support in a soft substrate (the snowshoe effect). A low L/H ratio corresponds to a more convex shell that maintains the commissure above the sediment surface and should be an adaptation for survival in an area with a high sedimentation rate because it would allow the commissure to stay above the accumulating sediment. (Others have suggested that such a bowl-shaped shell would be adapted to a fluid substrate by keeping the commissure from sinking below the sediment—the iceberg effect.) High alation could be an adaptation either to soft substrate by increasing the surface area of the shell for support on the substrate (snowshoe effect) or it could increase stability relative to overturning in turbulent environments.

Alexander measured these three parameters, as well as length as a measure of size, for specimens of *Rafinesquina* from seven stratigraphically separated localities (Fig. 5.10). He interpreted the depositional environment of each of these localities largely in terms of variation in water depth and then attempted to explain the size and shape of specimens in terms of the inferred depositional conditions. Small, thin, and compressed specimens occur in the deeper-water, soft-substrate environments; large, inflated moderately convex shells are found in high-energy environments; specimens with high convexity occur in an area of rapid sedimentation; alation did not differ systematically with environment but did increase through time. Alexander interprets this to indicate that alation evolved as an adaptation that aided both in support on a fluid substrate and in stability in a turbulent environment. The field data thus support the theoretical interpretation.

Finally, Alexander investigated the effect of size and shape on stability in currents by testing brachiopod models in a flume. Fourteen models were prepared that included all of the common shapes found in the natural populations and a range of sizes (Fig. 5.11). The models were first placed on sand in the flume with their hinge line perpendicular to the current direction. The current velocity was gradually increased, and the velocity at which each model was overturned by the current was noted (Fig. 5.11). In this experiment all models overturned at velocities between 1 and 2 m/s, with no particular pattern in terms of size or shape. Another experiment was performed with the hinge parallel to the current direction. Many of the specimens were more stable in this orientation, with currents up to 3 m/s being required to overturn them. The alate and convex (or geniculate) alate forms were the most stable. This suggests that these morphologies could be adapted to life in strong currents provided that the orientation with the hinge parallel to current is the life orientation. More recent research by LaBarbera (1977) suggests that many brachiopods did (and do) live in this position.

By far the most comon analytical approach in adaptive functional morphology has been to recognize a specific structure, describe it, and interpret it in terms of its adaptive significance with respect to both the life habits of the organism and the parameters of the environment. This analysis may be at the natural history, observational level, or may include modeling and experimentation. The studies of McGhee and Alexander are examples of this general

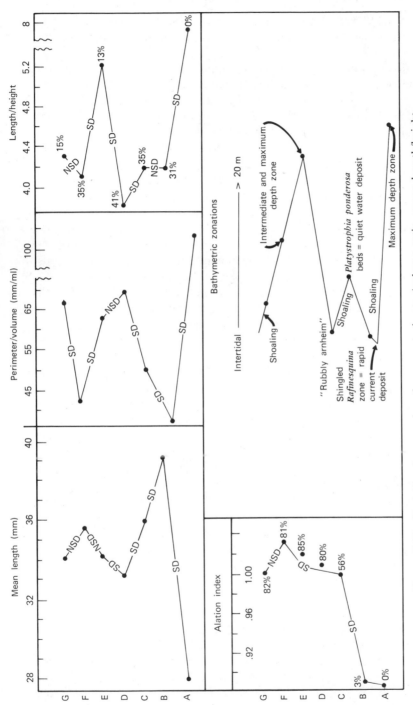

Figure 5.10 Variation in mean length, mean perimeter/volume ratio, mean length/height ratio, and alation index for *Rafinesquina alternata* among the stratigraphically successive samples A through G. The percentage values for length/height ratio and alation index indicate the frequency of geniculate specimens and alate (index > 1.00) specimens, respectively. NSD and SD indicate no significant difference and significant difference between successive means based on the student *t*-test at the 0.05 probability level. After Alexander (1975), Journal of Paleontology, Society of Economic Paleontologists and Mineralogists.

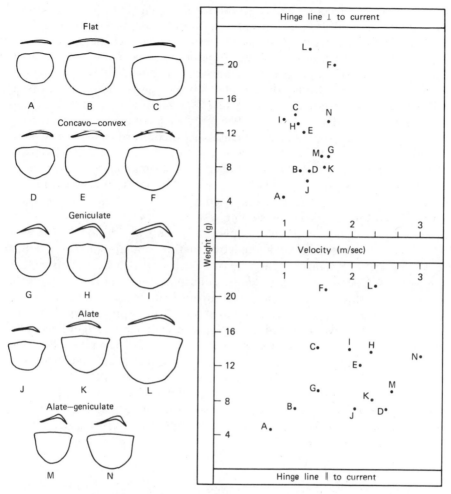

Figure 5.11 Shape and size of models of *Rafinesquina alternata* and current velocities at which they overturned. In the hinge line perpendicular orientation, the anterior margin faced upstream. After Alexander (1975), Journal of Paleontology, Society of Economic Paleontologists and Mineralogists.

approach. A number of others are presented below to indicate the range of approaches that has been used, and additional examples are given in Table 5.3.

The complexity of the skeleton and the close relationship between the skeleton and the body clearly indicate that the morphology of the vertebrate skeleton is functional, and functional relationships are often much more apparent in vertebrates than in invertebrates. The feeding method and the nature of the food used by an extinct bivalve or brachiopod is highly speculative for example, but jaw structure and teeth can reveal much about the type of food and feeding method used by vertebrates. Interpretations such as this are a routine part of the description of vertebrate fossils; more challenging to interpret are unusual structures that have no modern counterpart. The dinosaurs include a number of examples of such unusual structures, and one of the most unusual is the series of bony dermal plates projecting along the back of the well-known genus *Stegosaurus* (Fig. 5.12). The interpretation of the function of these plates by Farlow et al. (1976) is an example of the engineering approach incorporating both experimental and theoretical analysis. The common interpretation of *Stegosaurus* plates has been that they served a defensive function as armor, or that they functioned as a sexual display. Farlow et al. (1976), on the basis of the similarity between the plates and cooling fins on manufactured heat exchangers, proposed that their primary function was probably as a forced convection cooling system.

To evaluate their hypothesis, Farlow et al. constructed models of *Stegosaurus* that they tested in a wind tunnel by a method similar to that used to test mechanical heat exchangers. The rate of heat loss was measured on models with various types of fins in order to determine the "paradigm" for heat exchange fins on dinosaurs. They constructed scale models from aluminum cylinders to represent the body and thin metal fins to represent the plates on the back. Heaters were embedded within the cylinder and thermocouples were attached to the model to determine its temperature. The rate of heat loss was determined by monitoring the temperature of the model and of the surrounding air. Three types of fins were tested: paired, continuous; paired, interrupted; and staggered, interrupted (the latter type is what is actually found in *Stegosaurus*). The rate of heat loss was then determined for each of the three models and also for a model without fins.

Figure 5.12   The dinosaur *Stegosaurus*.

When oriented parallel to the wind, each of the finned models increased the rate of heat transfer relative to the unfinned model by about 35%. The continuous finned model was least efficient in heat loss per unit area because the continuous fin had a larger surface area. The three models differed considerably when oriented perpendicular to the wind direction. Paired continuous fins increased heat transfer by only about 15%; paired interrupted fins increased heat transfer by 27%; and staggered interrupted fins increased heat transfer by 33%. Thus the staggered interrupted condition is almost equally effective in either orientation, clearly an advantage to a large dinosaur with little opportunity to get out of the sun. The fact that *Stegosaurus* has a plate arrangement close to the paradigm strongly suggests that the plates did indeed function as cooling fins.

For the cooling fins to be effective, the dinosaur had to have an effective method of heat transfer from the body to the plates. In the models this was accomplished by the high conductivity in the metal, but in the animal this was probably accomplished by the circulation of blood through the plates, as indicated by the highly vesicular nature of the bone. Farlow et al. calculated that to work effectively as a heat exchanger, a wind of at least 4.5 m/s (10 mi/h) would be required. *Stegosaurus* might thus be used as a crude paleowind velocity detector! *Stegosaurus* appears to have lived in a fairly open, savannah-like area where moderate to strong winds might be expected.

In recent years considerable controversy has arisen over the possible endothermic or warm-blooded nature of the dinosaurs (Bakker, 1975). A heat-loss mechanism might be an advantage to a cold-blooded animal to allow it to dissipate metabolic heat. It would be even more advantageous to a warm-blooded animal that controls its temperature within narrow limits.

Further work by Buffrenil et al. (1986) has strengthened the thermoregulatory function of the *Stegosaurus* plates in two ways. The first is a step-by-step consideration of other possible functions, very much in the style of the paradigmatic method described previously. The second is a detailed examination of the internal structure of the plates by means of serial sections and thin sections. These confirm the light porous and vascular nature of the plates and strengthen the comparison with the osteoderms of present-day alligators (Seidel, 1979). The data available favor the thermoregulatory explanation, but are not sufficient to establish whether heat gain or heat loss was a more important function—an interesting conclusion that would contribute greatly to the question of the extent to which *Stegosaurus*, and dinosaurs in general, were warm blooded. In a similar study of the sailback pelycosaurs, such as *Dimetrodon* and *Edaphosaurus*, Haack (1986) concluded through a mathematical energy exchange analysis that the high dorsal "sail" was effective in gathering heat, and thus could increase the rate at which the organism could warm up in the morning. The sail was of much less value for dumping excess heat, however.

The analysis of cephalopod form in general and of specific structures such as shape, septa, and ornamentation has been intensive and has involved the use of flume studies, models, and mathematical analysis. A good example of the use of models and experimentation similar to Alexander's (1975) study of brachiopods is the study of streamlining in cephalopod shells by Chamberlain (1976). Ammonites and nautiloids are presumed to have been active swimmers such as *Nautilus*, the one living nautiloid genus. Differences in shell morphology suggest that fossil cephalopods may have differed widely in their swimming ability. The problem, then, is to determine which shell features might have been adaptations for more efficient swimming. Chamberlain attempted to answer this question by testing models of ammonoid and nautiloid shells in a flume that was designed for testing streamlining in ship hull design. While moving plexiglass models of cephalopods through the water at different speeds, he photographed the pattern of water motion around the specimens.

The streamlining and thus swimming efficiency was determined from the drag force, DF, on the model:

$$D_F = \tfrac{1}{2}\rho V^2 A C_D$$

where ρ is the density of the fluid (water), V the velocity of the specimen, A the cross-sectional area of the specimen (body size), and C_D the drag coefficient. The drag coefficient is dependent on the body shape or streamlining and thus is a measure of streamlining. The lower the value of C_D for a specimen, the lower will be the drag force and thus the greater the swimming potential—a species that spends a great deal of time swimming or that must swim rapidly would benefit from the lowest possible C_D value.

The study of flow patterns around the models showed that the water flowing past the leading edge of the model first flowed smoothly over the surface (in hydrodynamic terminology the flow was "attached" to the shell) but then broke up into turbulent eddies as the flow passed over the umbilicus. Further turbulence occurred at the trailing edge of the shell, especially as the water passed the aperture (Fig. 5.13). To show the quantitative relationship between the drag coefficient and shell shape, the shell shape was described by the parameters S, W, D, and F. Three of these (S, W, and D) are the same parameters as those used by Raup (1966b) in his quantitative description of coiled shells, as described earlier. S is a measure of cross-sectional shape of the whorl (Fig. 5.14). An S value of 1 indicates a circular or equidimensional cross section; lower S values indicate a more compressed cross section. W is a measure of the whorl expansion rate or the ratio of the diameter of successive volutions. Shells with large W values are more globose than those with small values. D is a measure of the relative distance of the whorl from the coiling axis. The larger the D value, the larger will be the umbilicus (the depression around the coiling axis). Because the ammonites and nautiloids studied are planispirally coiled, the translation rate $T = O$. F is a measure of the position of the umbilical shoulder, or the position of the greatest width of the shell relative to the diameter of the shell.

In general, the C_D value increases with increasing values for W. Shells with a large W value have a larger cross-sectional area in the direction of motion

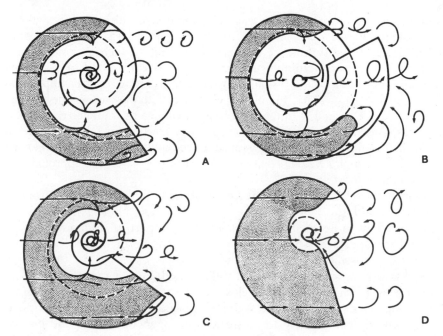

Figure 5.13 Sketch of flow lines around shells and shell models: (*A*) lytoceratid shell model; (*B*) serpenticonic shell model; (C) widely umbilicate oxyconic shell model; (*D*) *Nautilus pompilius* shell. The area of boundary layer attachment is shown by stippling. The dashed lines mark the umbilical shoulder (widest point in cross section) on the outer whorl. After Chamberlain (1976), copyright The Palaeontological Association.

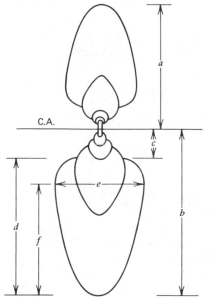

Figure 5.14 Definition of geometric parameters of shell form. $W = (b/a)^2$; $D = (c/b)$; $S = (e/d)$; $F = f(b + a)$; C.A. is coiling axis. After Chamberlain (1976), copyright The Palaeontological Association.

and thus the drag force is increased. The value of C_D also increases with an increasing value of D. A large D corresponds to a large umbilicus, which results in increased turbulence because the current "detaches" from the shell surface as it passes the umbilicus. Small S values correspond to more compressed shells, which have a smaller cross section, are more streamlined, and thus have lower C_D values. Finally, C_D decreases with increasing F because the current stays attached to shells with large F values for a longer relative distance as it passes the shell. Combining these characteristics, the most streamlined shell would be compressed with a reduced umbilicus, a low whorl expansion rate, and high F value. The lowest measured C_D value was approximately 0.1 and the highest about 1.0.

One might suspect that the soft tissues of the organism extending from the aperture would affect the streamlining, but Chamberlain concluded that the effect is likely to be small unless a considerable part of the body extended beyond the aperture. This is not the case for modern *Nautilus*, nor was it probable for fossil cephalopods either.

The lowest C_D value measured is still about 10 times larger than that for a fish or squids, which come much closer to the most streamlined, fusiform or cigar shape. This points out the limitation placed on the swimming efficiency of shelled cephalopods by the historical-phylogenetic factor—the possible shape and thus streamlining is limited by the internal growth program. The shell can approach a maximum efficiency within the historical-phylogenetic constraints, but the relatively large cross-sectional area, umbilicus, and large aperture all result in a drag coefficient much larger than that of the more efficient fusiform shape.

Another factor limiting the swimming ability of the shelled cephalopods is the relative positions of the swimming thrust and drag force (Fig. 5.15). The thrust generated as water is expelled from the hyponome acts on the lower part of the shell, but the drag force acts over the entire shell. These opposite but displaced forces produce a rotational moment. The resulting rocking motion can be observed in the swimming motion of modern *Nautilus*. An in-

Figure 5.15   Effect of drag and thrust on shell orientation during movement. The shell is shown moving from right to left. Drag (D) acts to the right from the center of dynamic pressure. Thrust (J) acts to the left in the direction of motion from the point of application by the hyponome at the ventral margin of the aperture. These forces cause a clockwise rotation, shown by the curved arrows. After Chamberlain (1976), copyright The Palaeontological Association.

crease in thrust force to increase the swimming speed increases shell rotation, and thus an upper limit to swimming velocity is established.

Another vigorously pursued question about the functional morphology of cephalopods has been how shell shape and septal pattern determine the strength of the shell, and thus its capacity to withstand crushing as the animal moves up and down in the water column during its life or from day to day. The cephalopod is unique in that much of the shell interior is empty, with the body itself confined to the body chamber. Thus the empty chambers could collapse if the organism submerged to a depth where the water pressure exceeded the shell strength. If the crushing strength of a fossil cephalopod could be estimated from its morphology, it would indicate the maximum depth to which the cephalopod could have submerged, and thus would indicate the maximum water depth if the animal lived throughout the full depth range.

This question has been investigated by a variety of approaches. For example, models of cephalopods have been constructed and the crushing pressure experimentally determined (Saunders and Wehman, 1977). The crushing pressure has also been established for living *Nautilus* (Ward et al., 1980). It is difficult to predict from living *Nautilus* what the crushing pressure of fossil cephalopods would have been because of significant differences in the general shell shape and in the septa (Chamberlain and Chamberlain, 1986). The values derived from *Nautilus*, however, do provide one point on the strength versus morphology curve that can be developed further through an engineering analysis of the septum as an arched to saddle-shaped surface with a smooth (nautiloid) to highly corrugated (ammonoid) attachment margin against the shell wall.

An engineering analysis of the septum must first establish the relative magnitude of stresses applied to the septum across the shell and back through the body chamber onto the last-formed septum. The second step is a rigorous description of the septum shape. An important starting point for this has been the membrane model, in which the shape of the septum is compared to that surface with minimum area that would have as its outer edge the boundary with the wall—the suture line. Analog models have been constructed with soap bubbles on a wire hoop bent to conform to the suture line (Seilacher, 1975). The extent to which the septum and the soap bubble surface are dissimilar then indicates the extent to which the shape of the septum is determined by external stresses (Bayer, 1977a,b).

The integration of the results of these observational, experimental, model, and engineering approaches has led to an increasingly complex explanation of the septum as a structural element of the cephalopod shell. Nevertheless, the bathymetric interpretation of shell morphology has not yet been fully resolved (Chamberlain and Chamberlain, 1986).

As stated at the beginning of this chapter, adaptive functional morphology has tended to focus on unusual structures—on the extreme forms in the total form spectrum. One justification for this is that the analysis of the extreme cases helps to clarify the functional attributes for the whole range of forms that, in themselves, might be so "normal" that they are nondiagnostic. A more satisfactory approach, however, is to use the full range of morphologies as the analytical unit. The advantage of this method is that comparisons within the full spectrum can be made, and co-occurring change in diverse aspects of the morphology as the organism maintains a developmental history through time can be taken into account. This approach has seldom been carried out, but the comprehensive studies by Seilacher (1979, 1984, 1985) on irregular echinoids and on pelecypods are good examples of this approach and its potential.

6

Ichnology

Body fossils have been, historically, the focus of interest in paleontology and the major source of data for the paleontologist interested in determining ancient environments. This concentration on body fossils is logical: They are easily seen, identified, and compared to one another and to living relatives, and they possess morphological, microstructural, and chemical skeletal features that can be studied and interpreted in detail. Trace fossils, on the other hand, are relatively lacking in these attributes and so have played a much smaller role in paleoecology until the last decade or two. Even this limited study has shown, however, that they yield much information about many aspects of paleontology, and that they provide valuable information about the depositional environment for both the paleoecologist and the sedimentologist.

Ichnology is the study of the effects of biological agents on sediments. The resulting biogenic sedimentary structures (*lebensspuren* or traces) include a wide range of features formed by the activity of organisms during their life, but exclude the remains of the organisms themselves. Those preserved in the fossil record are ichnofossils or trace fossils. The most thoroughly studied categories of trace fossils, and the ones that will occupy most of our attention in this chapter, are tracks, trails, and burrows in soft sediment, and borings and other evidence of bioerosion. *Bioturbation* is a general term applied to the disruptive effect of organisms on primary sedimentary features, but is most commonly used when the resulting biogenic structures are pervasive and/or impart a generally stirred appearance to the sediment. Coprolites, fecal pellets, and similar excremental fossils are grouped in the broad category of trace fossils but have not been as intensively studied. Biostratification structures include a range of effects on sediment texture, structure, and chemistry caused by organisms living within the sediment, and stromatolites, a lamination formed primarily by sediment trapping and binding by filamentous algae and bacteria.

Trace fossils are of particular value because they are found in many sedimentary rocks that are devoid of body fossils, and thus they provide in those instances the only record of life. In addition, they supply virtually the only record of all the organisms that have little or no mineralized skeleton and are thus unlikely to be preserved as body fossils. For example, they have been important in establishing much of our knowledge of Precambrian life, of the existence and abundance of deep-burrowing infauna in the Early Paleozoic (Sheehan and Schiefelbein, 1984; Miller and Byers, 1984), and through their footprints, of ancient terrestrial vertebrates (Lockley, 1986). In sediments originally containing both skeletal material and traces, diagenesis may destroy the skeletal material by solution or recrystallization, but at the same time enhance the traces.

In addition, and of special value in paleontology, trace fossils are almost always preserved in place. Skeletal parts may be transported for a considerable distance from the life site before burial, or be reworked from older to younger strata, but the sedimentary record or trace of an organism can be redeposited under only very unusual circumstances. Although trace fossils are not reworked, however, they may penetrate to a considerable depth within the sediment, so that the trace fossils at a particular horizon may be a temporally mixed assemblage formed by organisms burrowing into the sediments to different depths during and after deposition of that horizon, and consequently living in very different microhabitats as well. Finally, trace fossils are important to the sedimentologist because they record the activity of organisms in deposition and penecontemporaneous modification of sediments.

THE TRACE FOSSIL RECORD

Our knowledge of trace fossils is strongly biased by aspects of their construction and by processes of diagenesis. Dwelling structures, which have a distinct form and borders, are in some cases lined with mineral different than the adjacent sediment. They may be left open after the death of the organism to be filled later by a different sediment and are thus more likely to be preserved in a recognizable form than a feeding trace constructed as an organism moves over or through the sediment. Traces in firm-to-hard substrate should have a better chance of preservation than those in soft sediment because in the latter case distinct burrow margins will not be formed. In addition, tracks and trails on the sediment surface should generally have a much lower chance of preservation than those formed within sediment. Diagenesis plays an important role in preservation of traces because as an organism burrows into the sediment it may change the local chemistry in a number of ways—most commonly by bringing oxygenated water down into otherwise anoxic sediments and by concentrating organic material within the sediment. In addition, porosity and permeability of the sediment may be strongly modified by the organism so that diagenetic fluids will be channeled by the trace through the sediment. Because diagenesis is commonly controlled in these ways by traces within the sediment, trace fossils are made evident by effects such as selective cementation, chertification in chalks (Bromley and Ekdale, 1984a), dolomitization in limestone, and pyrite rims (Byers and Stasko, 1978).

Modern traces are largely known from tracks and trails, which are easily observed; in contrast, trace fossils are most commonly observed in vertical section in sedimentary rocks and thus are more likely to be infaunal (Seilacher, 1957). Although infaunal traces have become much better known because of the widespread use of box cores, the difference in our knowledge of modern traces and of trace fossils is still significant. An example of this is provided by analysis of traces recorded in photographs of the deep-sea floor by Kitchell et al. (1978), in which they reported a much lower abundance of grazing traces than would be expected on the basis of their abundance as trace fossils in deep-water sedimentary rocks. Ekdale (1980) has pointed out, however, that photographs of the sediment surface will not provide a true measure of these largely infaunal traces.

Most of the trace fossil literature is based on outcrop study, and relatively little work has been done with cores. Recognition and analysis of trace fossils in cores is difficult because of the small size of the core, the often unfamiliar appearance of traces fossils as exposed on the core surface, and because weathering has not enhanced lithologic differences to make the trace fossil stand out from the surrounding sediment. On the other hand, a continuous and unweathered core contains valuable information not commonly available from the outcrop. Chamberlain (1978) and Ekdale et al. (1984) have presented comprehensive surveys of trace fossils as exposed in cores.

Estimation of traces fossil abundance is an important aspect of their study because it may provide evidence of the abundance of trace-making organisms and of the rate of deposition as it limited the time available for the organisms to make traces. Droser and Bottjer (1986) have proposed a semiquantitative scale of abundance of bioturbation that should help to standardize estimates by different workers (Fig. 6.1).

The literature on ichnology is voluminous and widely scattered, and in this chapter we can provide only a survey of the aspects of ichnology most important in paleoenvironmental reconstruction. Several publications provide a more comprehensive survey or more detailed treatment of specific topics. The second edition of the volume on trace fossils and problematica of the *Treatise on Invertebrate Paleontology* (Häntzschel, 1975) is the basic treatment of trace fossils and the primary resource for identifying them. Ekdale et al. (1984) and Frey and Pemberton (1985) provide recent and comprehensive surveys of ichnology. Curran (1985) has compiled case studies of trace fossils from a wide range of environments and geologic ages. The discussion by Golubic et al. (1984) provides an introduction to microboring traces. Farlow (1987) and Lockely (1986) give current treatments of vertebrate traces and their paleoeco-

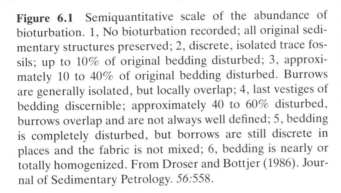

Figure 6.1 Semiquantitative scale of the abundance of bioturbation. 1, No bioturbation recorded; all original sedimentary structures preserved; 2, discrete, isolated trace fossils; up to 10% of original bedding disturbed; 3, approximately 10 to 40% of original bedding disturbed. Burrows are generally isolated, but locally overlap; 4, last vestiges of bedding discernible; approximately 40 to 60% disturbed, burrows overlap and are not always well defined; 5, bedding is completely disturbed, but borrows are still discrete in places and the fabric is not mixed; 6, bedding is nearly or totally homogenized. From Droser and Bottjer (1986). Journal of Sedimentary Petrology. *56:*558.

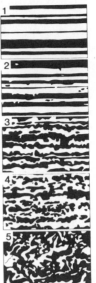

logic value. Among recent publications on stromatolites, Monty (1977) gives a historical development and brief ecological overview, and Monty (1981) reviews studies of stromatolites in a wide range of geologic settings and ages.

TRACE FOSSIL CLASSIFICATION

Trace fossils have been categorized (1) as morphologic–biologic entities, (2) by their position in the sediment, and (3) according to the behavior that formed them. Each classification yields specific and useful information for trace fossil analysis.

The morphologic–biologic classification is descriptive, based on the morphologic characteristics of the fossil. The nomenclature conforms to the rules of zoological nomenclature (Häntzschel, 1975) but is largely confined to the generic and specific levels. A biologic classification relating a trace fossil to the organism that made it is difficult for several reasons: (1) Identification of the causative organism is rarely possible because the animal is seldom found together with the trace; (2) most trace-forming organisms do not leave distinctive traces and, in fact, the same trace may be made by several organisms; and (3) an organism may form traces that are very different under different environmental conditions, in different sediments, or at different seasons or stages of life. Consequently, the biologic classification identifies morphologically distinct form taxa and does not connote particular causative organisms. Differences in organism, behavior, and sediment open the way for numerous slight morphologic variants to be recognized by species names, but the generic level is generally used in environmental interpretations, and the specific level only in detailed local and taxonomic studies.

The basic terminologies used for classifying traces according to their position relative to the original depositional surface and bedding surfaces as they formed and have been preserved, proposed by Seilacher (1964) and Martinsson (1970), are illustrated in Fig. 6.2. According to Seilacher's classification, a trace entirely within a bed is a full relief trace, a trace at a bed boundary is a semirelief trace, a trace at the upper surface of the bed of interest is an epirelief trace, and a trace at the lower surface is a hyporelief trace. According to Martinsson's classification, a trace entirely within the bed is an endichnion (pl. = endichnia); on the upper surface of the bed, an epichnion; on the lower surface of the bed, an hypichnion; and outside the bed of interest, an exichnion. The important genetic aspect of the location of the trace fossil relative to bedding is whether the trace is "predepositional" or "postdepositional." This distinction has been most commonly used in describing traces in beds above and below a marked discontinuity in deposition. In sequences of interbedded mudstone and turbidite, for example, traces along the sole of a turbidite may have been formed within the underlying mudstone but were exposed as the turbidity current scoured the surface, and then were filled by turbidite sand. These traces, relative to the turbidite, were predepositional. Correspondingly, subsequent burrows down into this turbidite sand are postdepositional. These, and more complex classifications, are presented in fuller detail by Frey and Pemberton (1985).

The value of this classification approach is illustrated by the problems of classification and interpretation of hyporelief traces, which are common at the base of coarse-grained limestone and sandstone in contact with underlying

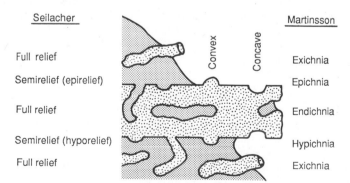

Figure 6.2 Terminologies for the preservational classification of trace fossils by Seilacher (1964) and Martinsson (1970). From Ekdale et al. (1984). Society of Economic Paleontologists and Mineralogists. Short course No. *15:* 22.

TABLE 6.1   Behavioral Categories of Traces

I. Dwelling Structures (Domichnia)

The organism constructs a dwelling by burrowing into soft sediment or by boring into hard rock, skeletal material, or wood. In contrast to feeding burrows, the trace has a relatively fixed and permanent form, may conform closely to the shape of the inhabitant, and may be lined by sand grains, shell fragments, or organic material. The organism derives its nourishment from outside the dwelling rather than from the substrate itself.

II. Feeding Burrows

A. MINING (FODINICHNIA)

The organism moves through the substrate, processing the sediment more or less in bulk in search of food. The traces are distinguished from dwelling burrows by having a changing and impermanent outline and unlined walls. If the burrow is immediately back-filled as the organism mines the sediment and shifts its burrow, a distinctive "spreite" structure, as in *Zoophycos,* may be formed (see Fig. 6.6). Probing and mining structures radiating from a central location are common types of fodinichnia.

B. GRAZING (PASCICHNIA)

In contrast to fodinichnia, formed as an organism mines or probes within the sediment, pascichnia are formed in the process of feeding on the surface or at a level within the sediment. If the organism carries out an efficient search procedure, the trace may be highly structured, resembling the closely spaced pattern of a lawn mower or a tightly wound spiral. Meshlike burrow systems and highly structured traces that may serve both dwelling and farming functions comprise the subcategory of graphoglyptid traces [Seilacher (1977b); agrichnia of Ekdale et al. (1984)].

III. Locomotion (repichnia)

The organism travels on or through the sediment. No purpose may be evident but may be as diverse as idle wandering, searching for mate or food, or migration to a more suitable location. Vertebrate traces generally fall into this category. Locomotion traces are characterized by the absence of a highly structured pattern, as in grazing or mining traces and by a stronger linear aspect. Many are probably nondiagnostic feeding traces.

IV. Resting Structures (cubichnia)

The mobile animal temporarily comes to rest and more or less buries itself in the sediment. The reason is generally unknown, but the organism may be resting, hiding from a predator, or it may be a predator itself hiding and awaiting prey. The trace is characterized by its temporary use, its shape as a depression on the substrate, and the outline of the organism with details of the ventral morphology.

V. Escape Traces (fugichnia)

Traces formed by the movement of an organism as it attempts to work its way upward through rapidly deposited sediment.

mudstone. This classification emphasizes the important question of whether (1) the trace formed on or in the muddy seafloor before deposition of the sandstone and was filled as sand was spread across a surface trail or an exhumed burrow (predepositional), or (2) the trace formed by an organism burrowing down to the underlying sediment after or during deposition (postdepositional or syn-depositional).

The behavioral classification is based on the assumption that specific environmental conditions lead to specific behavioral responses, regardless of the kind of animal. Consequently, if we know how behavior was determined by environmental conditions, and if we could recognize from the trace fossil the behavior that was responsible for it, we could interpret the trace fossil to determine the depositional environment. The fact that the same animal may make different traces under different conditions, that different animals may make similar traces under similar conditions, and that trace fossil assemblages in different lithofacies are of different behavioral composition support this approach.

The fundamental behavioral categories of traces are dwelling structures (domichnia), feeding burrows (fodinichnia), grazing traces (pascichnia), locomotion traces (repichnia), resting structures (cubichnia), and escape traces (fugichnia) (Table 6.1).

Many traces can easily be assigned to one or the other of these categories. The web-lined trapdoor spider burrow (Fig. 6.3), for example, is strictly a domichnion. In many other cases, however, the behavioral significance of a trace is not evident or the trace is not the result of a single activity but of a combination of activities, such as locomotion and feeding or dwelling and feeding. To what extent, for example, is a locomotion trace actually a record of nonsystematic feeding? Some of the potential behavioral combinations, as well as some of the other behavioral categories that have been proposed, are presented in Fig. 6.4.

ENVIRONMENT AND BEHAVIOR

The behavioral interpretation of trace fossils has been most commonly in terms of bathymetry. The reason for this is that water depth is a major component of most reconstructions of marine paleoenvironments. In addition, other environmental parameters that directly affect behavior, such as environmental stress, predation, water energy, food supply, and sediment characteristics, are difficult to identify individually, but are correlated in general with bathymetry. Salinity and climate do not appear to be major determinants of behavior.

Environmental Stress and Predation

Most organisms in both subaerial and subaqueous environments live at or very near the sediment surface because that is the site of maximum flux of

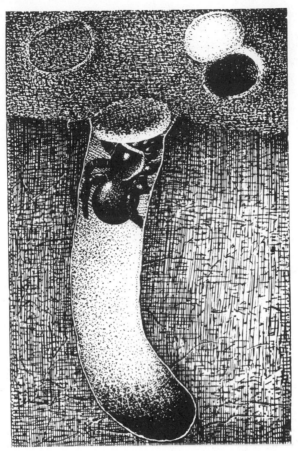

Figure 6.3 Trapdoor spider lurking at door of its web-lined dwelling. Upper right, burrow with door open; upper left, burrow with door closed. Burrow is inconspicuous because outer side of door is covered with dirt. From Kaston (1978).

resources for respiration and nourishment: The substrate provides support for both plants and animals and nutrients for rooted plants, and is thus a surface of both production and accumulation of organic material. The overlying medium contains oxygen for respiration and, in aqueous environments, nutrients in solution and in suspension as phytoplankton, zooplankton, and organic detritus in various stages of decomposition. The underlying sediment is a zone of accumulation of the organic material, but lacks light for photosynthesis and commonly is low in oxygen for respiration.

  In general, environmental variability is high above the substrate, decreases markedly just below the surface, and decreases less rapidly with increasing depth, approaching stable, invariant conditions asymptotically (Fig. 6.5). Thus the essential survival strategy for many animals is to utilize resources at

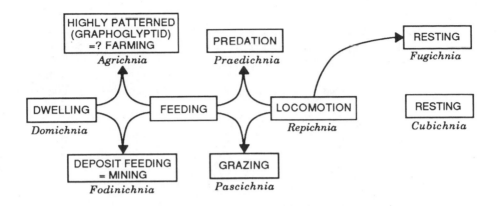

Figure 6.4 Activity (in blocks) and categories of resulting trace (in italics). After Ekdale (1985) Palaeogeography, Palaeoclimatology, Palaeoecology. *50:* 65.

and above the substrate and live in the more equable habitat below the substrate.

A dwelling burrow provides protection for the inhabitant from both predators and physical conditions above the substrate. Predation must be an important determinant for the burrow dweller, but is difficult to evaluate because generally predator and prey organisms are not both preserved, so the interaction between them cannot be established. For example, the deep burrow of many bivalves, such as *Panopea,* is at least in part to avoid predation, as indicated by the fact that both modern and Tertiary deep-burrowing bivalves are less likely to be preyed upon by naticid gastropods than shallow-burrowing bivalves. Because such biologic interactions are generally indeterminate, traces are commonly interpreted only in terms of the physical environment. In that respect a major function of a dwelling burrow must be for survival in fluctuating conditions where the organism can wait out unfavorable intervals, whether they are predictable and periodic ranging from diurnal tidal to seasonal, or are unpredictable and irregular as in the case of infrequent storms or extreme temperatures.

In terrestrial environments, several kinds of seasonal climatic change cause vertebrates to burrow. A few mammals and many reptiles and amphibians burrow during the cold season to survive by hibernation, permitting them to live at much higher latitudes than the winter climate would otherwise allow. Dry, hot summer weather is equally intolerable for many amphibians and

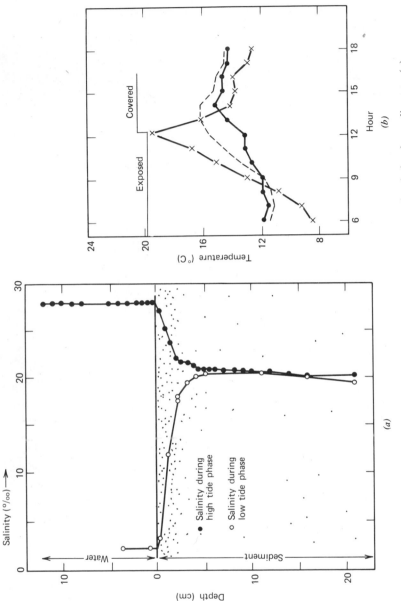

Figure 6.5 Plots of salinity and temperature ranges versus depth in the sediment. (*a*) Salinity profiles at high and low tide of the sediment and the immediately overlying water in the Pocasset River estuary, Massachusetts. After Sanders et al. (1965). (*b*) Temperature variation at depths of 1 cm (crossed line), 10-cm (dashed line), and 20-cm (dotted line) depths in intertidal sand, July 3, 1963, in Tomales Bay, California. After Johnson (1965b).

lungfish, and they retreat into burrows to aestivate. Thus the existence in the fossil record of hibernation and aestivation burrows could indicate seasonality of temperature or rainfall in the ancient climate, and potentially the depth of these burrows could be mapped as an indication of climatic gradient and thus of paleolatitude. Aestivation burrows indicating strongly seasonal terrestrial climate characterized by summer period of drought have been described (Berman, 1976; Olson and Bolles, 1975) and have been used to infer lateral climatic gradients and temporal climatic changes during Permian time in the southwestern United States (Olson and Vaughn, 1970).

In marine environments, variability is broadly correlated with water depth and distance from land. The daily-to-seasonal fluctuations of temperature and salinity in the water column that are caused by climatic fluctuations in the overlying atmosphere are greatest at the surface and decrease rapidly with depth to the thermocline at a few hundred meters or less, below which conditions are relatively stable. Fluctuations in salinity, turbidity, and sedimentation also are correlated in general with proximity to land, but in more detail they depend on variations in freshwater runoff from the land, and the magnitude of these fluctuations depends on the local geography and bathymetry as they restrict mixing of the coastal water with the open ocean water. An indication that dwelling burrows are used by the benthic organisms to avoid fluctuations in the environment above the substrate is the general decrease in burrow depth with increasing water depth and distance from shore in open marine settings. The same relationship of burrow depth and environmental variability is even more evident in estuarine environments, where gradients of change are steeper at any site, and fluctuations increase up-estuary, away from more equable marine conditions. In the Pocachet River estuary, with variability as indicated in Fig. 6.5a, epifaunal taxa are poorly represented and are largely restricted to the most marine end of the estuary. Each infaunal taxon has a preferred depth within the sediment and those that burrow most deeply are taxa with the least tolerance for rapid salinity fluctuations (Sanders et al., 1965).

An example of burrow depth correlated with environmental stress in one locations is provided by the mantis shrimp *Squilla* in Narragansett Bay. It constructs a broadly U shaped burrow to a depth of 15 to 20 cm in the summer, but an approximately vertical tube as much as 4.1 m deep in the winter. Myers (1979) postulated that the deep winter burrow is a means of avoiding the cold bay water. (This is, in addition, an example of diverse trace morphology by a single species, and of the great depth of sediment mixing and reworking that may occur.)

Food, Water Energy, and Sedimentation

Food for trace-forming organisms is available in, on, or above the sediment. Although many organisms feeding from the water column will form dwelling burrows, not all dwelling burrows reflect suspension feeding because the inhabitant may be a predator (as in the trapdoor spider of Fig. 6.3) or may come out of its burrow to graze on the substrate. In addition, many suspension feeders, such as shallow-burrowing bivalves, do not establish a permanent domicile and thus they cause a general bioturbation rather than a distinct dwelling burrow.

Feeding burrows are formed as the animal works through the sediment for food. The animal may mine the sediment, extract the food, and expel the sediment out of the burrow and onto the substrate to form a mound at the burrow opening and an ever-larger burrow that serves also as a dwelling. *Ophiomorpha* is a good example. On the other hand, as the animal mines the sediment it may backfill the burrow with the sediment it has worked over, so that an open dwelling burrow with fixed walls is not formed, as exemplified by *Teichichnus* or *Zoophycos*.

Organisms grazing on the substrate or along nutrient-rich horizons within the sediment move constantly as they search for food, leaving a trace through or on the sediment. The resulting feeding traces differ widely in pattern and symmetry as a result of the search strategy used by the particular animal. A basic tenet in interpreting feeding traces has been that evolution has been in the direction of optimal feeding efficiency. That is, the feeding strategy should be to gain the maximum nourishment with the minimum expenditure of energy. The problem is like that of mowing a lawn or of searching for a lifeboat with shipwreck survivors in the middle of the ocean—how to cover all the area while minimizing overlap. If this efficiency hypothesis is correct, morphology of the feeding trace depends on abundance and patchiness of food and location of the food, whether on or in the sediment. As a broad generality, food is most abundant in shallow water near the continental margin because organic productivity is determined by the contribution of nutrients and organic material from the land, by the localization of upwelling and nutrient recycling along the continental margin, and by the restriction of benthic plants to the photic zone of the shallow shelf. Overlaid on this general pattern of decreasing organic material from shore to the deep ocean floor, of course, are local differences due to geographic and oceanographic conditions. The proportions of the available food in suspension, on the substrate, or in the sediment depend primarily on the *water energy*. (Water energy is a poorly defined term, but qualitatively describes the sum total of water motion or turbulence due to current and wave action.) Where water energy is high, the bulk of the available food is in suspension, little is on the substrate because it is frequently resuspended, and an intermediate amount accumulates in the sediment as a result of occasional intervals of rapid deposition. Consequently, suspension-feeding organisms should predominate in areas of high water energy because the efficient feeding strategy would be to utilize the relatively abundant food in the water column, and dwelling structures should be the predominant trace fossil. At the other extreme, on the deep ocean floor, both organic productivity and water energy are low, so that the total amount of food will be less and will largely accumulate on the substrate. In addition, because the sedimentation rate is low, little of the food will have a chance to be deeply buried before being used by organisms at or near the substrate. Consequently, grazing and shallow mining structures formed by animals reworking the small but consistently present food should be the predominant trace fossil.

The amount of food present within the sediment depends on the rates of (1) organic productivity and accumulation, and (2) sedimentation. If the sedimentation rate is sufficiently high, some of the organic matter may be buried before it can be consumed at the surface. Some organic matter on the substrate may also be buried by burrowing organisms as they dump excavated material on top of the organic-rich surface. If fluctuations in sedimentation result in deposition of coarser, less-organic-rich sediment on top of finer, more organic-rich sediment, animals may burrow down through the surface sediment in order to feed from the buried, more nutritive horizon. If food is abundant, no systematic search strategy is necessary because food is available with little effort whichever way the animal turns. If food is scarce, however, a systematic search pattern with a high order of symmetry should be most efficient.

These general relationships between environmental conditions and both dwelling and feeding behavior should lead to parallel trends of trace morphology. In shallow water, with unstable conditions, high water energy, and high productivity, the dominant trace should be dwelling burrows of suspension feeders. In deeper water, with more stable conditions but with lower water energy, sedimentation rate, and productivity, dwelling burrows of suspension feeders should decrease in abundance relative to traces of both surface and in-sediment deposit feeders, and with decreasing productivity and sedimentation rates, more efficient systematic feeding patterns become increasingly advantageous.

This behavioral progression is broadly correlated with bathymetry. It is not determined by water depth in itself, but by the several factors described that change systematically with depth, but that may change along other environmental gradients as well. For example, low water energy and surface deposition of food resources may occur in shallow but protected bay environments as well as in deep water.

The general relationships between behavior and environment may not be valid at extreme levels of the environmental factors. For example, high water energy increases suspended food supply and so should promote suspension feeding animals in general. At the same time, it increases sediment movement, which is disadvantageous for these largely sessile animals, and increases turbidity due to suspended sediment, which inhibits suspension-feeding animals by clogging the food-gathering mechanism. As another example, abundant organic content on the seafloor should favor deposit feeders, but organic content in itself is not important in determining the abundance of deposit feeders, for much of it may be refractory and of little or no nutritive value. In addition, too much organic material, even if usable, may decrease the oxygen content within the sediment and thus inhibit deposit feeding (Bader, 1954).

Sediment characteristics will also modify this general model of environment and behavior because (1) the ease with which an animal can move through the sediment will determine the shape of its burrow, the relative rates of deposition of organic material and thus of burrows, within the sediment, and (2) bearing strength will determine the suitability of the substrate for benthic organisms: In very soft, fluid sediment, all benthic organisms may be excluded unless they have special morphologic adaptations that prevent them from sinking into the substrate (Thayer, 1975a).

The net result is that these relationships between dwelling and feeding behavior and water energy and trophic resources provide valid paleoenvironmental criteria but only as generalities. Analysis of behavior and trace morphology must take into account the full range of environmental factors.

Invertebrate tracks and trails that are linear, wandering, or erratic in plan view are commonly attributed to locomotion, but the behavior cannot be determined because the trace morphology is not diagnostic. Vertebrate footprints, however, prove to be very valuable in providing both biologic and paleoenvironmental information (Lockley, 1986). Analysis of assemblages of footprints and tracks can tell something about the herding and social structure of the animals in addition to their size and locomotion. Dinosaur footprints have been interpreted to establish such aspects of the paleoenvironment as location of shoreline, water depth, current direction, sediment consistency, and depositional and climatic cyclicity.

Oxygen

The abundance, size, and depth of burrowing are also closely linked to the oxygen content in the ocean. In an oxic environment, an abundant shelled and soft-bodied benthos is expected and the sediment is consequently thoroughly burrowed. With decreasing oxygen levels, in a dysoxic environment, the infauna becomes progressively less diverse and smaller; and first shelled and than soft-bodied infauna disappear. In an anoxic environment, no burrowing organisms are present (Rhoads and Morse, 1971; Thompson et al., 1985). Several trace fossils have been recognized as particularly diagnostic of oxygen content. *Chondrites* is found in a wide range of environments but seems to be the most tolerant of dysoxic conditions and is commonly the dominant or only trace fossil present in dark organic-rich sedimentary rocks (Bromley and Ekdale, 1984b). *Zoophycos, Thalassinoides,* and *Planolites* are other common traces that appear to be tolerant of low oxygen levels. That is, at lowest oxygen levels, only *Chondrites* might be present, but with progressively increased oxygen level, *Zoophycos, Thalassinoides,* and *Planolites* would be present. A comparable progression of trace fossil from *Planolites* to *Chondrites* would be found with increasing depth in the sediment because oxygen normally decreases rapidly downward in marine sediments and is absent at depths of more than a few centimeters. Thus most burrows are within anoxic sediment and the burrower survives by circulating oxygenated water down into the burrow. Because oxygen is consumed by reacting with the reduced sediment along the burrow wall, the normal oxygen level of the overlying ocean will be more difficult to maintain at the bottom of a deep burrow than of a shallower burrow, resulting in an oxygen-determined ecologic stratification or tiering of trace fossils (Figs. 6.6 and 6.7; Ekdale, 1977; Seilacher, 1978; Bromley and Ekdale, 1986). This stratification is important because it may result in complex crosscutting relationships that would reflect the oxygen content in the environment and fluctuations of the content. Under uniform conditions and slow sedimentation, the deepest burrowers would be the last ones active in a horizon and would cut across and perhaps even obliterate the earlier formed but subsequently buried shallower traces. If the oxygen content in the overlying water were to change, however, the depth of the different traces should shift. If oxygen decreases, shallower tiers should disappear and the deeper tiers would be at shallower depths. Conversely, if only *Chondrites* were present, but then the oxygen content increased, the depth of activity of the *Chondrites* animal should move downward and a *Zoophycos* tier should become established with its traces cutting across and obliterating the early shallow *Chondrites*. Savrda and Bottjer (1986) have developed from this logic a model of tiering and crosscutting relationships that reflects variation in oxygen content of ancient bottom waters.

Figure 6.6 Tiered structure of the trace fossil suite in chalk (Upper Cretaceous), northern Jylland, Denmark. From shallowest to deepest (left to right) trace fossils are *Planolites*, *Thalassinoides*, *Zoophycos*, large *Chondrites*, and small *Chondrites*. Scale equals 1 cm. From Bromley and Ekdale (1986), Composite ichnofacies and tiering of burrows: Geol. Mag., Cambridge Univ. Press.

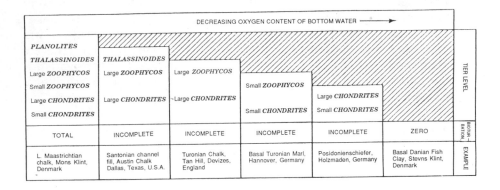

Figure 6.7 Tiering relations of trace fossils, primarily oxygen-controlled, in selected examples of Mesozoic marine strata that represent oxic to anoxic depositional environments. From Bromley and Ekdale (1984), Science 224:873. Copyright 1984 by the AAAS.

TRACE FOSSIL ANALYSIS AND ENVIRONMENTAL INTERPRETATION

Distribution Patterns and Ichnofacies

Environmental interpretation of trace fossils is based on two sources of information. One is the distribution of trace fossils in known geologic settings. The other source is the distribution of modern traces. The latter approach has the advantage of combining organism and trace within a known environment, and of being able to observe the organism and its behavior as it forms a trace. It has the disadvantage of not incorporating the effects of geologic time and taphonomic processes, as well as the difficulty of viewing the sediments and traces in three dimensions. The study of trace fossils within a known ancient environment of deposition and stratigraphic framework is commonly handicapped by lack of knowledge about the animal or its behavior, but it is aided by being able to work within the context of both time and space and by the diagenetic enhancement of the traces.

Figure 6.8 is a representative summary of the data that can be derived from these sources. It illustrates the strength of this approach, and also unfortunately, the wide environmental range for many common trace fossils. This fact has been demonstrated repeatedly, for example by the presence in terrestrial habitats of traces that were otherwise thought diagnostic of a marine habitat. Another well-documented example is *Ophiomorpha,* a trace fossil once considered diagnostic of the littoral and inner sublittoral but subsequently recognized to have a very broad bathymetric distribution (Frey et al., 1978). These examples, however, emphasize that part of the difficulty has been caused by relating trace fossils in the past to such a generalized environmental characteristic as water depth rather than to the specific and more immediate parameters. A case in point, as discussed previously, is the diagnostic value of *Chondrites* as an indicator of dysoxic conditions rather than of water depth.

Another approach in the interpretation of trace fossils has been the description of ichnofacies, assemblages of trace fossils diagnostic of particular bathymetric settings (Fig. 6.9; Seilacher, 1967). Each facies is named after a typical trace fossil and is characterized by the common occurrence of a number of other trace fossils. It is fundamentally established on behavioral characteristics of the traces. Seilacher (1978), for example, has distinguished among the ichnofacies the zone of suspension feeders (*Skolithos* and *Glossifungites* ichnofacies), the zone of generalized sediment feeders (*Cruziana* ichnofacies), the zone of churners (*Zoophycos* ichnofacies), and the zone of systematic grazers and farmers (*Nereites* ichnofacies). In addition to these, the *Teredolites* ichnofacies has been described to comprise the borings found in wood (Bromley et al., 1984) and the *Trypanites* ichnofacies (Bromley, 1972), to comprise the borings found in hard lithified substrates. These two facies are thus less determined by a definable suite of environmental characteristics than by a specific substrate. The *Nereites, Zoophycos, Cruziana,* and *Skolithos* ichnofacies are typical of marine soft sediment. The *Glossifungites* ichnofacies has a much more limited distribution, in firm but lithified sediment; it has been described in detail by Pemberton and Frey (1985). The *Scoyenia* ichnofacies, consisting of terrestrial trace fossils, has not been well characterized in either behavioral terms or by generic composition, but largely by contrast to the otherwise marine ichnofacies (Frey et al., 1984). Terrestrial trace fossils are not well known, and many in fact closely resemble marine trace fossils. With further study the *Scoyenia* ichnofacies can probably be subdivided into assemblages characteristic of specific nonmarine habitats.

The ichnofacies model has proven to be very robust because behavior can be linked to a number of depth-correlated environmental parameters. However, as discussed previously, depth inferences for the ichnofacies clearly should not be too rigorously applied because the primary environmental control on trace fossil distribution is by these other parameters and not depth itself. Nevertheless, these ichnofacies have been recognized in many studies and fall within the correct gradient of water depth. In fact, the ichnofacies model has become so well established that if it cannot be recognized, the explanation of differences between the observed trace fossil distribution and

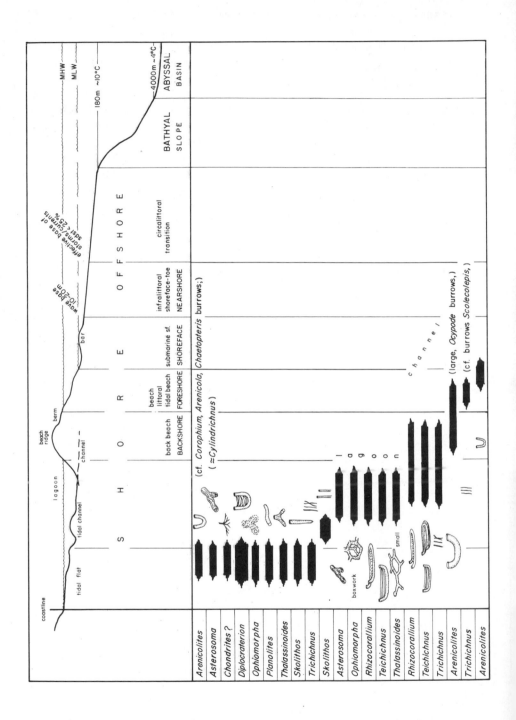

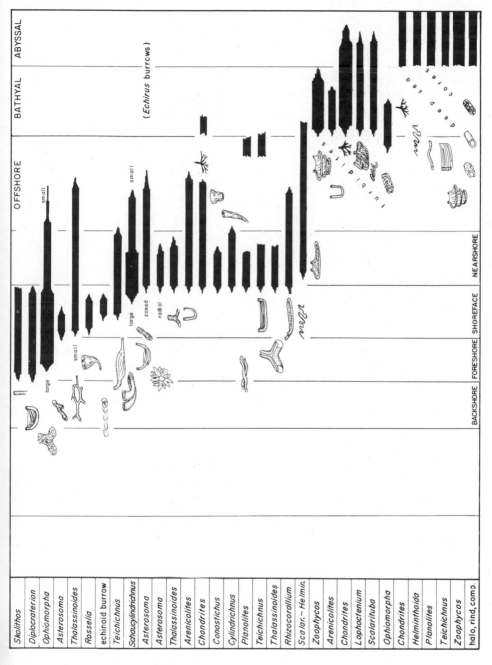

Figure 6.8 Environmental distribution of trace fossils commonly found in cores. After Chamberlain (1978), Trace Fossil Concepts, Society of Economic Paleontologists and Mineralogists Short Course.

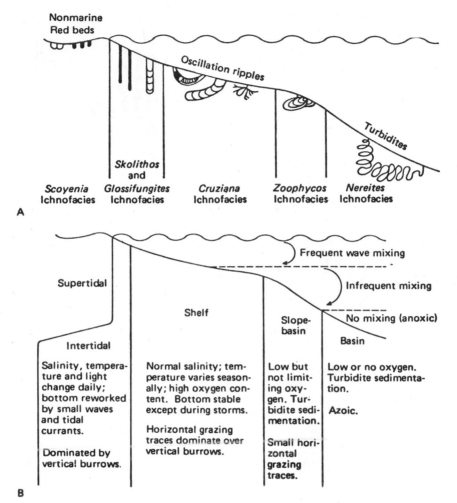

Figure 6.9 Terrestrial to offshore profile showing generalized distribution of trace fossil morphology, ichnofacies, and important environmental parameters. After Rhoads (1975).

the ichnofacies model may comprise a significant aspect of the analysis and interpretation.

Trace Fossil Abundance

Much of the analysis of trace fossils has been focused on relating particular trace fossils, assemblages, and causative behaviors to specific environmental conditions. The abundance of trace fossils is also important. Because the rate of bioturbation observed on the seafloor greatly exceeds the normal low average rate of sedimentation, marine sediments are generally thoroughly churned. Even though sedimentation is generally an episodic process whose rate may on occasion far exceed the "normal," burrowing may be so rapid and extend so deeply that even in areas of rapid sedimentation, intensive bioturbation is the rule throughout the sedimentary column.

In environmental terms, too, bioturbation is nearly ubiquitous. Along common environmental gradients of increasing stress and decreasing diversity, soft-bodied organisms, and thus traces, tend to be present even though shelled invertebrates are not able to live. In places where trace fossils are absent, either deposition must have been very rapid or the environment was so exceedingly unfavorable that all benthic organisms were excluded. Examples of lack of trace fossils being a result of rapid sedimentation would be beach deposits and turbidities. On beaches, although net sedimentation may be slow relative to rate of burrowing, reworking and redeposition are continuous, so that even if traces are formed, they are soon destroyed. Turbidities are deposited episodically and rapidly, and generally in deep water where burrowing is less common than in shallow water, so that burrowing will not be able to work down into the bed before it is buried. Two examples of trace abundance being controlled by limiting environmental conditions are evaporites and diatomites. Evaporites are deposited in an environment that is outside the tolerance for benthic organisms. The reason for the absence of trace fossils in laminated diatomites is not immediately evident because the abundance of organic material should be advantageous as a food source for benthic organisms. The abundant organic material probably depletes available oxygen in and on the surface of the substrate, and in addition, the diatomaceous ooze probably does not provide a substrate firm enough for macroorganisms.

Absence of oxygen is the most common explanation for the lack of bioturbation. The principal locations at present where burrowing is diminished because of reduced oxygen is within silled basins and in the oxygen minimum zone (OMZ) on the open seafloor. In the silled basins of the southern California borderland, for example, water circulation is limited below sill depth. Consequently, available oxygen is depleted, macrobenthic organisms are lacking, and the microbenthic assemblage is of low diversity and consists of dwarfed and deformed individuals along with a normal planktonic and nektonic assemblage. As a result, the sediments are generally laminated and trace fossils are absent. When the water below sill level occasionally turns over and is replaced by normal oxygenated seawater, a normal benthic fauna flourishes and the sediments are burrowed, but then conditions gradually revert as the oxygen level decreases (Harman, 1964). In the OMZ off the California coast, burrowing is present where the oxygen level in the overlying water is as low as 0.3 mL/L (Thompson et al., 1985). Absence of a burrowing fauna is predicted only at oxygen values lower than 0.1 mL/L.

Both the presence and absence of biogenic sedimentary structures are of paleoenvironmental significance. The distribution patterns of trace fossils will increase in value for paleoenvironmental reconstruction, as we are able not only to describe them but to explain them in behavioral–environmental reconstruction. Yet we need to know more about the effects on the trace fossil record of evolutionary changes in trace-making organisms and in the evolution of behavior through geologic time (Seilacher, 1977a).

TRACE FOSSILS AS CLUES TO SEDIMENTATION

The morphology, abundance, and relation to bedding of trace fossils are controlled by the characteristics of the sediment in which they form. Consequently, trace fossils may be of great value in describing depositional processes and primary sedimentary characteristics. When human beings first walked on the moon, views of their footprints in lunar soil were immediately transmitted to earth, and the depth of the impressions and the apparent cohesiveness of the soil in these vivid pictures formed the basis for speculation about the nature of the sediment, its depositional history, and the lunar environment. Trace fossils in the geologic record offer similar interpretive possibilities. Thayer (1983) and McCall and Tevesz (1982) provide broad surveys of the interactions between organisms and sediments.

Sediment Strength

If the nature of a sediment when it was deposited can be determined, that information will provide important clues about the depositional water energy, about subaerial exposure events or specific physical–chemical conditions that might have altered the strength of the sediment after deposition, and about the substrate as an important environmental parameter for the benthic biota. Sediment may range from soft, flocculent, and easily eroded to firm, with high bearing strength and strong resistance to erosion. Sediment consistency is not easily determined from sedimentary features of the rock, but can be, potentially, from the trace fossils in the sedimentary rock. At one extreme, a lithified substrate is indicated by borings. Sediment with high cohesiveness and bearing strength is indicated by borrows that are well preserved in original shape and outline and may have remained open during and after the life of the inhabitant even though they were unlined. Somewhat less cohesive sediment is suggested by burrows that are similarly well preserved but were able to maintain their form only because they were lined with organic matter, pellets, shell fragments, or sand grains. Substrate that was soft and fluid is indicated by indistinct, collapsed, and deformed burrows and a general swirled wispy bioturbation.

In addition to providing information about the sediment strength, trace fossils provide information about the subsequent history of the rock. The extent to which trace fossils are changed from their original shapes is a useful measure of sediment compaction. (Does the fact that trace fossils are seldom described as having been compacted or flattened indicate that the bulk of sediment compaction occurs so near the surface and so soon after deposition that it precedes most burrows?)

Rate of Sedimentation

The abundance of trace fossils is determined (1) by the rate of trace formation, which is controlled by the complex of environmental factors that determine the abundance of organisms and the level of their activity, and (2) by the rate of sedimentation. Thus an increase in trace fossil abundance may indicate either a decrease in depositional rate or an improvement in the environment and consequent increase in the abundance and level of activity of trace-making animals. Although trace fossil abundance is commonly used to infer sedimentation rate, the relative importance of these two independent variables cannot be determined without additional information.

Hardgrounds and Depositional Hiatuses

Under some conditions, marine substrate may become a lithified hardground, either by submarine cementation or by meteoric cementation during brief periods of subaerial exposure. In either case, no erosion may have occurred or be evident and the hardground may be discernible only from the borings extending down from the surface. The traces are recognized as borings because they cut sharply across all components of the rock, including grains and cement, body fossils, other borings, and prelithification burrows and bioturbation. Because hardgrounds form in areas of fluctuating depositional rate, nondeposition, and local scour, a complex detailed stratigraphy may develop, including both burrowed surfaces and bored surfaces. The chronology of overlapping generations of trace fossils is commonly difficult to unravel, but may be the only means of determining the sequence of depositional events. Intensive burrowing or boring may reduce a stratum to an apparent rubble of clasts of the underlying rock type in a matrix of the overlying rock type (Fig. 6.10). As a result, a surface of nondeposition may appear to be an eroded, unconformable contact, and suggests a much greater hiatus in the stratigraphic record than actually occurred. This may be particularly difficult to recognize in cores, with the limited exposure they provide.

A surface of this type has been described at the top of the Middle Eocene Stone City Formation of east-central Texas by Stanton and Warme (1971). The Stone City Formation, overlying deltaic sandstone and overlain by marine mudstone, was deposited in a shallow, delta-margin setting during the initial phase of the last major transgressive–regressive cycle of the Middle Eocene in the area. Historically, detailed correlation of Tertiary strata in the Gulf coast region has been difficult in outcrop because of marked lateral facies changes. Consequently, regional disconformities were sought as possible time horizons to establish an improved stratigraphic framework. The upper surface of the Stone City Formation was interpreted as such a regional

Figure 6.10 Intensively burrowed surface, Eocene, Alabama, showing two generations of burrow (1, early; 2, late) and very irregular resulting surface.

disconformity by Stenzel (1935) because he observed that the surface was uneven and because he believed that there were clasts lying on it composed of the underlying sediment. Therefore, he concluded that the surface formed during a period of subaerial exposure, lithification, and erosion was of sufficient magnitude to be of regional extent and valuable for correlation. Analysis of these features in terms of trace fossils, however, leads to the alternative conclusion that apparent clasts are the lithified fillings of burrows or borings that are more resistant to weathering than is the surrounding ground mass. The surface may have been a hardground or perhaps was burrowed intensively during a period of slow to nondeposition, but the apparent lithoclasts and surface unevenness appear to be only the normal effects of biologic activity, either burrowing or boring, or both.

APPLICATIONS IN SEDIMENTOLOGY

The potential contribution of trace fossils in sedimentologic analysis is illustrated in a study of lower Eocene, Wilcox strata at a locality in east-central Texas (Warme and Stanton, 1971). The strata consist of laminated unfossiliferous claystone overlain by several 0.6- to 1.0-m beds of muddy sandstone (Fig. 6.11). Regional stratigraphic data suggest that the sediments were deposited in a deltaic setting. (Extensive deposits of lignite that apparently

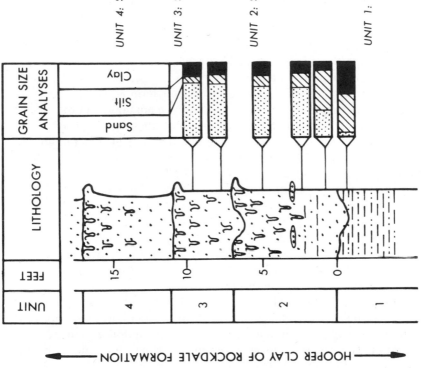

UNIT 4: Sandstone; very fine; cross-bedded near base; burrows in upper part; weathered and poorly exposed.

UNIT 3: Sandstone; very fine; basal contact erosional; cross-bedded in lower part with abundant clay flakes along bedding surfaces, grading upward to massive, burrowed; *Ophiomorpha* burrows abundant, increasing upward.

UNIT 2: Sandstone; very fine; cross-bedded, grading upward to low angle or flat-bedded laminations, massive in upper half; basal contact erosional; plant impressions along clay partings in lower half, *Ophiomorpha* burrows, clay lined, in middle and upper parts and increasing upward in abundance.

UNIT 1: Claystone, siltstone and very fine sandstone, interlaminated; bedding irregular, lenticular, rippled; plant impressions rare.

Figure 6.11 Lithologic column in the Hooper Clay Member of the Eocene Rockdale Formation, Texas. Units are individual pulses of sedimentation. During intervening time intervals, units were burrowed from upper surface, and then scoured as the overlying unit was deposited. After Warme and Stanton (1971), reprinted from Trace Fossils: A Field Guide to Selected Localities in Pennsylvania, Permian, Cretaceous, and Tertiary Rocks of Texas and Related Papers, B. F. Perkins, Editor. By permission of the School of Geoscience, Louisiana State University. Copyright 1971 by the School of Geoscience, Louisiana State University.

accumulated in a deltaic swamp are only slightly higher in the section and are mined within a few miles of the outcrop.) Each sandstone bed is cross-stratified in the lower part but grades upward into increasingly bioturbated sandstone containing distinct *Ophiomorpha* burrows. The boundary between the cross-bedded and burrowed parts of each bed is very irregular, whereas the upper contact of each bed is an uneven scoured surface with 10 to 20 cm of local relief. The sandstone beds were deposited as crevasse splays built into the shallow water of an interdistributary or interdeltaic bay. The absence of trace-forming organisms in the underlying delta-fringe claystone can be explained in several ways: rapid sedimentation and high turbidity, a soft fluid substrate, and variable but generally low salinity. Each sand bed was deposited rapidly as an individual pulse associated with stormflood and overlevee river flow. The rapidity of deposition of each sand unit is indicated by scour of the underlying depositional surface and by the stratification. Following each of these depositional events, "callianassid" crustaceans and perhaps other infauna colonized the sandy seafloor and began to burrow into the sediment. The burrowing obliterated completely all primary sedimentary structures in the upper part of each bed, and only the last generation of dwelling burrows were preserved as abundant *Ophiomorpha*. A very shallow water setting is inferred from the scoured uneven upper surface cutting across the *Ophiomorpha* at the top of each bed. The trace fossils are useful in reconstructing the sedimentation events in two ways: They document the episodic pattern of deposition by accentuating the bedding contacts, and they allow an estimate of the time interval between deposition of individual beds by considering the depth to which burrowing had extended before the next scour-and-sand deposition event terminated the burrowing in that bed. The depth of burrowing is, of course, difficult to determine exactly because the subsequent scouring event removed an unknown amount from the top of the bed. Wetzel and Aigner (1986) have proposed that the depth of scour can be determined if the trace fossils occur in a tiered arrangement. This is an interesting possibility, but it depends on an expectable depth distribution of the tiering. They offer several examples in which depths of the different tiers are apparently rather uniform within a stratigraphic interval and area. In these cases, then, depth of scour can be determined quite precisely by the extent to which the upper part of the sequence is missing. Except in well-studied situations, however, the tier depths may be difficult to establish because depth of burrowing depends not only on time, as indicated above, but on sediment characteristics (Ekdale, 1985) and distribution gradient of oxygen in the sediment (Savrda and Bottjer, 1986).

Sedimentation in shallow marine environments is commonly very much like that in this Eocene example, consisting of infrequent but rapid pulses of sedimentation, perhaps with initial scour and relatively longer intervals of little deposition. Thus the trace-forming organisms living within the sediment may be repeatedly forced to shift their positions up or down to reestablish themselves at the optimal depth below the sediment surface. This activity may generate distinctive sequences of overlapping generations of traces similar to

Figure 6.12 Burrows in the Pliocene Etchegoin Formation, California. Dark sandstone is medium grained, well sorted, thin bedded to laminated. It was deposited in beach to nearshore conditions. Matchbook in each photograph provides scale. (*a*) General view of outcrop; (*b*) close-up view. Light-colored burrowed sediment is muddy sandstone.

those caused by fluctuations in oxygen content as proposed by Savrda and Bottjer (1986). As an example, erosional and depositional events during the Pliocene in central California are illustrated in Fig. 6.12. Deposition in a high-energy, probably shoreface environment is suggested by the regional stratigraphic setting and the texture and sedimentary structures. The episodic nature of scour and deposition can be recognized only from the trace fossils. The presence of stable surfaces from which the underlying sediment was intensively burrowed indicates that gross deposition was slow. Sometimes, burrowing was abruptly terminated by subsequent deposition, although no evidence of scouring on the existing surface can be seen. At other times, as in the Eocene example, a period of little or no deposition and development of an intensively burrowed surface was terminated by erosion of the upper part of the burrowed zone and followed by rapid redeposition, leaving only the lowest parts of burrows intact.

The ability of benthic organisms to move up or down during sedimentation or scour to maintain their preferred position at or below the depositional surface differs from one species to another (Kranz, 1974). Among benthic marine organisms, attached epifauna obviously have no migration ability, whereas vagile infauna should be well adapted for vertical migration. If deposition is very rapid, the upward migration path becomes an escape structure, or

Figure 6.12 (Continued)

fugichnion (Fig. 6.13). In addition to providing information about the depositional process, recognition of escape structures helps in sorting out original life
assemblages of body and trace fossils. At a specific stratigraphic horizon, the
fossil assemblage may include individuals that had lived at that sediment surface, along with individuals that migrated upward during times of sedimentation, and others that burrowed down from higher, younger depositional surfaces. The resulting mixed and overlapping fossil assemblage may only be
sorted out by trace fossil evidence of vertical differences in original position.

In the extreme case, active burrowing may create biogenic stratification. In
the most commonly described process, burrowers work through the sediment
to a depth of perhaps 20 to 30 cm, and cast the sediment from the lower limit
of burrowing onto the surface. The coarser sediment fraction and shells of
organisms living on and in the sediment settle passively to the lower limit of
burrowing. If the rate of sedimentation is slow, a layer of coarser sediment

Figure 6.13 Escape burrows in Pleistocene oolite lime grainstone, New Providence Island, Bahamas.

and shells would accumulate that combines many generations of individuals that had lived during a potentially wide range of external marine environmental conditions at the depositional surface 20 to 30 cm higher in section (Rhoads and Stanley, 1965). The process of biogenic stratification is complex, however, as indicated by downward as well as upward movement of sediment during burrowing (Powell, 1977).

ORGANISMS AS AGENTS OF SEDIMENTATION

The role of organisms in modifying primary sedimentary structures and the sediments themselves may be profound. Much of the discussion so far has focused on traces formed by the activity of organisms, implying that primary sedimentary structures were being destroyed concurrently. The activities of organisms may not leave distinctive traces but may be more pervasive and have major sedimentological effects. Because they are primarily of sedimentologic interest, they are discussed only briefly.

Stratification

Shell-lag layers formed by burrowing organisms have been recognized in modern sediments, and described in the preceding section. This is a process that may be common but has seldom been recognized in fossil shell concentrations. More widely recognized are the sediment binding and trapping effects of organisms. The foremost example is the deposition and stratification of carbonate sediments as stromatolites through the action of filamentous blue-green algae or cyanobacteria. Sediments are trapped and held by a wide range of other mat-forming organisms as well, however. For example, the excellent preservation in finest detail of fossils in the Jurassic Solnhofen limestone (discussed briefly in Chapter 7) and Posidonienschiefer in Germany has been explained by the presence of algal/fungal/bacterial mats that covered decaying carcasses and kept them in their original articulated condition despite bottom currents and perhaps scavengers (Keupp, 1977; Kauffman, 1981).

Stratification may also result from biological sorting and redeposition of a previously poorly sorted sediment. An example of this process has been described at Mugu Lagoon in California, where as burrowing organisms move through the sediment and redeposit the excavated material onto the surface, tidal currents sort out the finer fraction of the excavated sediment and carry it onto the higher tide flats. As a result, the sediment in the bay and lower tide flats becomes progressively coarser and better sorted, and as the higher tide flats and marsh prograders over them, the resulting stratification is the typical fining-upward sequence of bay to intertidal to marsh sediments (Warme, 1967).

The tubes of burrowing organisms are commonly believed to retard erosion on the seafloor. For example, Fager (1964) has described how a marine tube-building polychaete worm and a small anemone colonized a sandy seafloor and stabilized the sediment against erosion by their baffling effect on the wave surge. In fact, however, such burrow tubes may have a destabilizing effect on the substrate. In studies using the tube-building polychaete worm *Owenia*, Eckman et al. (1981) found that with worm tubes present, the water velocity at which scour began was less than 25% of that velocity when tubes were absent. The explanation for the common observation of more stable conditions when tubes are present is that a blanketing mat formed by benthic diatoms, filamentous algae, and animals is enhanced by the sediment burrowing.

Texture

Sorting can be improved by organisms resuspending sediment for subsequent winnowing by waves and tidal currents, as described at Mugu Lagoon. The opposite effect, of decreasing sorting by burrows cutting across and mixing beds of different textures, is probably more common. Evidence for this mixing consists of the filling or lining of burrows that is texturally different from the surrounding sediment, and from the swirled and churned mixture of different textures in thoroughly bioturbated sediments. Many poorly sorted massive mud–sand strata are probably the result of thorough bioturbation of original thin distinct beds of mud and sand.

Sediment Destruction and Construction

Burrowing and boring both destroy and create sediments. Carbonate sediment may be dissolved by the acidic gastric fluids as it passes through the gut of sediment swallowers, so that grain-size reduction and net loss occur. Mayor (1924) estimated that holothurians swallowing carbonate sediment in Pago Pago harbor dissolve about 400 g of sediment per square meter each year. Mechanical abrasion in the guts of organisms may also be an important process of sediment destruction. Biting, scraping, and boring by a wide range of animals, fungi, and algae break down and recycle large amounts of loose sediment and rock (Warme, 1977). This bioerosion may disagregate the substrate, resulting in sediment of the same, finer, or coarser grain size. In general, net loss of rock results from bioerosion, either from acids used by the bioeroder, or by subsequent enhanced solution of fine debris. An indication of the magnitude of this bioerosion is the estimate by MacGeachy and Stearn (1976) that on average, 20% of the volume of the skeleton of the coral *Montastrea annularis* that grows on the reefs of Barbados is destroyed by the boring activity of macrooganisms. This is largely caused by sponges, but also by bivalves, sipunculids, polychaete worms, and barnacles. Boring by clionid sponges is largely by plucking of fine sand- to silt-size particles. Bioerosion by the pedunculate barnacle *Lithotrya* on the underside of skeletal grainstone beach rock ledges in Puerto Rico produces carbonate mud from the grainstone (Fig. 6.14), but at the same time, the boring weakens the beach rock so that it breaks readily into cobble-to-boulder-size clasts (Ahr and Stanton, 1973).

Bioerosion by microorganisms, algae, bacteria, and fungi, is even more pervasive than that by macroorganisms, resulting in the micritization that is endemic in carbonate sediments (Golubic et al. 1984). Although the effects of algae are limited to the photic zone, the effects of fungi and bacteria are found in the deep ocean, limiting the potential value of microborings in the fossil record as bathymetric indicators.

Chemical Effects

Organisms modify the chemistry of sediments, removing organic material as they feed within the sediment, and adding organic material as fecal pellets, or organic burrow linings, and as carcasses when they die. The clay minerals

Figure 6.14 Fragment of beachrock, composed of skeletal lime grainstone, from Isla Icacos, Puerto Rico. Sample is in original position, with borings of *Lithotrya* extending upward from lower surface. After Ahr and Stanton (1973), Journal of Sedimentary Petrology, Society of Economic Paleontologists and Mineralogists.

swallowed by crustaceans may be modified in the gut and excreted as fecal pellets of a different mineralogy (Pryor, 1975).

The more general effect of organisms is to create within the sediments a well-mixed and open chemical system. As they burrow they oxygenate the sediment by pumping water down into the burrow system and by casting sediment onto the surface, where it can be oxidized. In the process, however, they bury organic material lying on the surface. The net effect may be difficult to predict, but chemical reactions and changes within the sediment are promoted and maintained by the constant stirring and mixing, long after deposition (Aller, 1982b).

7

Taphonomy

Taphonomy is the study of the postmortem history of fossils. It is important because one of the major problems in paleoecology is how to explain the range of abundance of fossils that is common in sedimentary rocks. Most strata are only sparsely fossiliferous, whereas most fossils occur in infrequent concentrations. Do these differences in abundance reflect environmental differences, or preservational, taphonomic differences? The same question applies, in more detail, to the differences in abundance of particular taxa—abundances that might be determined by taphonomic, preservational characteristics because of factors such as life habit, morphology, and skeletal characteristics. An understanding of taphonomic processes is also essential to explain why fossils are present at all, considering that one finds so few organisms lying on the present-day seafloor waiting to be buried and incorporated into the fossil record. Thus taphonomy is a fundamental aspect of paleoecology, and is important for three reasons: (1) it helps in understanding how the fossil assemblage differs from the original community from which it was derived; (2) it provides insight into the depositional and postdepositional environment; and (3) it helps to establish the temporal significance of a fossil assemblage—whether it was formed during an instant in time or through the mixing and accumulation of numerous, temporally distinct communities over a long period of time. The role of taphonomy in paleoecology is diagrammed in Fig. 1.4. The taphonomic processes that operate on the organisms of an ecosystem from the time of death until they are finally collected by the paleontologist are listed in the box labeled "Fossilization." The record of these processes and their effects form the "Taphonomic Data" (Fig. 1.4), that paleontologists hope to be able to use in their interpretation of the fossil record. The components of taphonomy are (1) *necrolysis,* which deals with the decomposition and disaggregation of the organism at the time of death; (2) *biostratinomy,* which deals with the sedimentologic processes

223

that incorporate the potential fossils into the sediment to become part of the stratigraphic record; and (3) *fossil diagenesis,* which deals with the chemical processes that may dissolve and alter the fossil.

Taphonomic processes are essentially negative in that they create a fossil assemblage that is very different from the original community, and thus they severely limit our ability to determine what the original community or the depositional environment was like. Paleontologists have long been aware of this problem as they worked with the death assemblage or thanatocoenosis to get to the life assemblage or biocoenosis. Two major sources of confusion were recognized: A fossil assemblage may contain specimens that were added to or removed from the life assemblage, causing mixing in both time and space; and the fossil assemblage is normally only a small part of the original community because of loss of potential fossils by a variety of processes. Much of taphonomic research has focused on these processes in order to be able to recognize if and how they affected the fossil assemblage, and to be able to assess the taphonomic bias built into the fossil record. This work has been valuable because it has systematized the problems we face in interpreting the fossil record. Two studies have been particularly noteworthy in the development of the topic. Fagerstrom (1964) categorized fossil assemblages on the basis of their content relative to the original death assemblage (and community from which it was derived) into unaltered (the original death assemblage), transported (the winnowed-out and redeposited component), residual (the untransported remnant), and mixed (the original death assemblage with both additions and subtractions). Then he described and tested criteria by which the nature of the fossil assemblage could be established. Lawrence (1968), through a detailed analysis of a fossil oyster bank and comparison with a modern oyster bank, documented the taphonomic loss and added the hopeful note that some of the loss based on the presence or absence of macrofossils could be reduced if the more subtle information in the fossil assemblage was also used. Examples of this information, which he labeled as redundant, would be borings left by nonpreserved gastropods, siliceous spicules of otherwise soft-bodied sponges, and the xenomorphic imprint on the base plate of barnacles of the nonpreserved substrate shell. In contrast to studies that have emphasized how much information has been lost from the fossil record, other studies have emphasized the positive conclusions that can be derived from taphonomic data. As indicated in the box labeled "Taphonomic Data" in Fig. 1.4, if the results of taphonomic processes can be recognized in the fossil record, the environmental conditions specified by these processes can also be recognized.

TAPHONOMIC PROCESSES

Necrolysis

Necrolysis encompasses the processes acting on the organism upon its death and soon afterward, but excluding processes of transportation and sedimentation, which fall within the realm of biostratinomy, and of chemical alteration, which is the subject of fossil diagenesis. Thus it is concerned largely with the processes of decomposition and disaggregation.

It is evident from the sparsity of fossils in sedimentary rocks that most organisms are not preserved, and thus that the transformation of an organism into a fossil depends on unlikely preservational circumstances. Requisites for preservation are (1) a skeleton that is resistant to decay and diagenetic change, (2) rapid and gentle burial before all evidence of the organism has been destroyed, or (3) unusual conditions that slow down the rate of destructive processes. The first of these requisites deals directly with necrolysis because decay of soft tissue is the usual and normal initial step in taphonomy, and once the soft tissue is gone, skeletal elements are easily scattered. Rapid burial is important to move the organism into the reducing environment within the sediment, thus further reducing decomposition, and at the same time, placing the organism below the range of destruction of scavenging and boring organisms and of breakage and abrasion by water motion. Amanoxic depositional settings has been the explanation proposed for most instances of unusually complete preservation of soft-bodied organisms or of the soft-tissue parts of skeletonized organisms. Decomposition that destroys the soft parts of skeletonized organisms, removes soft-body organisms entirely, and promotes the disaggregation of multielement skeletons is the normal circumstance.

Lawrence (1968) estimates that 7 to 67% of the species in modern marine benthic communities are soft bodied and have little potential for preservation. R. G. Johnson (1964) and Schopf (1978) both suggest that on average only 30% of the species are likely to be incorporated in the fossil record. Comparison of the total benthic macrofauna fauna with its skeletonized and potentially fossilized component in 195 samples on the southern California shelf illustrates the extent of nonpreservation and the resulting biased fossil record (Table 7.1; Stanton, 1976). At both the generic and specific levels, only about 25% of the fauna is likely to be represented as fossils. Nonpreservable crustaceans and "worms" make up the bulk of the fauna in number of taxa, number of individuals, and biomass. Among the echinoderms, only a clypeasteroid echinoid has a good chance of being preserved. Brittle stars, although a major component of the shelf fauna, and potentially preserved, leave only a scattering of tiny skeletal elements upon death and decomposition, and these have been essentially unrecognized and unreported in microfaunal studies of the modern sediments or of Cenozoic sediments of the same region. The pattern of taxonomic preservation typified by this example from the southern California shelf strongly affects the fossil record of the fauna not only in terms of taxonomic composition but also in characteristics such as life habit, guild, trophic classification, and diversity. This is most evident in a trophic characterization of the fauna because the total fauna largely utilizes food found within the sediment, but the preservable and potentially fossilizeable part of the fauna feeds largely from the substrate (the sediment surface) or the overlying water. Thus inferences about the trophic characteristics of the original fauna would be poorly recorded in the preserved fossil assemblage (Stanton, 1976).

TABLE 7.1   Preservable Component of Benthic Invertebrate Macrofauna, southern California shelf

	Total		Shelled	
	Genera	Species	Genera	Species
PHYLUM				
Coelenterata	8	11	—	—
Bryozoa	1	1	—	—
Brachiopoda	2	2	2	2
Platyhelminthes	4	9		
Rynchocoela (Nemerteans)	6	12	—	—
Nematoda	1	1	—	—
Annelida	212	402	—	—
Sipunculida	4	6	—	—
Echiurida	3	4	—	—
Phoronida	3	5	—	—
Arthropoda	167	266	—	—
Mollusca	139	277	130	271
Class				
Gastropoda	67	132	60	218
Amphineura	1	1	1	1
Pelecypoda	67	138	67	138
Scaphopoda	3	5	2	4
Cephalopoda	1	1	—	—
Echinodermata	29	40	1	1
Total	159	1,036	133	274

Two other examples of the value to paleoecology of the study of taphonomic processes are by Plotnick (1986) dealing with anthropods and by Meyer and Meyer (1986) dealing with crinoids. Plotnick (1986) has explored the factors that limit so severely the fossilization potential for arthropods. To do this, he buried shrimp in marine sediment and then monitored the taphonomic processes that resulted in the disappearance of specimens. Bacterial decomposition was important, but destruction by scavengers and disturbance by burrowing infauna were more important. This conclusion has two important implications for the fossil record: The first is that rapid burial is necessary to minimize each of these destructive processes. The second is that abundance of fossil arthropods may be inversely proportional to the abundance of deep infauna that would disturb the fragile skeleton. This may well explain the relative abundance of arthropods in the early Paleozoic, and their scarcity since then, as depth of burrowing increased (Ausich and Bottjer, 1985).

A common observation in Paleozoic limestones is that whereas pelmatozoan (crinoid and blastoid) columnals may be exceedingly abundant, calices may be rare. Thus it is well-nigh impossible to estimate how many individuals are represented by the columnals, although a few instances of unusual preservation give some hint of the ratio of stem to calyx (Lane, 1973). Several mechanisms have been postulated for the apparent scarcity of calices: They disarticulate quickly and so cannot be recognized; they do not disarticulate quickly, but with decay become bouyant and drift away; or they are nipped off the column by predatory fish. Meyer and Meyer (1986) have investigated this aspect of echinoderm necrolysis by studying the modern crinoids on the Great Barrier Reef, Australia. They have observed that (1) crinoids disarticulate within a few days after death, except for the calyx, which may survive intact long enough to become abraded; (2) the crinoidal parts are deposited essentially where they lived with very little winnowing of the smaller parts; and (3) there is no evidence of bouyant transport of body parts or intact calices. These observations on modern crinoids imply for the paleoecologist that the preservation of calices does not require quiet water, although preservation of the calyx together with column, arms, and so on, does indicate very rapid burial or else very quiet water and absence of scavengers. Although not observed, the removal of calices by predators remains a possible explanation for the relative scarcity of calices in fossil occurrences.

Skeletonized planktonic microfossils are well represented in the fossil record in spite of severe postmortem loss because of the great number of living individuals. Their small size is an advantage while living because it reduces their settling velocity and thus makes it easier for them to remain in suspension, but conversely, the small size retards settlement to the bottom and increases the likelihood that they will dissolve or be eaten while still in the water column. In fact, many of these organisms would not reach the seafloor if they were not eaten and their test incorporated into fecal pellets with larger size and greater settling velocity (Schrader, 1971).

Marine vertebrates are seldom preserved intact because they tend to float during decomposition, at which stage they are prime targets for scavengers that destroy and disarticulate the skeleton. Even in the absence of scavengers, the skeleton will disintegrate and fall to the seafloor bit by bit. Schäfer (1962, 1972) has described the necrolysis of marine vertebrates thoroughly and in detail; his study remains the basic reference on the necrolysis of marine vertebrates.

Terrestrial organisms have a very poor chance of preservation because they generally die in areas of slow or no deposition and thus are commonly decomposed or are devoured and scattered by scavengers before they have a chance to be buried (Fig. 7.1). It is important to both the archaeologist and paleontologist to understand the necrolysis of terrestrial vertebrates—for the archaeologist, to recognize and distinguish between physical processes and the effects of human and nonhuman predators when analyzing archaeological material. As a result, studies have been carried out on the progressive decomposition, disarticulation, and disappearance of carcasses in the absence of scavengers and carnivores, on the effects of nonhuman carnivores

and scavengers, and on the utilization of animals in present-day aboriginal human societies that is perhaps comparable to utilization as recorded at archaeological sites. Behrensmeyer and Hill (1980), Fisher (1984), and Shipman (1981) provide an introduction to these aspects of vertebrate necrolysis.

The fossil record of terrestrial plants, as for animals, is largely in lake, fluvial, or particularly deltaic deposits, because these are the sites of active nonmarine sedimentation. Several studies have documented the relative preservability and transportability of the leaves and woody parts of plants (e.g., Scheihing and Pfefferkorn, 1984; Spicer and Greer 1986; Gastaldo et al., 1987). In modern deltaic sediments in Mobile Bay, Alabama, for example, plant accumulations consist of material that lived at or near the site or that was transported in by river either as suspended or bedload from the upper deltaic plain or from more distant upland locations. Thus the relative abundance and degree of preservation of the taxa in this modern peat deposit can be explained in large part sedimentologically by considering the plant parts as sedimentary particles that are transported, fragmented, and destroyed within the water current (Gastaldo et al., 1987).

Pollen, being small, abundant, and having an outer surface that is very resistant to decomposition, has a better chance of preservation than do other plant parts. The interpretation of a pollen assemblage is difficult, however, because this good preservability, combined with the ease of transport, means that most pollen assemblages are a mixture of pollen derived from plants with very different pollen-producing capabilities and transported both by water currents and wind, from perhaps different directions and probably from diverse habitats, and for different distances. Delcourt and Delcourt (1980) and Lowe and Walker (1984) provide an introduction to the taphonomic considerations in pollen analysis.

The probability that a marine organism, even with a skeleton or shell, will be preserved is low because burial is not rapid enough to prevent disarticulation, physical attrition, biological destruction, and chemical solution. Burial depends largely on the rate of deposition, but estimates of these rates in the literature are of dubious value in the context of taphonomic processes. Sedimentation rate is generally calculated by dividing sediment thickness by the interval of time in which the sediment accumulated. This average rate, which may range from negative values (episodes of scour) to a rate several orders of magnitude greater than the long-term average rate, however, has little to do

---

Figure 7.1 Taphonomic information loss for terrestrial vertebrates. The sequence shows the transition of a whole animal to fossilized fragments of bone. *Top to bottom,* a living herbivore; the carcass of the herbivore; disarticulation and destruction of the carcass by predators; further destruction by trampling; cracking and splitting of bones through weathering; invasion by plant roots; burial in sediments; fossilization; displacement and breakage of fossils by faulting; exposure by subsequent erosion. Drawing by David Bichell. From Shipman (1981), Life History of a Fossil. Reprinted by Permission, Harvard Univ. Press. Copyright 1981 by the President and Fellows of Harvard College.

with the short-term or instantaneous rate, which is paramount for an organism while living or if it is to be buried and preserved. Nevertheless, published rates of sedimentation, aside from at local sites of rapid sedimentation, are low enough that with a reasonable productivity of shell-bearing organisms, the seafloor should be paved with shells (Schindel, 1980; Kidwell, 1982; Behrensmeyer and Schindel, 1983). The fact that this is not the case is explained in part by biological processes—shells may be destroyed by predators and dissolved and mechanically weakened by organisms boring into the shell for nourishment or to construct a dwelling. (Clionid sponges are particularly important and active in the latter role.) The bulk of the shells must disappear, however, because of solution while lying exposed on the seafloor or only shallowly buried. The uppermost few centimeters of sediment forms a chemically active zone because of the abundance of decaying organic matter, the mixing by burrowing organisms, and the diffusion of material between the sediment below and the overlying water mass (Aller, 1982b).

The resulting loss of shell material is illustrated by the recruitment, growth, death, and disappearance of mollusks in bays along the Texas coast during a study lasting several years (Cummins et al., 1986b). Because both recruitment and mortality were abrupt in this strongly seasonal environment, discrete populations could be recognized and followed from recruitment through life and then into the death assemblages by sampling at 6-week intervals. *Mulinia lateralis,* a common pelecypod of the Texas bays, is a typical example of this cycle (Fig. 7.2). Within a year after a recruitment event, that population of *Mulinia* is essentially gone from the death assemblage. Thus, under these normal conditions none of the potential fossils is preserved, and rapid burial below the chemically active zone during occasional storms seems to be the only way to account for the shells abundantly preserved in the underlying sediment. One might suppose that this poor preservation of the *Mulinia* population is explained by the onshore coastal location. It has been proposed that the poor body-fossil record from marginal marine strata is the result of abundant production of plant material there that, upon decay, forms CO_2, lowers the pH of the bay and interstitial pore waters, and enhances shell solution. However, for probably the same reason, dead shells are also surprisingly scarce on the present-day seafloor, irrespective of water depth, latitude, or geographical setting. Decay of organic material and oxidation of sulfide minerals in the sediment throughout the oceans is sufficient to dissolve all the shells produced unless they are rapidly buried (Boudreau, 1987).

Biostratinomy

Biostratinomy encompasses the sedimentological processes that form the fossil assemblage from the original community. These include the physical and biological processes of transportation and deposition and postdeposition compaction. Of major importance are the fluid dynamic forces that act on the potential fossil as if it were a sedimentary particle. For marine organisms, these forces are

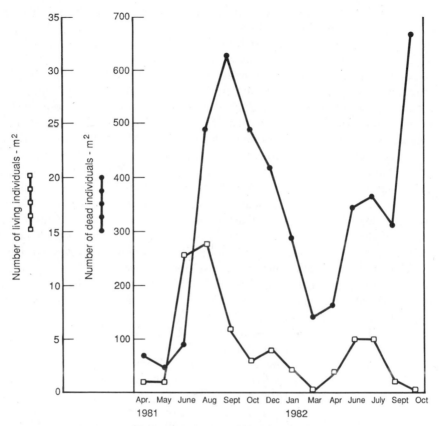

Figure 7.2 Abundances of live and dead *Mulinia lateralis* in Laguna Madre. Samples collected at approximately 6-week intervals. Abundance of dead shells is far greater than of live individuals following a recruitment episode because dead abundance is cumulative, whereas the live abundance is the number alive at a particular time. After Cummins et al., (1986b), Palaeogeography, Palaeoclimatology, Palaeoecology, *52:* 3.

the result of water motion. In the nonmarine realm, they are the result of water motion in lakes and rivers and of the flow of air over the earth's surface. These forces are an integral part of the environment, and their effects are much broader than the taphonomic ones that are the focus of this discussion. They affect, for example, the trophic structure of the community by determining the location of trophic resources in the marine environment, that is, whether organic matter is winnowed out of the substrate and maintained in suspension, or settles out of suspension in quiet-water settings. They help to determine behavior of organisms and thus resulting trace fossils. They are particularly important in determining the adaptive functional morphology of organisms. Biostratinomy is important, then, in explaining the nature of the fossil assemblage as determined by these forces, and in providing information about the fluid-dynamic conditions of the paleoenvironment by interpreting the resultant characteristics of the individual fossils and the fossil assemblage. These conditions have typically been deduced from lithologic features, but fossils may be more informative because they form a population of largely in situ clasts that have predictable initial skeletal characteristics and size distributions. In contrast, terrigenous clasts have unpredictable initial properties that were determined by provenance, weathering, and transportation. Although biostratinomy has in the past dealt primarily with physical, sedimentological processes, biological processes should not be overlooked because they may result in similar effects on, for example, fragmentation, sorting, transportation, or bedding.

The characteristics of a fossil assemblage are those of the original biocoenosis, as subsequently modified by numerous processes during the life of the individual organisms, through death, burial, sedimentation, diagenesis, exposure, and weathering. To extract from paleontologic data the biostratinomic aspects of the fossil assemblage, we must be able to isolate the effects of biostratinomic processes from those of other taphonomic processes and from the original characteristics of the life assemblage. For example, to interpret the size-frequency distribution of a species in a fossil assemblage in terms of the sorting that may have been caused by a certain amount of wave action, and perhaps deduce water depth from this, we would ideally start with a model for the size-frequency distribution of the species when alive, determined solely by its population dynamics. Then the differences between this model and the actual observed size-frequency distribution could be analyzed in terms of taphonomic processes, perhaps particularly the water energy of the paleoenvironment. This logical sequence is seldom explicitly stated, and the starting model used as a basis of comparison is generally not presented in any detail. For example, it has been commonly assumed that a well-sorted population of fossils is due to wave or current action and therefore to transportation (Boucot, 1953; Fagerstrom, 1964). However, it could also be due to population dynamics—the result of a single pulse of recruitment followed by low postlarval mortality and a catastrophic mortality event. As another example, the interpretation of fossil orientation only rarely proceeds from a rigorous description of the expected or probable orientation of the living organisms (Toots, 1965). However, an analysis that does not specify the presumed original orientation and then compares the fossil orientation to it lacks a logical basis for interpretation and may well be erroneous if not essentially meaningless. In both these cases, a null hypothesis against which the data can be tested is needed.

The mechanics and effects of fluid motion on skeletal material must be known in order to determine from the fossils this parameter of the paleoenvironment, but these have been described only by the general concept of "water energy." Present understanding of the effects of water motion on skeletal material is derived largely from tumbling-mill experiments and from flume studies in which the water flow, substrate texture, and bed form are related to the movement and stable orientations of skeletal particles of different size and shape. These studies have clearly demonstrated the effects of unidirectional flow in sorting and orienting skeletal material, but they do not provide satisfactory models for oscillating water movement due to wave action in offshore settings or for the turbulence of the breaker zone. The characteristics of the fossil population or assemblage that are most likely to be affected by water energy are skeletal condition, texture, orientation, and depositional fabric (Table 7.2).

Skeletal Condition

Skeletons and shells will be disarticulated, broken, and abraded as they and the surrounding sediment are moved by flowing water, whether this is during transportation for some distance or during agitation that is more or less in place. The extent of disarticulation, breakage, and abrasion is generally corre-

---

TABLE 7.2   Taphonomic Criteria

Characteristic and Interpretation

SKELETAL CONDITION

Disarticulation: Biologic causes of disarticulation probably dominant; articulated skeletons reflect no agitation and, more important, no disturbing biota

Fragmentation: Considered diagnostic of transportation; probably caused primarily by biologic agents (predation, scavenging, bioturbation), may result from postdepositional compaction

Abrasion: Rounding of edges and loss of fine sculptural detail predominantly reflects water energy; may also be caused by solution and microborers

TEXTURE

Size-frequency distribution (SFD) of individual species is strongly controlled by population dynamics of live population, postmortem predation, and solution; water energy is probably a minor factor. SFD of total assemblage is indicative of water energy; differential sorting of body parts indicates water energy and hydrodynamic equivalence, differential compactional crushing, predation, or solution/ decomposition

ORIENTATION

Orientation is diagnostic primarily of water energy, but also of biologic agents (predation, scavenging, bioturbation)

FABRIC

The accumulation of fossils reflects rate of production versus loss and sedimentation, redeposition by water energy, and mixing and concentration by bioturbation

---

lated with the magnitude and duration of the movement the skeleton has undergone during one or more depositional cycles, and is commonly expressed as the amount of transportation. The simple correlation with distance of transport may lead to erroneous conclusions, however. Living mollusks are scoured from the shallow seafloor off the Texas coast during hurricanes and quickly redeposited while still alive and in perfect condition on the beach or even across the barrier island into the bay, into a setting very different from that in which they had been living and for which they might be considered diagnostic. In contrast, the pelecypod *Donax* lives at the water's edge, but dead shells on the beach are disarticulated and badly worn by constant washing to and fro in the surf, without having undergone any transportation. Despite such complications, skeletal condition is a generally useful criterion for determining if the individuals in an assemblage have been transported or are preserved where they lived (Fagerstrom, 1964).

Articulation.   Skeletons are preserved in complete and articulated condition only under unusual conditions of rapid burial or of no physical or biological disturbing forces. In many marine invertebrate assemblages, individuals that are also in living position or still attached to the substrate confirm this conclusion. Normally, progressive decay and decomposition of connective tissue after death is inevitably followed by disarticulation, as Schäfer (1962, 1972) has described in detail for marine organisms as they lie on a tidal flat or float in the water. Articulated fossils preserved in sediment that on sedimentologic grounds was deposited very slowly require a setting that was quiet to minimize physical scattering, and that was inhospitable for predators, scavengers, and bioturbators. Anoxia is the usual explanation for such abiotic conditions. Disarticulation of crinoids, as described by Meyer and Meyer (1986) and discussed under "Necrolysis," is a good example of the necrolysis of a particular organism and its consequent accumulation as sedimentary particles. A classic example at the facies rather than individual level is provided by the excellent preservation of entire skeletons of fish and other organisms in the Jurassic Solnhofen Limestone of southwestern Germany as described by Keupp (1977). The Solnhofen Limestone was deposited in a broad lagoon separated from the open marine Tethys Sea of southern Europe by a more or less continuous reefal sill. Marine organisms lived in the lagoon during times of open connection between lagoon and ocean, but with subsequent isolation of the lagoon from the open ocean and the establishment of hypersaline conditions, the marine organisms in the lagoon died. Their excellent preservation is due to the absence in the stratified anoxic bottom water of benthic scavengers or burrowers that would have disturbed their carcasses. In addition, a coccoid blue-green algal mat blanketed and protected them both from bottom currents and from benthic organisms that would reappear when improved connection with the open ocean was reestablished and the lagoonal water became reoxygenated. As another example, fish are preserved in excellent condition in laminated diatomaceous sediment because benthic organisms that would disturb the skeletons and sediment are apparently absent for two reasons: Because of high productivity and rapid deposition of diatoms, the organic content in the sediment is very high, and oxygen in the sediment and in the water immediately above the substrate is absent; and the diatom ooze has a very high water content and thus insufficient bearing strength to support large benthic scavengers.

Fragmentation. Shell fragmentation has commonly been considered a good indicator of water energy or transportation. Patterns of shell breakage presumably caused by physical processes have been studied by Hollmann (1968) from material collected on the sandy beaches of the northwest Dutch coast after a storm. Two examples are illustrated in Fig. 7.3. He concluded

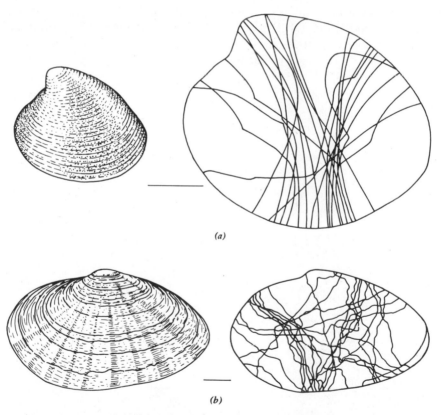

(a)

(b)

Figure 7.3 Fracture lines in two recent bivalve species from the northwest Dutch coast. Note the differences in fracture pattern and the correlated differences in shell morphology. Bar length: 1 cm. (*a*) *Venus gallina*, 20 examples (*b*) *Mya arenaria*, 25 examples. From R. Hollmann (1968), published in Paläontologische Zeitschrift, *42:* 226. E. Schweizerbart'sche Verlagsbuchhandlung Nägele u. Obermiller, Stuttgart.

that the morphology of shell fragments is independent of specific processes of breakage, but is controlled by shell solidity, concentric growth lines and seasonal rings, and details of sculpture. The shell microstructure of crystallites and organic matrix also controls shell strength and the size and shape of the particles into which it disaggregates (Ginsburg, 1956).

In contrast to Hollmann's results and the common notion of physical processes of breakage, little firm evidence is available to support this assumption. In experimental studies of the factors that control their preservation, shells have been tumbled in barrels with sediment of different grain size and sorting, and resulting size decrease and weight loss monitored (Chave, 1964; Driscoll and Weltin, 1973). The results are largely explained by abrasion, as discussed more fully in that section. Fragmentation occurs primarily in these studies after the shell has been seriously weakened by abrasion.

Skeletal breakage by predators is common. Many crustaceans, for example, feed on mollusks by chipping away the shell margin of pelecypods or the outer lip of gastropods in order to reach into the shell far enough to grasp and feed on the body (Fig. 7.4; Papp et al., 1947). Numerous marine vertebrates crush invertebrate shells as they feed. Skates and rays, for example, crush pelecypods with their massive flat-surfaced teeth, ducks crush pelecypods and

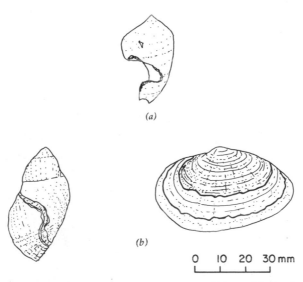

(a)

(b)

0   10   20   30 mm

Figure 7.4 Shell fragments produced by crab predation: (*a*) fatal predation on gastropod; (*b*) nonfatal predation on gastropod and bivalve and subsequent repair. From R. Hollmann (1968), published in Paläontologische Zeitschrift, *42*:227. E. Schweizerbart'sche Verlangsbuchhandlung Nägele u. Obermiller, Stuttgart.

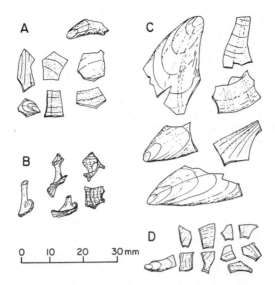

Figure 7.5 Shell fragments produced by bird predation on *Mytilus edulis* and *Littorina littorea: (A)* mussel fragments; *(B) Littorina* fragments from eider duck excreta; *(C)* mussel fragments; *(D)* mussel fragments from gull excreta. After Trewin and Welsh (1976). Palaeogeography, Palaeoclimatology, Palaeoecology, *19:*223.

gastropods in their gizzard (Fig. 7.5; Trewin and Welsh, 1976), and shore birds peck at the tests of sand dollars until they break away the dorsal surface. The resulting fragment shapes are controlled by the technique of the predator and by the morphology and shell strength of the prey. As a consequence, fragmentation by predation should be different from that from water energy, and should potentially be recognizable in the fossil record (compare Figs. 7.3, 7.4, and 7.5). A less dramatic but more pervasive biologic process is the boring by algae, worms, sponges, and a host of other organisms that weaken the shell and thus make it more susceptible to mechanical breakage (Warme, 1975). Papp et al. (1947) and Vermeij (1978) have proposed that morphologic features in gastropods, such as low spire, narrow aperture, varices, and aperture-rim thickening, are examples of coevolutionary adaptations that reduce the predation and shell breakage by crabs.

Because shells under usual conditions of slow sedimentation are almost certainly separated, widely scattered, and selectively destroyed by predators and scavengers, fragmentation as well as disarticulation cannot be simply equated with water energy or amount of transportation, as Cadée (1968) has demonstrated for the modern organisms of the Ria de Arosa, Spain. Shells on the beach there are entirely disarticulated and fragmented, and the normal inference would be that this resulted from constant agitation and tumbling as the shells were transported onto the beach and then remained in the surf zone. However, shells found in midbay, on a muddy substrate far removed from wave agitation, are equally disarticulated and fragmented. Crustacean predators are largely responsible for fragmentation of the shells in midbay and probably for much of that of the beach material as well.

Shells may also break during sediment compaction, but these fragments should easily be recognized because they should be together in the rock, approximately in their original relative position, whereas fragments generated by transportation or biologic activity should be scattered. Breakage by compaction has not been widely reported in fossil assemblages. A shell floating in fine-grained sediment would probably not be broken because the sediment would flow around the fossil during compaction. Breakage is likely only when the grain size of the sediment is large relative to the shell and thus cannot flow around the shell during compaction or if the shell is in a coquinoid layer in which shells are in contact and form points of impact or leverage with one another. Several studies have established a quantitative basis for this phenomenon, however: Brenner and Einsele (1976) have measured shell strength and shown that it is correlated with morphologic parameters such as solidity, curvature, and microstructure; and Taylor and Layman (1972) have shown on the basis of limited data that shell strength is generally correlated with the size of the microstructural unit in the shell.

Abrasion.   The rounding of skeletal parts is generally explained by abrasion, and considered diagnostic of water energy. In contrast, a skeleton that is buried where it lived should have angular edges and well-preserved fine sculpture. Semiquantitative measures of abrasion rates for different kinds of marine invertebrate skeletons are available from experimental studies using tumbling barrels (Chave, 1964; Driscoll and Weltin, 1973) and field conditions (Driscoll, 1967, 1970). These studies show that rate of abrasion is determined by the size of the skeletal particle relative to that of the sediment and by skeletal microstructure (Fig. 7.6). As sediment texture increases, the rate of abrasion increases for all skeletal material, as demonstrated by the relative durability of large and small *Spisula* shells—the shells that are smaller relative to the sediment are more likely to be broken as well as abraded. The dense and fine-grained shells of the gastropod *Nerita* and pelecypods such as *Spisula* and *Mytilus* are most durable. Intermediate forms include corals, with hard but moderately porous structure, and oysters, with coarsely crystalline structure but with the individual crystallites separated by organic material. The least durable skeletons are the very porous and organic-rich echinoderms and algae. These data support the concept that if shells are to be preserved in original, unabraded condition, they should not have been transported. It is also important that they be buried quickly because chemical solution and boring by algae and fungi and other organisms will round the edges and give the shell a superficial appearance of abrasion (Perkins and Tsentas, 1976).

Although the correlation of abrasion with amount of transportation is generally valid, as noted earlier a storm can easily move shells quickly and deposit them, still alive, in pristine condition in very different habitats than where

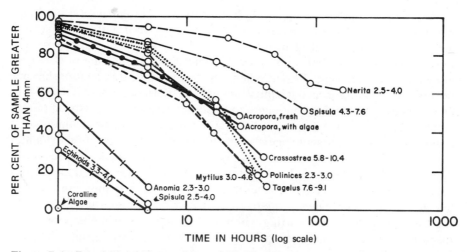

Figure 7.6 Durability of invertebrate skeletal material in a tumbling barrel also containing chert pebbles. Numbers following each name are initial size range in centimeters. After Chave (1964).

they had lived, and shells may be severely abraded at the site where they had lived, without any significant transportation.

Texture and Sorting

Three factors determine the susceptibility of fossils to transportation: (1) size—in general, smaller fossils will be more readily moved, (2) density—less dense and more bouyant fossils should be more readily moved, and (3) shape—rounded shapes are more readily rolled on the seafloor and thin flat shapes are more readily put into and maintained in suspension. These general relations are difficult to quantify, however, because of the great variability in shell shape among invertebrates and of bone shape among vertebrates. A particular level of water energy should winnow out the most easily transported individuals of a taxon or skeletal parts of an individual and leave behind a residual fossil assemblage. This assemblage will be determined by size and sorting attributes of each taxon present, and should be in hydraulic equilibrium with the associated sediment.

The texture and sorting of a fossil assemblage depend on the original size-frequency distributions of the populations in the life assemblage, and on the effects of preservational and other size-selective processes in addition to that of fluid flow. The size-frequency distribution in a living population, discussed in detail in Chapter 8, depends on the rates of recruitment (birth), growth, and mortality through time, and thus permits a wide range of initial size-frequency distributions. Although the size-frequency distribution of the original population in a death assemblage cannot be described rigorously, many studies have assumed that it should be negatively skewed, with a predominance of small individuals because of high juvenile mortality, and that this negatively skewed initial distribution would be modified by current sorting toward a normal distribution because of preferential winnowing of the small individuals (Fagerstrom, 1964). When first formulated, this relationship appeared to be useful for recognizing winnowed or transported assemblages, but field studies of natural assemblages and computer simulations as well (Craig and Oertel, 1966) have demonstrated that the size-frequency distribution of a population in a fossil assemblage may be the result of the population dynamics of the living population, and of a range of factors such as predation, diagenetic susceptibility, and strength to resist crushing during sediment compaction. Commonly, the size range of individual species populations in both live and fossil assemblages is relatively narrow with juveniles notably underrepresented, whereas the sorting within the whole assemblage is poor. Thus the total assemblage suggests lack of winnowing and sorting, and the size-frequency distributions for each taxon are best interpreted in terms of its population dynamics and the preservational potential of the dead shells within the resulting size range. Cummins et al. (1986a) have proposed that the spatial distributions of species can be used as an indication of transportation. They found that the distributions of species on a shallow bay floor covaried more strongly than would be expected from natural recruitment patterns. The explanation was that wave action had redistributed the shells into a uniform distribution. Their shuffling was not necessarily over any great distance, but it was enough to create postmortem patterns of distribution and co-occurrence that differed significantly from the life patterns.

Different parts of a skeleton do not have the same probability of being incorporated into the fossil assemblage because of taphonomic processes that preferentially destroy or remove some parts more than others. In general, smaller and more fragile skeletal parts are relatively more susceptible to diagenesis, winnowing, fragmentation, and abrasion. These taphonomic processes begin immediately after death, as discussed previously in the section on necrolysis. The differential preservation of skeletal parts of vertebrates has been applied to fossil assemblages in order to determine the relative impact of predators and physical processes on the fossil assemblage so that the original biota and the depositional setting could be reconstructed (Voorhies, 1969; Behrensmeyer, 1975).

Right–left sorting of pelecypod shells on the beach due to shape differences has been described by many workers. The extent of the sorting, attributed largely to the mirror-image asymmetry of the two valves, is indicated by data collected by Lever (1958) from the Dutch North Sea coast (Table 7.3). Because of a longshore component, waves move onto the beach at an angle but then flow off the beach directly down the slope. The two valves are affected differently by this directional aspect of the water flow: The valve that is oriented with relatively gently sloping surfaces facing both directions of water flow will remain on the beach; the valve that is not will be rotated and moved

TABLE 7.3   Percentages of Left and Right Valves of Bivalves at Three Locations on the Dutch North Sea Coast, July–August 1955[a]

	Texel			Den Helder			Petten		
	%L	%R	Total	%L	%R	Total	%L	%R	Total
Spisula subtruncata	39.7	60.3	1,616	65.3	34.7	1,948	52.0	48.0	1,681
Spisula solida	44.3	55.7	149	60.7	39.3	56	45.6	54.4	57
Macoma balthica	47.9	52.1	328	43.7	56.3	213	48.3	51.7	149
Donax vittatus	41.0	59.0	178	63.0	37.0	370	46.3	53.7	121

[a]From Lever (1958).

continuously with each wave and be preferentially abraded and transported (Fig. 7.7; Behrens and Watson, 1969). The differences in gross morphology of the two valves are of primary importance; differences in fragility, weight, ornamentation, and in hinge structures projecting out from the valve are also important. In a detailed study of right–left sorting of bivalves on the Georgia coast, Frey and Henderson (1987) have documented that sorting takes place both on the beach and the shallow shelf, and that sorting and shell movement are different during storm and nonstorm periods. Because of these complexities, the interpretation of right–left sorting depends on knowing the location of the fossils within the original paleogeography—whether they accumulated, for example, on the beach, offshore, or in the tidal inlets.

Differential preservation may also be determined by selective predation on a particular part of the prey. Naticid gastropods, for example, feed on bivalves by boring a hole through the shell to get to the tissue of the animal. The hole is commonly situated near the umbo, and will be preferentially on one valve rather than the other. The unbored valve, being stronger, will more likely be preserved in the fossil record. Birds, too, may cause differential preservation of bivalve shells. The oyster-catcher (*Haematopus ostralegus*) feeds on clams exposed during low tide by breaking the shells with its beak. Results of a study on the north coast of Northern Ireland indicated that the left valves of individuals of *Venerupis, Spisula,* and *Mytilus* were preferentially destroyed (Carter, 1974), so that about two-thirds of the remaining valves were right valves.

Orientation

Each organism living on or in the substrate has a specific and preferred orientation. This orientation is rarely preserved unless the organism is firmly attached to the substrate or lives and dies well below the sediment surface so that it cannot be reoriented after death. If the life orientation is preserved in the fossil record, it indicates that the fossil was a member of the original

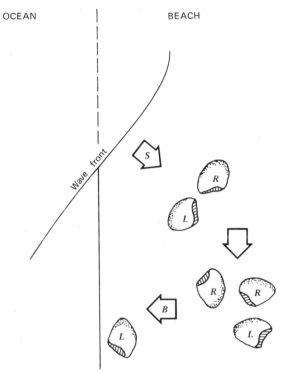

Figure 7.7 Differential sorting mechanism. The stippled portion of each valve is the broad, gently sloping edge that faces the current when the valve is in a stable orientation. In this position the valve is moved only by strong currents. However, when the steep-sided beak (vertically lined) faces a current the valve is unstable and may be rolled or rotated and transported by a relatively weaker current. Thus the oblique swash (*S*) approaching from the right as one faces the sea orients the valves so that the *R* valve is relatively stable with respect to the backwash (*B*), but the *L* valves tend to be carried back down the foreshore. Furthermore, as a result of backwash orientation the beak of the *R* valve points into the next oblique swash which will thus be able to carry it further up the foreshore. Oblique waves approaching from the left would sort the valves in the opposite manner. After Behrens and Watson (1969), Journal of Sedimentary Petrology, Society of Economic Paleontologists and Mineralogists.

community at that place and that disturbing and transporting processes were absent. In addition, life orientation may tell about the environmental parameters to which the organism was responding. Living oysters of the species *Crassostrea virginica,* for example, are preferentially oriented so that the plane along which the two valves meet, the plane of commissure, is approximately parallel to the tidal currents flowing over the oyster bank (Lawrence, 1971b; Frey et al., 1987). Measurements of the plane of commissure of fossil oysters, then, should indicate the direction of current flow in the ancient

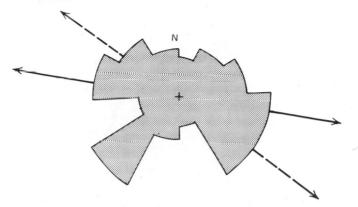

Figure 7.8 Orientation of plane of commissure of 86 Oligocene oysters. These are in a bank that formed in a channel. The dashed arrow is orientation of channel. The solid arrow is vector mean of oyster orientations. After Lawrence (1971b), Journal of Paleontology, Society of Economic Paleontologists and Mineralogists.

environment. This has been applied in a study by Lawrence (1971b) of the Oligocene oyster *Crassostrea gigantissima* at a site in North Carolina where the oysters had formed a bank within a tidal channel. The orientation of the tidal channel could be determined from independent geologic evidence, and the direction of current flow based on the oysters closely approximates the channel direction (Fig. 7.8).

Commonly, however, original orientation is not preserved, and the orientation of skeletal parts within the fossil assemblage is the result of fluid flow before burial and of bioturbation after burial. Numerous examples have been described from the fossil record of bedding planes bearing oriented fossils. The usual way to display these data is by plotting the azimuth of a linear axis of the fossil on a rose diagram. Reyment (1971) describes the quantitative statistical procedures for analyzing this type of data. In some cases, features of the fossils or traces they made on the substrate indicate clearly the direction of flow. For example, the starfish illustrated in Fig. 7.9 has been shoved toward the top of the slab, as indicated by the bent-under upper arm tip and the downward bend of the right arm. From the starfish alone, one might conclude that the starfish had been deformed by a current flowing from top to bottom, but the drag marks leave no doubt as to the current direction (Seilacher, 1960).

Interpretation of oriented fossils has been strengthened by flume studies and field observations of modern-day material. Futterer (1978a,b) provides a comprehensive survey of the results of flume studies. The potential for estimating of current velocity from shell orientation is demonstrated by the orientations of *Scrobicularia* valves at different velocities (Fig. 7.10). The interpretation becomes more difficult, however, when several shells are near each other on the seafloor, creating eddies and thus indefinite orientation patterns

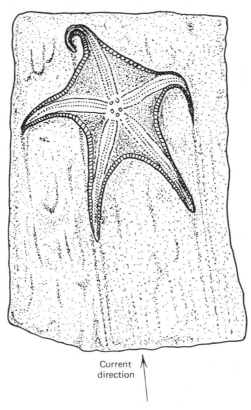

Current
direction

Figure 7.9   Drag marks on substrate and bending of starfish indicate current direction
was from bottom to top of illustration. After Seilacher (1960), copyright Hessisches
Landesamt für Bodenforschung, Wiesbaden.

(Fig. 7.11). Nagle (1967) has established orientation criteria from flume studies (Fig. 7.12, for example) and applied these to field observations in a study
of the fossils in Devonian strata of Pennsylvania. Typical orientations of the
fossils are illustrated in Fig. 7.13 for two brachiopods, an elongate bivalve and
pelmatozoan stem segments. The bimodal rose diagrams indicate that the
fossils were oriented by wave action rather than a unidirectional current, with
the predominant wave movement from south-southeast to north-northwest.
    Alexander (1986) has integrated laboratory and field observations on
brachiopod orientation to arrive at useful paleoenvironmental criteria. His first
step was to determine from field observations what the life orientation of the
brachiopod was. Flume studies then predicted the stability and reorientation
patterns of different brachiopod morphologies under a range of current conditions. By combining these results, it is possible to recognize in a fossil assemblage the extent of disorientation and the probable paleocurrent conditions.

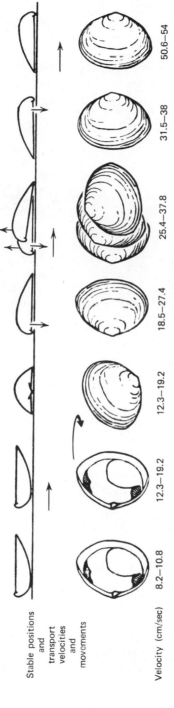

Figure 7.10 Behavior of single valves of the bivalve *Scrobicularia plana* in a water current at different velocities. The velocity is measured 1 cm above the substrate. The arrows indicate the velocity ranges and patterns of motion for shell transport. From E. Futterer (1978a), Published in Neues Jahrbuch für Geologie u. Paläontologie, Abhandlungen, 1978; E. Schweizerbart'sche Verlagsbuchhandlung Nägele u. Obermiller, Stuttgart.

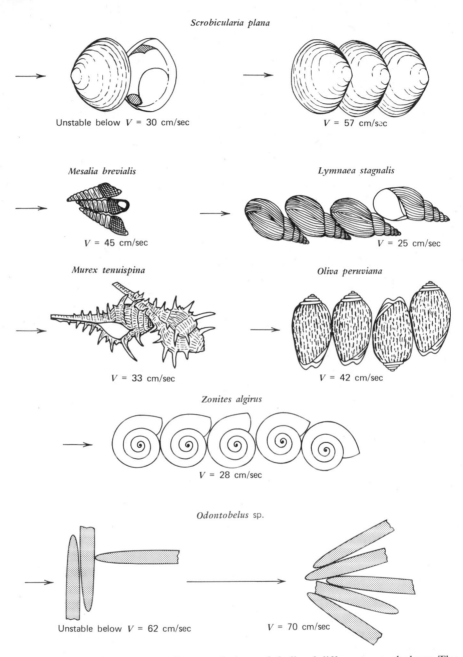

Scrobicularia plana

Unstable below V = 30 cm/sec

V = 57 cm/sec

Mesalia brevialis

V = 45 cm/sec

Lymnaea stagnalis

V = 25 cm/sec

Murex tenuispina

V = 33 cm/sec

Oliva peruviana

V = 42 cm/sec

Zonites algirus

V = 28 cm/sec

Odontobelus sp.

Unstable below V = 62 cm/sec

V = 70 cm/sec

Figure 7.11 Stable patterns of accumulations of shells of different morphology. The arrow indicates current direction. The current velocity in each case is that at which the figured pattern is attained. The velocity is measured 1 cm above the substrate. After Futterer (1978a). From E. Futterer (1978a), Published in Neues Jahrbuch für Geologie u. Paläontologie, Abhandlungen, 1978; E. Schweizerbart'sche Verlagsbuchhandlung Nägele u. Obermiller, Stuttgart.

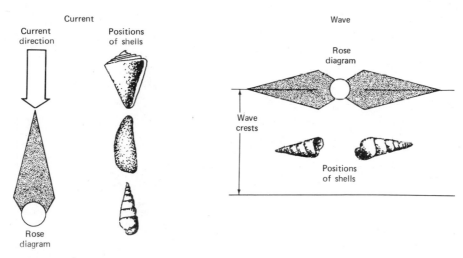

Figure 7.12 Diagnostic orientations of bivalve and gastropods by current and wave energy. After Nagle (1967), Journal of Sedimentary Petrology, Society of Economic Paleontologists and Mineralogists.

Orientation of fossils is generally attributed to wave or current action. The absence of orientation, however, does not necessarily indicate quiet-water conditions because initially oriented shells may be scattered by predators and scavengers after deposition, by scour of the sediment and disorientation during burial, and by bioturbation. Finally, compaction of the enclosing sediments will tend to rotate inclined fossils into a more horizontal position that might resemble a current-formed planar orientation.

Fabric

Fabric is the arrangement of fossils within the sediment. The concentration of fossils more or less densely into layers reflects primarily the balance between net accumulation rate (productivity minus taphonomic loss) and sedimentation rate. The magnitude of these parameters is difficult to determine because they are not independent—sedimentation rate affects both productivity and taphonomic loss, productivity may affect sedimentation rate and so on. Thus it may not be easy to establish whether a sparsely fossiliferous horizon is the result of low productivity, rapid sedimentation, high taphonomic loss, or indefinite interactions of the three. Fossil-rich horizons most commonly reflect the concentrating effect of water energy, either by transporting and depositing the fossils or by winnowing out the sediment and leaving a residual concentration of fossils behind.

Biogenic bedding, caused by burrowing organisms reworking sediment on to the surface and shells in the sediment then settling gradually to the lowest level of burrowing, has not been widely recognized in the fossil record.

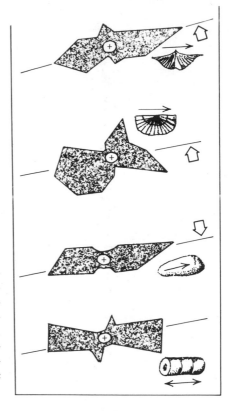

Figure 7.13 Orientations of two brachiopods, an elongate bivalve, and pelmatozoan stem fragments in the Devonian Mahantango Formation, Pennsylvania. Arrow by each fossil is direction measured. Double arrow is inferred current direction. After Nagle (1967), Journal of Sedimentary Petrology, Society of Economic Paleontologists and Mineralogists.

Biogenic bedding should be common because bioturbation is so pervasive. On the other hand, most burrowers are too small to be effective in turning over the sediment, and the process should have a negative feedback effect because the increasing shell concentration should inhibit the infauna and preclude then the formation of a dense sediment-free shell layer. In this case, however, a diagnostic graded sequence of downward increasing shell density that should be recognizable in the sediment and biogenic bedding should be conspicuous when present. In contrast to its effect in producing biogenic bedding, bioturbation probably more commonly disrupts and mixes discrete fossil horizons, destroying fine bedding and temporal resolution as well.

Fossil Diagenesis

Diagenesis serves to both destroy and preserve fossils. Permineralization of plant material, for example, is an important means of preservation. Similarly, the early alteration of the high-magnesium calcite skeleton of echinoderms results in a skeleton of low-magnesium calcite that appears to be particularly resistant to further alteration.

Differential solution of skeleton material has been widely described and is perhaps the most notable example of diagenetic bias of the fossil record. The most common example of this phenomenon is solution of aragonitic fossils and preservation of the more stable calcitic ones in an assemblage. This solutional bias imposed on the fossil record is well documented by comparison of the faunas of living and fossil oyster banks. Although a modern oyster bank is dominated by oysters, the number of associated taxa may be very large. For example, Wells (1961) reported that a fauna of 303 species occurred in the *Crassostrea virginica* banks of North Carolina. Most of these are soft bodied and not likely to be preserved unless they leave a recognizable structure such as borings, or form a xenomorphic structure—an impression on an encrusting surface (Voigt, 1979). Of those taxa that are reasonably common and have mineralized skeletons, however, loss of the aragonitic species would remove all of the gastropods and one-third of the pelecypods (Table 7.4). This is the case in fossil oyster banks as well—taxa other than oysters are rare and found only with careful search.

Solution loss is controlled by morphology, microstructure, and other aspects of the fossil as well as by mineralogy. Among foraminifers, for example, agglutinated forms are much less likely to be preserved than calcareous forms, and among the calcareous forms, those with thinner walls and larger pores are less likely to survive (Smith, 1987). Flessa and Brown (1983) have studied the relation of shell form on dissolution by putting a wide range of shells into 4% acetic acid and monitoring the progressive shell loss on the different shell shapes and sizes. Their major conclusion is that solution rate is correlated with the surface area-to-weight ratio and with porosity and otherwise low shell

TABLE 7.4   Effect of Aragonite Loss upon Preservation of the Recent Minimal Oyster Community[a,b]

Taxa	Number of Species	A	C	A + C
Coelenterata	1	—	1	—
Bryozoa	3	—	3	—
Mollusca				
Gastropoda	9	9	—	—
Pelecypoda	13	4	3	6
Arthropoda				
Crustacea	4	—	4	—
Totals	30	13	11	6
Percentage of total community	38	16	14	8

[a]After Lawrence (1968).
[b]A, completely aragonitic skeletons; C, entirely calcitic skeletons; calcitic skeletons with very minor aragonite, as in crassostreids; A + C, appreciable aragonite plus calcite in skeletons.

density. The complexities of shell dissolution have also been examined by suspending shells on deep-sea moorings in the Drake Passage for 52 days (Henrich and Wefer, 1986). They found that organic coatings, intraskeletal pore space, size and shape of crystallites, mineralogy, particle size, and the presence of internal sediment within the shell must all be considered.

Complete loss of fossils in a sediment by solution is probably common but can seldom be established. McCarthy (1977) has described a Permian beach sand that is at first glance devoid of fossils and apparently never contained any. However, scattered fossiliferous concretions indicate that the sandstone, like modern beach sand, did contain shells that were dissolved by unsaturated water flushing through the unconsolidated sediment except where early cementation had protected them. In lithified rock, solution of the fossil will leave a mold where the fossil had been, but in unconsolidated sediments, as in this Permian example, solution was followed by compaction, and any evidence of the fossils originally present was obliterated. The preferential preservation of fossils is widely recognized for other kinds of fossils as well. For example, Blome and Albert (1985) describe an analogous situation in which microfossils are better preserved because of early cementation resulting in concretions. Commonly, of course, the concretion forms because the fossil is a local nucleus, and one should not assume in this case that fossils were otherwise widely distributed but were not preserved where concretions are lacking.

If not only the fossil but much of the enclosing rock itself is dissolved, a very poor fossil record will be preserved. Bandel and Knitter (1983) have described a lower Jurassic example at Untersturmig, Bavaria, where the stratigraphic thickness of limestone beds within the Posidonia shale has been reduced to one-fortieth of the original, and the original fauna is entirely lost except within residual limestone nodules or as impressions on bedding surfaces. Cephalopod shells, which were aragonite, are completely gone and their original presence is recorded only by an impression on the bedding plane, but calcitic oysters that had encrusted the cephalopod shells are still preserved in their original position on the impression of the cephalopod.

The complexity of diagenetic effects on the fossil assemblage is illustrated by the comprehensive analysis of the Miocene Leitha Limestone of southeastern Austria by Dullo (1983). The richly fossiliferous limestone can be subdivided into several biofacies on the basis of the relative abundance of mollusks, corals, red algae, bryozoans, and foraminifers. These biofacies reflect differences in original paleogeographic setting and bathymetry, but the fossil content as preserved is largely the result of differences in diagenetic history. Details of these histories are presented in Fig. 7.14. In general, calcitic mollusks such as oysters and pectinids are well preserved in all the facies. Aragonitic mollusks with well-preserved primary microstructure are found only in the basinal marly facies. Aragonitic shells occur as molds or have been replaced by calcite in both the chalk and well-cemented limestone facies. Aragonitic fossils are absent as a result of leaching in the carbonate sand facies. These differences in preservation are determined by the extent to which each

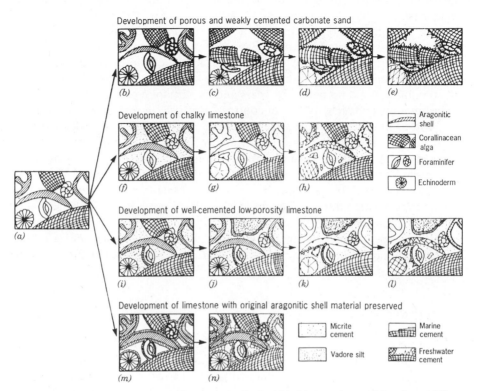

Figure 7.14 . Preservational differences in the Leitha Limestone resulting from different diagenetic histories. (*a*) original nonlithified sediment; (*b*) thin marine cements (Mg calcite) near the surface of the seafloor; (*c*) vadose leaching: removal of all aragonitic shells, compaction features, and brecciation; (*d*) vadose cementation with meniscus cements, dripstone cements, and rim cements with rhombohedral crystals; after the vadose cementation new marine cements sometimes can form because of the relative change in the sea level; (*e*) freshwater phreatic crystals are of lesser importance; (*f*) marine phreatic cementation by micritic Mg calcites; (*g*) leaching within the freshwater phreatic environment. Water passes through the sediment with a high velocity. Leaching occurs without coeval calcite growth; (*h*) only a few parts exhibit complete cementation due to reduced water circulation; (*i*) corresponds to (*f*); (*j*) during the vadose leaching vugs are formed which are later filled by vadose silt; (*k*) freshwater phreatic leaching together with synchronous; (*l*) replacement of the aragonite shells by calcite, some molds may be preserved; (*m*) Marine aragonite and Mg-calcite cements; (*n*) stagnant freshwater phreatic environment: most of the primary structures are preserved. Freshwater calcite crystals occur together with aragonite. From Dullo (1983). Facies *8*:54.

facies was subjected to marine phreatic, freshwater phreatic, or vadose diagenesis.

The time interval represented by a fossil assemblage must also be kept in mind. The fossil assemblage may have accumulated slowly by means of the year-by-year preservation of some small fraction of the community. In this case, the assemblage would represent a *time-averaged* sampling of a sequence of communities over a period of years and of perhaps a considerable range of environments (Fürsich, 1978). In contrast, because preservation is in general so poor, as indicated, for example, by the paucity of skeletons of dead organisms accumulating on the modern seafloor, the fossil record may well be formed by the occasional chance preservation of a single community of living organisms with only a relatively small contribution of older, dead shells. In this case the assemblage might be a fairly reasonable representation of the community existing during an event of short duration rather than the accumulation of meager sampling during a longer time interval. Examples from across the spectrum between the extremes of time averaging and a catastrophic event can be found in the fossil record, and must be taken into account in its interpretation.

APPLICATIONS

The process described in the preceding section and criteria based on them provide important information that can be used in determining both the depositional environment and the extent to which a fossil assemblage is in place and is derived from the original community. Ideally, these taphonomic criteria are applied in conjunction with sedimentologic data within the regional stratigraphic framework because this provides a strong limitation on the range of possible environmental conditions. For example, if the assemblage is preserved in a very fine grained sediment, extensive transport or winnowing of a fossil assemblage is not likely to have occurred. Similarly, a fossil assemblage in a conglomerate bed that clearly required considerable energy for its deposition is probably transported.

Evaluation of the Quality of the Fossil Record

Permian Bivalves of the Park City Formation, Wyoming—Taphonomic Imprint on an Invertebrate Assemblage

Boyd and Newell (1972) have described in detail the taphonomic and diagenetic processes that have produced the characteristics of the fossil assemblage in a thin bivalve-rich limestone. More than a dozen species of bivalves dominate the assemblage; bryozoans, gastropods, scaphopods, trilobites, crinoid columnals, nautiloids, and brachiopods are also present. Many of the fossils are silicified, others are preserved as sand casts or coarsely crystalline calcite. Formation of molds of the fossils was an important step in the complex

TABLE 7.5   Articulated and Single Valves of Most Common Bivalve Species in Samples from a Fossiliferous Bed in the Park City Formation (Permian, Wyoming)[a]

Species	Left Valves	Right Valves	Articulated Specimens
Kaibabella sp.	757	747	2
Scaphellina bradyi	557	575	1
Oriocrassatella elongata	390	364	2
Schizodus sp.	225	213	6
Astartella aueri	91	100	1
Nuculopsis sp.	66	68	16

[a]From Boyd and Newell (1972).

diagenetic history. The siliceous replacements and sand-cast fillings of these molds faithfully preseve fine details of the original shells.

The bivalves are largely disarticulated (Table 7.5) and scattered, and the abundance of paired specimens of *Nuculopsis* sp. is attributed to its numerous interlocking hinge teeth, which helped to keep the valves together after death. The numbers of right and left valves for each of the most common bivalve species are not significantly different (Table 7.5), and the size-frequency distribution of each bivalve species is wide and generally normally distributed. Valves of two of the most common bivalves, *Scaphellina bradyi* and *Oriocrassatella elongata,* are commonly broken, and the breaks occur in all parts of the shells rather than being concentrated in the thin and relatively fragile parts, as one might expect. Despite the high frequency of broken shells, abrasion is slight and fracture surfaces are generally sharp and angular. The shells appear to be randomly oriented in terms of concave-up or concave-down, flat-lying or inclined, and azimuth alignment.

Boyd and Newell conclude from these data that the bivalves were disarticulated, scattered, and broken on the seafloor, but that current or wave energy was too slight to be the causative agent, as indicated by the poor size sorting, slight right–left sorting, lack of abrasion, and preferred orientation. Boyd and Newell proposed that the fossil characteristics were the result of predators, mainly fish that disturbed and scattered the shells lying on the seafloor as they searched for and crushed living bivalves. Arcuate "bites" along the edges of some shells and the common pattern of breakage across the hinge area indicated a predator large enough to crush the whole articulated living bivalve. Holocephalan fish are likely candidates because their teeth are common in the Park City Formation.

Eocene Mollusks of the Stone City Formation, Texas—Taphonomic Analysis as Part of Community Reconstruction

The fauna of the Main Glauconite bed of the Stone City Formation was studied to attempt to reconstruct the original life community (Stanton and

Nelson, 1980). To analyze the community, the extent to which the relative abundance of species in the assemblage was the result of transportation or winnowing had to be determined.

The Main Glauconite bed is 1.5 m thick and consists predominantly of ovoidal glauconite grains 0.5 mm long and 0.2 mm in diameter that are authigenic replacements of fecal pellets (glauconite is used in a general sense; chamosite is probably the actual composition). The pelleted sediment was thoroughly bioturbated so that original primary sedimentary structures are not preserved. The matrix of the glauconite sandstone is made up of approximately equal proportions of clay and fine to very fine, well sorted, angular quartz sand. The basal contact of the bed is sharp, burrowed, and slightly irregular. The upper contact is also irregular because of intensive bioturbation. Continuous bioturbation during deposition formed a patchy churned mixture throughout the bed of pellet and matrix. Abundant body fossils, comprising 96 genera and 120 species in a sample of 6616 individuals are primarily gastropods and bivalves; bryozoans, octocorals, fish otoliths, scaphopods, crustaceans, and nautiloids are minor elements.

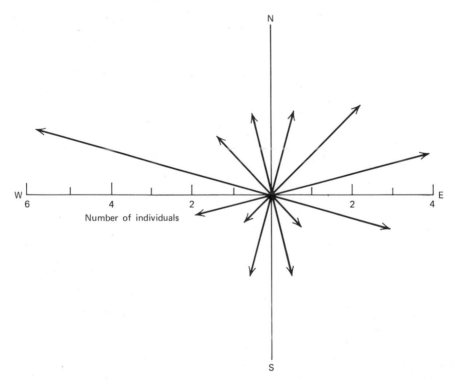

Figure 7.15 Rose diagram of polarity (bearing of anterior end) of 30 scaphopods from the Main Glauconite bed, Stone City Formation, Texas, Eocene. After Nelson (1975).

Eight characteristics of the fauna are used in the taphonomic analysis:

1. Fragile skeletons and fine sculptural detail are generally excellently preserved.
2. Less than 1.0% of the bivalves are articulated.
3. No fossils are found in a recognizable living position.
4. Many of the fossils are fragmented, but rounded and worn shells suggestive of abrasion are rare.
5. The total assemblage is poorly sorted, with shell size ranging from microscopic to several inches in length.
6. There is no preferred orientation among the scaphopods (Fig. 7.15).
7. Bivalves are predominantly inclined or concave-down, but a plot of the inclinations of the shells does not suggest any preferred orientation (Fig. 7.16).
8. The right–left ratios for only 5 of the 21 species are significantly different at the 95% level using the chi-square test. *Cubitostrea, Crassostrea,* and *Anomia,* three species of "oysters," are represented by an overabundance of the right valve, which is smaller and thinner, and so could be transported more easily than the thicker and heavier left valve. *Notocorbula* and *Vokesula,* two corbulids, are represented, in contrast, by an overabundance of the heavier, thicker right valve.

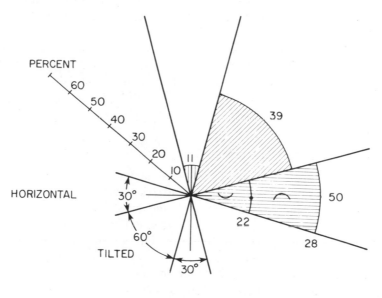

Figure 7.16 Orientation of bivalves in the Main Glauconite bed, Stone City Formation, Texas, Eocene. Radial distance in each sector is percentage of shell with vertical, tilted, horizontal concave-up, or concave-down orientation. After Nelson (1975).

Current-formed sedimentary structures such as crossbeds and ripples are absent. Terrigenous sand is a minor component of the sediment and is not coarser than fine sand. The glauconite grains were originally fecal pellets and probably not durable enough to survive more than local transportation.

The interpretation of these data is at first contradictory. The scarcity of articulation and the abundance of fragmented shells suggest water agitation sufficient to break and separate the shells. The right–left proportions suggest water currents that were strong enough to winnow out the more easily transported valves of *Notocorbula* and *Vokesula,* and to bring in the lighter valves of *Cubitostrea, Crassostrea,* and *Anomia,* but were not strong enough to affect the more robust species. In contrast, the poor sorting of the assemblage and the absence of shell orientation, of current-formed sedimentary structures, of shell abrasion, and of fine terrigenous matrix between the autochthanous glauconite pellets argue for very low water energy.

The resolution of these divergent interpretations must take into account, as in the preceding Permian example, the effects of biologic processes in the creation of the fossil assemblage. Evidence of predation by crustaceans is abundant: Both the outer lips of gastropods and the ventral margins of bivalves are chipped and many individuals bear multiple scars, indicating repeated unsuccessful attack. Thus much of the fragmentation may be the results of predation rather than of agitation by water movement. Predators would also have been effective in disarticulating and disorienting the shells. The intensive bioturbation probably also disoriented and perhaps disarticulated the shells. We conclude that water energy during deposition of the Main Glauconite bed was low. The lack of sedimentary structures and fossil orientation is negative evidence for the same conclusion, but these features may have been present but subsequently obliterated by bioturbation. The magnitude of the currents that were present could potentially be estimated from the results of flume experiments like those of Futter previously described (Fig. 7.10), or Alexander (1986).

Paleoenvironmental Reconstruction

Because reconstruction of the paleoenvironment based on taphonomy relies on the preservational character of the fossils, the focus in this work is strongly on the sedimentologic or diagenetic characteristics of the environment. The example of the determination of the diagenetic history of the Leitha Limestone by Dullo is an example of the latter type of analysis; the analysis by Allison (1988) of the diagenetic conditions necessary for soft-part preservation is another. Allison emphasizes that unusual conditions are necessary to preserve soft-bodied organisms, and that the conventional anoxic explanation is not sufficient because anaerobic processes continue to destroy the soft tissue on the seafloor or after it has been buried. Instead, the essential key to soft-body preservation is a formation of diagenetic minerals that replace the tissue or shield it from destruction. The sedimentary geochemical environment for pyritization, phosphatization, and carbonate precipitation are de-

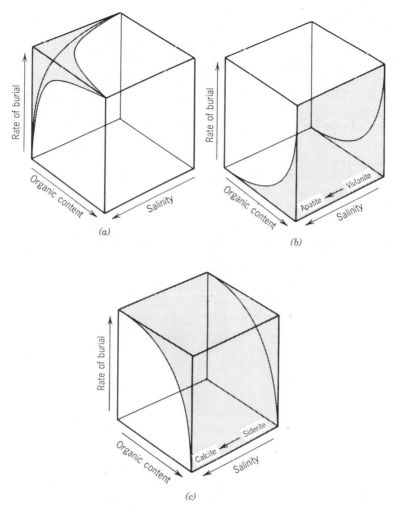

7.17 Depositional parameters (stippled area) required for: (*a*) pyritization of lightly skeletized and soft-bodied organisms [early diagenetic pyritization requires rapid (possibly catastrophic) burial, a low organic content, and the presence of sulfates]; (*b*) phosphatization of soft parts (early diagenetic phosphatization requires a low rate of burial and a high organic content); (*c*) preservation of soft parts within carbonates [field of exceptional preservation is characterized by rapid (possibly catastrophic) burial of organic-rich sequence]. From Allison (1988).

fined in Fig. 7.17. From this analysis, paleoenvironmental conditions in which one might expect to find exceptional soft-body preservation can be established, and at the same time, the presence of these diagenetic mineral and preservational artifacts provides information about the depositional as well as sedimentary-geochemical conditions.

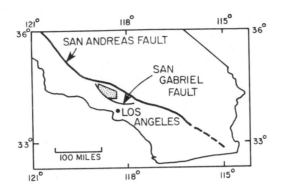

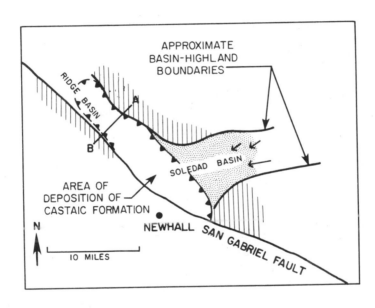

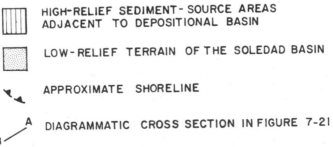

Figure 7.18 Location and local paleogeography of the Castaic Formation, southern California. Upper Miocene. After Stanton (1967).

The more common interpretation of taphonomic data deals with fossils as sedimentary particles and with accumulations of fossils as sedimentary rock units. The following example from the Castaic Formation deals primarily with the interpretation of fossils in place, transported, and in mixed assemblages. The example which follows that, from the Hauptmuschelkalk, relies on cyclic vertical changes in taxonomic composition, shell conditions, and fabric, as well as sedimentary features to infer a cyclic pattern of environmental variations. The example from the Arikaree Group of Nebraska focuses on the contribution of different body parts to the vertebrate assemblage, and preservation and orientation of the bones in order to infer transporting and depositional conditions.

Shell Beds in the Castaic Formation, Upper Miocene, of Southern California—Paleoenvironmental Interpretation of Taxonomic Composition and Distribution of Fossiliferous Horizons

The Castaic Formation consists of sedimentary rocks deposited in the Soledad and Ridge Basins, east of the San Gabriel Fault and along the margin of a transgressing sea, both on an open coast and within an embayment (Fig. 7.18; Stanton, 1967). It consists of three lateral facies (Fig. 7.19): (1) the lower part of the Violin Breccia, composed of poorly sorted sediments deposited along the western side of the basin, adjacent to the active San Gabriel Fault; (2) interbedded mud and sand deposited in midbasin; and (3) a complex of sand and gravel beds deposited as basal sediments along the eastern margin of the basin. The basal strata along the northeastern margin of the basin consist of lenticular and areally restricted sandstone and conglomerate beds that lap against the basal contact and extend as tongues from the contact a half mile or more out into the mudstone facies.

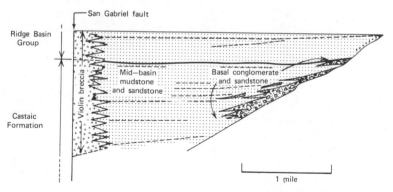

Figure 7.19 Diagrammatic cross section showing facies relationships within the northern part of the Castaic Formation, southern California, Upper Miocene. Subsequent deformation and erosion are not portrayed. After Stanton (1967).

The fauna in these basal strata consists primarily of mollusks with brachiopods, barnacles, echinoderms, and bryozoans as minor components. Two distinctive assemblages can be recognized. One is in the sandy beds adjacent to the basal contact, in very shallow water near shore. The other assemblage is in beds of sandstone and conglomerate that extend basinward from the basal contact and must have been located offshore as much as a half-mile or more.

The fossils in the nearshore assemblage are commonly in growth position, are whole, unworn, articulated if bivalves, and are in sediments like those in which their modern relatives live (Fig. 7.20). Living representatives of the taxa in the assemblage live on a sand or rocky substrate, intertidally or in shallow water: Some are restricted to the intertidal habitat; the rest range from the intertidal or base of the intertidal into water not more than a few tens of meters deep. As a whole, the assemblage appears to be essentially in living position in sediment deposited in water not more than about 10 m deep.

The assemblage in the offshore sandstone and conglomeratic sandstone beds contains the same taxa as in the nearshore sandstone, but the shells are generally disarticulated, broken, abraded, and occur in a matrix that is much coarser and less well sorted than the substrate in which present-day representa-

Figure 7.20 Articulated and single valves of *Lyropecten crassicardo* in shoreline basal pebbly sandstone in the Castaic Formation, southern California, Upper Miocene. Scale is 6 in. long. After Stanton (1967).

Figure 7.21 Shell fragments in offshore conglomerate of the Castaic Formation, southern California, Upper Miocene. After Stanton (1967), Short Contributions to California Geology, Special Report 92, California Division of Mines and Geology.

tives of the species normally live (Fig. 7.21). Fossils in the nearshore sediment are widely scattered, whereas those in the offshore beds are closely packed and commonly oriented concave-down. Several species that are restricted to the offshore beds are generally unbroken and unabraded and are not mixed with the disarticulated and broken shells, but are found in adjacent strata. The most abundant of these, *Turritella cooperi* and *Nemocardium centifilosum,* are most commonly found living today at depths of 60 to 100 m.

The relative proportion of right and left valves of the bivalve *Chione elsmerensis* is also useful in the reconstruction of the paleoenvironment. Individuals of this species lived in the shallow sublittoral environment, where storm wave action would have tended to transport the shells onto the beach or out into deeper water below wave base. Shells carried onto the beach would have been sorted by longshore currents. Few valves of *Chione elsmerensis* are found in the shoreline sand beds of the Castaic Formation, either because those remaining on the strand line were destroyed by abrasion, or because of the high intensity of solution of fossils in beach sands, as documented by McCarthy (1977). (The probability that fossil diagenesis was important is indicated by the relative predominance of calcitic fossils—the oyster *Cras-*

sostreatitan, the pecten *Lyropecten crassicardo*, and the sand dollar *Astrodapsis fernandoensis*, and the relative scarcity of aragonitic fossils.) Thus the specimens that are preserved, offshore, are the transported component from the beach. They would be right valves if the longshore current was running northward; left valves if it was running northward.

The reconstruction of the depositional environment depends on all the available paleontologic data, within the stratigraphic and sedimentologic framework. Normally, mud was deposited in midbasin and to within a few hundred feet of the shore. Sand and pebbly sand were deposited in water a few fathoms deep along the rocky shore at the basin edge. The abundance in living position in the shoreline sands of *Crassostrea* specimens as much as 46 cm long and *Lyropecten* specimens as much as 20 cm across indicates that conditions there were not greatly disturbed for long periods of time. At infrequent intervals, during large storms, the nearshore substrate and organisms were in large part eroded, mixed with pebbles and cobbles from along shore or newly introduced from the adjacent land, and redeposited as beds extending basinward for distances up to a mile and into water 60 to 100 m deep. Immediately afterward, a distinctive deeper-water fauna inhabited the offshore sandy substrate afforded by the sand and conglomerate tongues and persisted until this substrate was buried by the normally deposited mud. The preponderance of left valves of *Chione* in the sublittoral deposits indicates that there was a strong southward-flowing longshore current along the northwest margin of the basin during normal conditions.

Shell Beds in Lower Hauptmuschelkalk, Middle Triassic of Southwest Germany

The Lower Haupmuschelkalk in the vicinity of Crailsheim includes cyclic sequences grading from shale and marl at the base through dense limestone and shelly limestone to a shell bed at the top (Aigner, 1977). The shale and marl beds are 5 to 20 cm thick and the limestone beds are 5 to 10 cm thick (Fig. 7.22). The shale and marl beds contain few fossils, but these are generally articulated and scattered in the rock as single individuals. Body fossils are absent in the dense limestone overlying the shale-marl beds, but trace fossils are abundant, occuring along the lower surface of each limestone bed, and within it as dwelling traces and pellet-filled feeding traces and spreiten. Body fossils in a lime–mud matrix comprise as much as 50% of the shelly limestone. The fossils are commonly articulated and lithoclasts are rare. The fossils are more concentrated in the shell layer at the top of the limestone than in the underlying rock and may form a coquina. The top of the shell layer is an omission surface formed during an interval of nondeposition before the cyclic sedimentation was renewed with deposition of shale and marl.

The interpretation of the lithologic and paleontologic features by Aigner (1977) is illustrated in Fig. 7.23. The overall cyclic pattern is explained by changes in water salinity and temperature. Aigner considered the shell bed at the top of the cycle to be an essentially autochthonous storm deposit (shell-

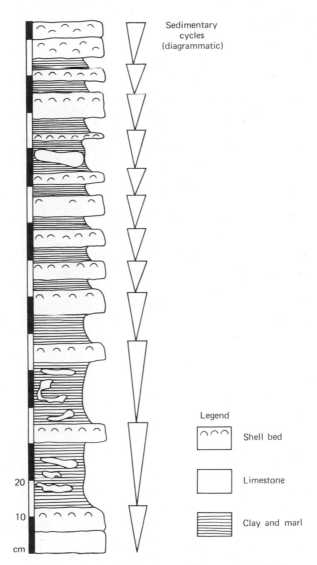

Sedimentary
cycles
(diagrammatic)

Legend

Shell bed

Limestone

Clay and marl

20

10

cm

Figure 7.22 Sedimentary cycles in the Lower Hauptmuschelkalk, southwest Germany, Middle Triassic. From T. Aigner (1977), Neues Jahrbuch für Geologie und Paläontologie, Abhandlungen, 1977; E. Schweizerbart'sche Verlagsbuchhandlung Nägele u. Obermiller, Stuttgart.

lag) concentrated from the underlying shelly lime mudstone. Three types of evidence suggest that the fauna is autochthonous: (1) bivalves, largely articulated, closed, unbroken, and unabraded; (2) scaphopods, presumably autochthonous because they would be difficult to transport; and (3) juvenile ceratite ammonites, which appear to have lived in the inferred depositional setting.

INTERPRETATION   LITHOLOGIC CHARACTERISTICS

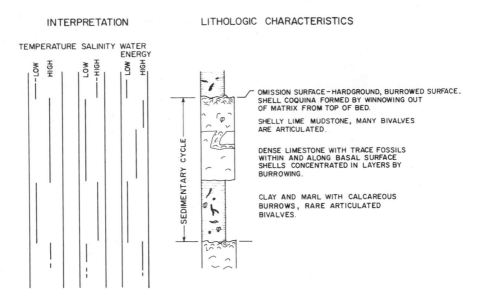

Figure 7.23 Detail of a sedimentary cycle in the Lower Hauptmuschelkalk, south-west Germany, Middle Triassic. Lithology of cycle and shell layer at top is caused by increasing water energy associated with changes in water characteristics and depth. From T. Aigner (1977), Neues Jahrbuch für Geologie und Paläontologie, Abhandlungen, 1977; E. Schweizerbart'sche Verlagsbuchhandlung Nägele u. Obermiller, Stuttgart.

The cyclic pattern of sediments culminating in the shell bed is attributed to changes in temperature and salinity. The increase in water energy at the end of the cycle is presumably the result of a storm. The position of the storm in the cycle may have been determined by the same climatic factors responsible for the change in salinity and temperature. The presence of the juvenile ceratites, which could be transported easily, the moderate degree of scaphopod lineation, and the general concave-down orientation of bivalve shells suggest that the water energy was moderate. Currents suspended and winnowed the lime mud sediment of the shelly seafloor and left a shell-lag pavement.

Mammal Assemblage of the Arikaree Group (Miocene), Nebraska—Depositional Setting Derived from Sorting, Preservation, and Orientation of Fossil Bones

Mammal bones are concentrated in a calcareous ashy sandstone several feet thick in localized outcrops in northwestern Nebraska. Hunt (1978) determined the depositional setting of the assemblage from fossil condition because these characteristics of the bones indicate current and wave action, and thus the amount of transportation and the degree of mixing of specimens from disparate habitats.

The assemblage consists of several hundred bones, 120 of which are identifiable. They come from a minimum of 13 or 14 individuals belonging to 11 species and 10 genera. None of the bones is articulated, although two individuals may have contributed several bones both to the assemblage as indicated by proximity and similarity of the bones each in size and condition. (Determining the number of individuals represented by fragments of body parts is a complex but significant problem. It was touched on briefly in the necrolysis section relative to estimating the number of crinoid individuals from a sample consisting of numerous columns but no calices. Badgley (1986) presents an in-depth analysis of the problem in her study of Miocene Mammalian assemblages. The orientation of the bones is described by three parameters: the bearing, in which the ends are not differentiated; the polarity, in which the azimuth of one end (the larger in this study) is measured; and the inclination. The rose diagram of the bearing of the long axis of 141 bones (Fig. 7.24) indicates a strong but diffuse northwest orientation and a lesser, broad maximum in the northeast quadrant. The plot of the azimuth of the large end of each bone provides a much clearer picture of northwest and northeast maxima (Fig. 7.25). The inclination of all the bones is low and not diagnostic. Flume studies and observations in nature have shown that elongate particles tend to be aligned either parallel or perpendicular to water movement (Fig. 7.13; Voorhies, 1969). The equal sizes of the northeast and southwest peaks in the polarity diagram suggest that current flow was normal to the northeast–southwest orientation. Hunt concluded that the predominant water flow was toward the northwest because flume studies by Voorhies have shown that bones aligned parallel to the flow direction have a strong tendency to lie with the large end downstream.

Bone abrasion was estimated qualitatively to be generally moderate and to decrease upward in the section, suggesting that the water energy and/or amount of transportation decreased with time. In addition, because the different parts of a skeleton, whether invertebrate or vertebrate, have different

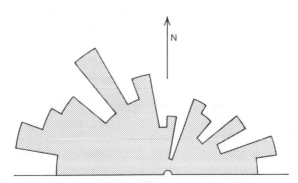

Figure 7.24 Bearing of long axis of 141 bones from Harper Quarry, Nebraska, Miocene. After Hunt (1978), Palaeogeography, Palaeoclimatology, Palaeoecology, *24*:11.

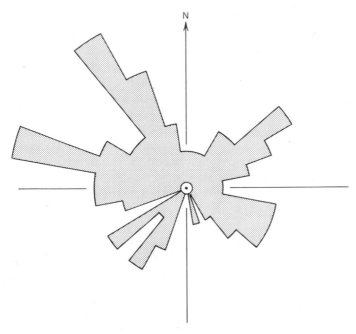

Figure 7.25 Rose diagram of polarity (bearing of large end) of 74 mammal bones from Harper Quarry, Nebraska, Miocene. After Hunt (1978), Palaeogeography, Palaeoclimatology, Palaeoecology *24:*22.

preservation and transportation potentials the fact that the bones of some kinds of animals are more abraded than those of others can be used to distinguish between autochthonous and introduced bones in the assemblage. Consequently, the proportion of skeletal parts in a fossil assemblage, as compared to the original proportion in the organism, is a clue to the extent of winnowing and transport, abrasion, and diagenesis. The proportions of skeletal parts in the vertebrate assemblage is portrayed in Fig. 7.26. Ungulate bones constitute 71% of the identifiable bones; 29% are carnivores. Complete skulls are absent in the assemblage, ribs are relatively abundant, and most ungulate bones are broken and fragmented.

The interpretation of the fossil assemblage by Hunt includes not only paleontologic data but sedimentologic and stratigraphic data as well. The orientation of bones in two perpendicular directions and the moderate-to-slight abrasion indicate that the bones were deposited in shallow water with generally weak currents. The relative abundance of easily transported vertebrae and ribs is also indicative of weak current action, because with stronger currents they would have been winnowed out of the assemblage. The decrease in abrasion upward in the section suggests decreasing water energy as the bed was being deposited. The conclusion that water energy was low is confirmed by the small amount of abrasion of small, compact bones. An explanation other than water agitation is necessary for the generally fragmented condition

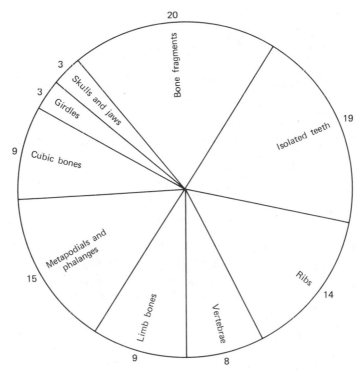

Figure 7.26 Percentages of skeletal elements in 198 mammal bones from Harper Quarry, Nebraska, Miocene. Cubic bones include podials (astragali, calcanea), sesamoids, sternebrae. Girdle elements include scapulae, innominates. After Hunt (1978), *Palaeogeography, Palaeoclimatology, Palaeoecology 24*:21.

of skulls, lower jaws, limb bones, and the other, more fragile, bones. The conclusion that water energy was low leads to the conclusion that much of this fragmentation was caused by carnivores.

Taphonomic Onshore–Offshore Gradient in the Pliocene Purisima Formation, California

Dense accumulations of fossils are uncommon, but of considerable interests because they represent unusual depositional conditions. Kidwell et al. (1986) have proposed that shell accumulations can be of biogenic sedimentologic, or diagenetic origin, and can be adequately described by their taxonomic composition, packing, geometry, and internal structure, and that these descriptive attributes can be interpreted in terms, respectively of ecology, hydrodynamics, topography of the depositional surface, and the ecologic and hydrodynamic history. This conceptual framework of descriptive and genetic aspects provides a basis for the expected environmental distribution of the different types of fossil accumulations. This classificatory and genetic scheme has been applied to the shell beds of the Purisima Formation of central California by Norris (1986).

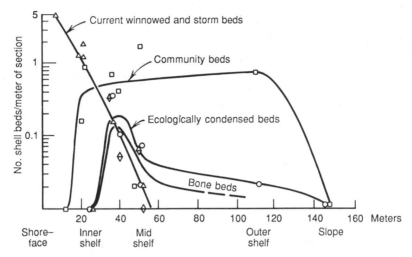

Figure 7.27 Onshore–offshore trends in genetic types of shell accumulations. Triangles, current-winnowed and storm beds; squares, community beds; circles, ecologically condensed beds; diamonds, bone beds. Log vertical scale used to separate curves. After Norris (1986), Palaios, *1:266*.

The Purisima formation contains marine shelf to nearshore strata ranging from offshore diatomaceous mudstone to nearshore shaly sandstone. The characteristics of the sediments and fossil accumulations along this environmental gradient can be studied in excellent outcrops in sea cliffs in central California. Norris (1986) has categorized the shell accumulations into genetic units of storm beds, current winnowed beds, bone beds, community beds, and ecologically condensed beds, with their bathymetric distributions illustrated in Fig. 7.27, and the characteristics listed in Table 7.6.

Whereas the examples from the Castaic formation and the Muschelkalk provide taphonomically based explanations of specific situations, the results from the Purisima Formation provide a general model for application and testing. The first interesting question is how general the Purisima model is. From our experience, it appears to be generally valid for the Neogene of California, but is it equally applicable to passive margin low-energy settings? [In this context Kidwell (1988) has suggested that, in fact, shell accumulations in different tectonic settings should be different.]

Taphofacies

Each one of the taphonomic processes has a more-or-less distinctive effect that is imprinted on individual fossils, the fossil assemblage, or on the shell accumulation. At the same time, each process is more or less effective within a specific habitat, and it should therefore be possible to characterize a habitat by distinctive and perhaps unique taphonomic attributes. This logic is the basis for interpreting the paleoenvironment and sedimentologic regime from taphonomic data, as illustrated by examples cited in this chapter. Brett and Baird (1986) have proposed that the co-occurring association of taphonomic characteristics of a habitat can be used to define taphofacies. This notion has interesting implications because it sets taphonomic analysis and interpretation within a sedimentary-stratigraphic framework. An example of this is illustrated in Fig. 7.28, in which they chart the taphonomic attributes of fossil accumulations as they are controlled by sedimentation rate and frequency and intensity of reworking. Schäfer (1962, 1972) anticipated the current interest in taphofacies when he observed the modern processes and effects and then postulated how they would be preserved in the rock record. The resulting taphofacies, which he termed taphocoenoses, provide an excellent starting point for an integrated analysis of taphonomy (Figs. 7.29 to 7.33).

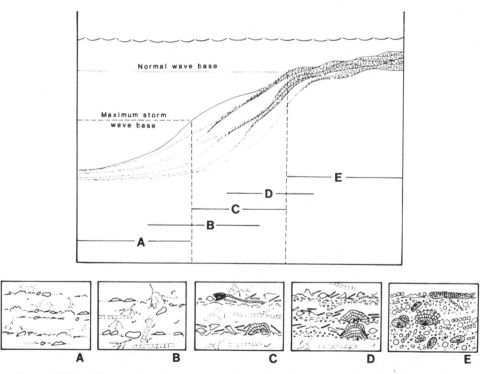

Figure 7.28 Diagram showing superposition of successive event deposits produced by storms of varying intensity. Lettered bars show expected depth ranges of different shell layer types: **A**, nondepositional surfaces smothered by thin distal mud layers; **B**, colonized soft-bottom surfaces buried by upslope winnowed muds; **C**, colonized winnowed pavements buried by thick layers of redeposited nearshore muds; **D**, rewinnowed and amalgamated shell beds smothered by thick mud layers; **E**, winnowed crinoidal grainstones with a few subtle internal burial horizons. From Miller et al. (1988), Palaios, 3:40.

TABLE 7.6 Characteristics of Shell-Bed Types[a]

Geometry	Thickness	Biofabric	Species Diversity	Sedimentary Structures	Interpretation
Storm Bed					
Lenses, stringers	1–7 cm; Thicker (to 30 cm) where amalgamated; laterally continuous over several tens of meters	Concave-down shells; shells unworn but may be fragmental; mud rip-up clasts, geopetal structures, and packstone; well-preserved fossils	Same fauna as in associated burrowed sediments	Hummocky crossbeds capped by bioturbated mudstone; basal erosion surface and gradational upper contact	Deposited during waning storm flows after erosion and winnowing of the seafloor
Current-Winnowed Bed					
Tubular to lensoidal	3–35 cm; up to 5 m where amalgamated; in lenses 2—4 m long	Shell poorly preserved with much rounded and fragmental debris; shells in a variety of orientations, many nest within each other; dense fragmental shell layers capped by whole shell lags; much amalgamation	Very high	Large-scale trough crossbeds and hummocky beds; lower contact erosional, upper contact gradational	Extensively amalgamated by migrating megaripples; low aggradation allowed repeated reworking of sediment and shells; shells deposited as lag behind bedforms

Bone bed

Tabular or with starved-ripple geometry	1–2 cm to as much as 50 cm	Pavements of gravel, phosphatic steinkerns, concretions, and bones; shells sometimes bored but in situ bivalves may be present; borings common	Moderate to high; includes crabs and vertebrates	Extensive bioturbation; sometimes small-scale tabular crossbeds	Record prolonged exposure and repeated burial and exhumation of hardparts; formed where sedimentation is very low or where the seafloor is eroded; most calcareous shell lost by bioercsion and dissolution

Community Bed

Lenses, clumps, or rarely tabular	10–30 cm thick, 1–3 m long	Articulated bivalves in life position but some disarticulated or randomly oriented; little broken shell debris	Low; often monospecific aggregations	Extensive bioturbation; some beds buried by storm deposits	Record colonization by gregarious fauna, or infauna that all burrowed to a common depth below a nonaggrading surface, or by colonization of mud firmgrounds

Ecologically Condensed Bed

Tubular, rarely starved-ripple geometry	1–45 cm thick; thickest where burrowed	Cemented clumps of barnacles, worm tubes, and bryozoans; much bored and encrusted shell; matrix-supported or shell-forming winnowed drifts; encrusters often in life position	Moderate to high; encrusting fauna of low species diversity (2–7 species), but host mollusk shells include many species	Extensive bioturbation; minor ripple lamination	Recolonization of exposed shell gravels by epizoans; could also record successive colonization of hardparts as they became available on a nonaggrading surface

[a]After Norris (1986).

Figure 7.29 Life assemblage—nonbedded biofacies. Characterized by skeletal material from an in-place community consisting largely of sessile benthic organisms such as corals, calareous algae, and sponges; bedding is lacking. Common in shallow agitated-water settings. A reef would exemplify this biofacies. (1) Living octocorals and scleractinian corals; (2) the calcareous alga *Lithothamnium;* (3) coral and shell fragments. After Schäfer (1962) Aktuo-Pälaontologie nach Studien in der Nordsee, Verlag Waldemar Kramer, Frankfurt am Main.

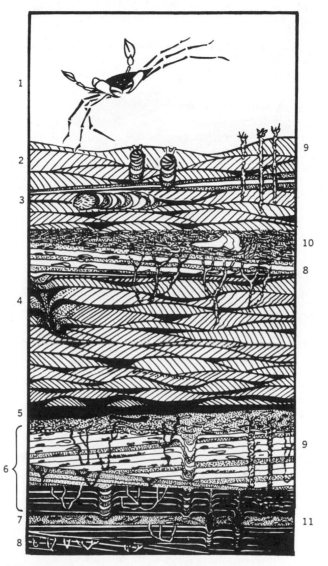

Figure 7.30 Life assemblage—active traction-deposition biofacies. Characterized by diverse short-duration benthic communities, by death assemblages and by scour surfaces. Forms in well agitated water and contains resulting distinctive bedding forms. Fossils derived primarily from vagile benthic organisms; fragmentation, reworking and transportation of hard parts is characteristic. (1) Swimming crab; (2) traces of pelecypods moving up to keep pace with sedimentation; (3) trace of heart urchin; (4) escape trace formed by polychaete buried by rapid sedimentation; (5) limb bone of bird; 6. escape traces formed by pelecypods buried by rapid sedimentation; (7) otoliths (fish ear bones); (8) worm burrows; (9) agglutinated worm tubes; 10. skull of marine mammal; 11. pelecypods in living position. After Schäfer (1962), Aktuo-Päalaontologie nach Studien in der Nordsee, Verlag Waldemar Kramer, Frankfurt am Main.

273

Figure 7.31 Death assemblage—active traction-deposition biofacies. Characterized by abundant diverse death assemblages, scour surfaces and bed forms indicative of high water energy. A benthic in-place community cannot become established because of the unstable substrate. Skeletal material is largely transported, reworked, winnowed, etc. (1) Crab claw parts; (2) limb bones of birds; (3) rib, skull and vertebra of marine mammal; (4) peat clast derived from coastal marsh by storm erosion. After Schäfer (1962), Aktuo-Pälaontologie nach Studien in der Nordsee, Verlag Waldemar Kramer, Frankfurt am Main.

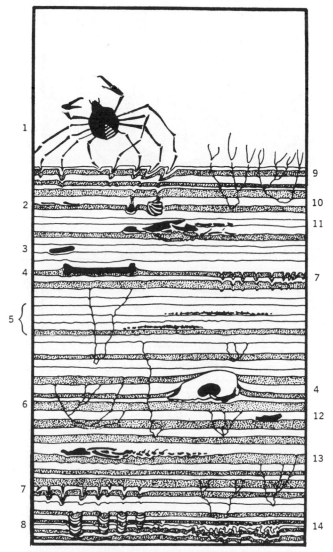

Figure 7.32 Life assemblage—suspension-deposition biofacies. Characterized by low energy deposition of sediment from suspension and a mixture of an in-place benthic community and remains from planktonic and nektonic organisms. The extent of bioturbation depends on the rate of sedimentation. Multi-element skeletons commonly intact because of quiet water and little scavenging activity. (1) Sea spider; (2) shells of the epiplanktonic barnacle *Lepas;* (3) *Nautilus* shell; (4) bone and skull of marine mammal; (5) star fish and brittle star skeletons; (6) worm tubes; (7) locomotion traces of crab (8) escape traces; (9) tubes of living worm; (10) pelecypods in living position; (11) crab remains; (12) coprolite; (13) fish skeleton; (14) resting trace. After Schäfer (1962) Aktuo-Pälaontologie nach Studien in der Nordsee, Verlag Waldemar Kramer, Frankfurt am Main.

Figure 7.33 Death assemblage—suspension-deposition biofacies. Characterized by low energy, deposition from suspension, lack of benthic community, and death assemblage of planktonic and nektonic organisms. Multi-element skeletons preserved intact because of quiet water and lack of benthic scavenging organisms, attributed to azoic conditions. (1) carcass of fish; (2) vertebra, tooth and skeleton of marine mammal; (3) coprolite; (4) plank-tonic gastropod; (5) *Nautilus* shell; (6) bird skull; (7) fish skeleton; (8) fish scales; (9) plates from the epiplanktonic barnacle *Lepas*. After Schäfer (1962) Aktuo-Pälaontologie nach Studien in der Nordsee, Verlag Waldemar Kramer, Frankfurt am Main.

8

Populations in Paleoecology

Organisms do not function as isolated individuals but are part of a population of many interacting members of the same species living together in an area. Members of a population interact with one another (especially in interbreeding and in competing for space and a common food source) and with populations of other species (especially in utilizing food resources and space). Interactions between populations are discussed more fully in Chapter 9. The population as a whole is affected by many physical aspects of the environment. To this point we have mostly been concerned with the impact of the environment on individual organisms, especially their distribution and morphology. In this chapter we consider the effects of the biological and the physical environment on population properties, such as (1) population growth rates and growth patterns, (2) age and size distribution of individuals within the population, (3) long-term variation in population size, (4) spatial distribution of population members, and (5) amount of morphological variability within the population. Because these properties are influenced by environmental parameters, we should be able to determine the nature of paleoenvironments by measuring these properties in fossil populations.

A simple definition of population is individuals of a species living together in an area. A more rigorous definition would state that the individuals live close enough to one another that each individual in the population has an equal opportunity of interbreeding with all members of the opposite sex within the group. Thus a population is a group of freely interbreeding organisms. In genetic terms, a population shares a common gene pool and there are no geographic or other barriers to gene flow. As is so often the case in ecology and paleoecology, boundaries are difficult to define. Where does one population stop and another begin? How complete must the isolation be between two areas before we should recognize members of the same species in the areas as belonging to different populations (e.g., corals on two adjacent reefs)? Should all individuals of a uniformly distributed, widespread species be considered as belonging to the same population (e.g., a planktic foraminifera species in the open seas)? In such a case, clearly all individuals do not have an equal opportunity to interbreed, but no natural boundary exists within the overall distribution. In practice the nature of the geologic record usually helps us with these boundary problems. We deal with only a sample of the population, and that sample, because of preservation and exposure of the strata, represents only a limited portion of the geographic distribution of the population. We discuss properties of the population, but we can really only measure properties of our limited sample and hope that it is representative. We define the boundaries of the population on the basis of the exposures or cores from which our samples came. We need not be concerned with boundaries because the boundary areas are conveniently not exposed or preserved.

As will be apparent in the ensuing discussion, some attributes of fossil populations are difficult to study because of the nature of the fossil record. A population consists of individuals living at a given instant in time. Ideally, a population should be studied by observing individuals living at the same time. Rarely does the fossil record allow us to do this. A great deal of the effort expended in studying fossil populations is in attempting to unravel or extract this instant-in-time view from the time- and space-averaged fossil record.

Ecologists have been very active in studying the properties of populations and the influence of environment on populations. Good reviews and general discussions of populations are given by MacArthur and Connell (1966), Odum (1971), Hutchinson (1978), Hedrick (1984), and May and Seger (1986). Paleoecologists have increasingly applied the approaches of population ecology, particularly the study of population structure of dynamics, in analyzing fossil populations. Useful general discussions of the application of population ecology to the fossil record are given by Hallam (1972) and Valentine (1973a). Much of the work on fossil populations has been done to interpret the structure and evolution of the population rather than to use population properties to interpret paleoenvironments. Our emphasis is on using the properties of fossil populations in interpreting ancient environments.

POPULATION GROWTH

The change in the size of a population is a fundamental attribute of the population that is strongly related to environment. Under favorable environmental conditions a population of a given species may increase rapidly in size to a high level. Under less favorable conditions the population may grow more slowly and remain relatively small. Variable environmental conditions will likely cause population size to vary. Unfortunately, population growth patterns on a short-term scale can seldom be studied in the fossil record because of the time-averaging effect. Invertebrate populations normally grow to full size in a few weeks or at most a few years. Most fossil assemblages probably accumulate over a period of many years so that short-term changes in population size are obliterated in the mixing process (Fürsich, 1978; Staff et al., 1986). The best opportunity to study short-term population growth in fossils would be in a section that was deposited very rapidly in an environment of low turbulence and minimal mixing. Nevertheless, the concept of population growth is important in helping us to understand many aspects of population and community structure. As we discuss in the next section, longer-term variation in population size may be influenced by short-term patterns.

Intuitively, what might one expect to be the pattern of change in population size? At least two types of patterns might be expected. If we assume that the population starts small, it might grow rapidly, perhaps even exponentially, to some value and then decrease abruptly. This pattern might be repeated over and over for a given population. This is a common pattern among modern populations and has been called the J-shaped pattern of population growth (Hedrick, 1984). It results from the presence of initially favorable but changeable conditions for the population. The changing parameter may be physical (e.g., temperature), chemical (e.g., salinity), or biological (e.g., food organisms). The cause for the abrupt decrease in population size may be density independent, that is, not dependent on the population size, as for example a sudden decrease in temperature; or the cause may be density dependent, that is, the overutilization of food resources. Many examples of the J-shaped population growth pattern are familiar to us from our daily experience—grasshoppers, mosquitoes, crab grass (Fig. 8.1).

The other pattern we might expect is for the population first to increase and then to level off gradually at some upper equilibrium value. This pattern is also common in nature and has been called the S- or sigmoidal-shaped pattern (Hedrick, 1984). Any given area can support only a certain-size population for an indefinite time period because some environmental factor, commonly the food source, is maintained at a certain level that places an upper limit to the population size. This upper population limit is called the carrying capacity of the environment. The control on the maximum size of the sigmoidal-shaped population growth pattern is thus density dependent, and the closer the population size approaches the carrying capacity (i.e., the greater the density), the slower will be the rate of increase in population size. In a perfectly balanced ecosystem the population ideally should remain at the carrying capacity unless some outside force disturbs the system. Examples of populations showing the sigmoidal-shaped pattern would include the bass in a balanced lake or pond, the deer population in a well-managed herd, or corals on a stable reef (Figs. 8.2 and 8.3).

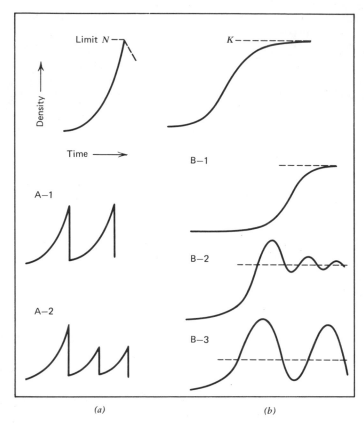

Figure 8.1 Population growth patterns (changing population density with time): (*a*) J-shaped pattern and variants; (*b*) sigmoidal or S-shaped pattern and variants. From Odum (1971), Fundamentals of ecology, 3rd ed., p. 184, copyright W. B. Saunders, Philadelphia.

    The most common patterns may be more complex than a simple J-shaped or sigmoidal-shaped one. Sometimes, observed patterns are a combination of the two basic patterns or even more complex patterns (Fig. 8.1). These may, in some cases, be caused by the population temporarily exceeding its carrying capacity. In other cases the carrying capacity itself may change as a result of environmental fluctuations. Density-independent factors may occasionally be superimposed on the sigmoidal-shaped pattern.
    Mathematical models have been developed to describe population growth [see Hutchinson (1978) for an interesting and entertaining discussion of the history of development of these models]. These are useful in allowing us to test more rigorously hypotheses that explain population growth by comparing the models with actual data. They also allow us to predict quantitatively the population growth rate and pattern.

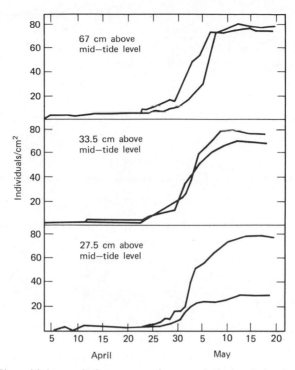

Figure 8.2 Sigmoidal population curves for populations of the barnacle *Balanus balanoides* on rocks in the intertidal zone of the Firth of Clyde, Scotland. The two curves in each box represent two samples at the same tide height. Reprinted from Effects of competition, predation by *Thais lapillus* and other factors on natural populations of the barnacle *Balanus balanoides*, by J. H. Connell, in Ecological Monographs, *31:* 65, by permission of Duke University Press. Copyright 1961 by the Ecological Society of America.

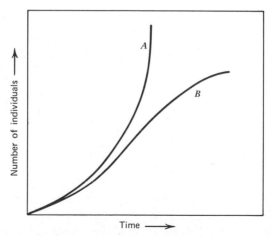

Figure 8.3 Exponential growth curve (*A*) and logistic equation (*B*) for modeling population growth.

The simplest mathematical model describes population growth with no upper limit to population size with the equation

$$\frac{dN}{dt} = rN \tag{8.1}$$

in which N is population size, t is time, and r is a rate constant that describes the rate of increase in population size (Fig. 8.3). In words, this model simply says that the rate of change in population size is equal to the size of the population times a population growth constant. An equation of this type should be familiar to geologists because it is identical to the radioactive decay equation used in age dating. In the decay equation N is the number of atoms of the radioactive element and the decay constant, λ, is used instead of r. λ is negative, indicating a decrease in the population size of radioactive atoms, whereas r is positive, indicating a growth in the number of organisms.

The r value is a function of the interaction between natality (birth rate) and mortality (death rate). Each population has a characteristic natality that is basically controlled by the physiology of the species but may vary depending on environmental conditions. Similarly, the population has a characteristic mortality that is also controlled by the organism's physiology and environmental factors. Clearly, the population will grow most rapidly under those conditions that maximize natality and minimize mortality. The maximum value of r for a given population is called the intrinsic rate of natural increase or the biotic potential of the population. The actual value of r for a given set of environmental conditions is normally less than the biotic potential, and the difference between the two is a measure of the environmental resistance to maximum growth rate (Odum, 1971).

The biotic potential varies greatly between organisms. Some species have an extremely high natality rate, and if the mortality rate is not too high will give high r values, hence rapid rates of population growth. Diatoms, bacteria, foraminifera, and mosquitoes would be examples of species of this type. Other species have much lower biotic potential and thus much slower population growth rates. Elephants, humans, and probably dinosaurs fall into this category.

Of course, the exponential growth model cannot explain the whole story of population growth, for it predicts populations of astronomical size within relatively short time periods. This model led Thomas Malthus to his predictions of disaster for the human race due to geometric population increase [see discussion in Hutchinson (1978)]. Clearly, something not accounted for in this model happens to restrain population growth. The exponential growth model is useful in introducing the concept of biotic potential and does fairly accurately depict the growth of many populations in their early stages. Some examples of the J-shaped pattern could be compared to the exponential growth pattern that is abruptly terminated at a certain point. The main shortcoming of the exponential growth model is that it does not take into account any factor limiting growth rate even though density-dependent factors will always operate and should be accounted for in the population growth model.

The logistic equation, which was originally developed by P. F. Verhulst (Hutchinson, 1978), takes into account the density-dependent factors and gives a reasonable approximation of the sigmoidal-shaped population growth pattern. In this model

$$\frac{dN}{dt} = rN \frac{K - N}{K} \tag{8.2}$$

in which K is the upper equilibrium limit or carrying capacity that, in theory, the population size approaches but does not exceed, and the other symbols are as defined previously (Fig. 8.3). As can be seen from the equation, as N approaches K, that is, as the population size nears the carrying capacity, the term $(K - N)$ approaches zero and thus the rate of population growth must approach zero. This models the effect of density-dependent factors acting on population growth. The logistic equation has been questioned in recent years as an appropriate model for describing population growth (May and Seger, 1986). It is a simplification of reality, but it is useful for visualizing the factors responsible for population growth patterns. More complete treatments of the nature of population growth and structure can be found in the biological literature (e.g., Murray, 1979).

POPULATION SIZE VARIATION

We have discussed how populations grow from a small starting size to a maximum. This can be viewed as a short-term fluctuation in population size. But population size may also vary on a time scale of years, decades, or longer. Many examples of population size variation have been described, especially in populations in terrestrial environments [see Odum (1971) and Hedrick (1984) for reviews]. Marine examples are also common (Coe, 1957), although they have not been as extensively described, perhaps because of the greater sampling difficulty.

Modern Examples

Among the examples from terrestrial environments are the lynx and snowshoe hare of the Arctic, which vary tremendously in number in 9- or 10-year cycles (Fig. 8.4). Every 9 or 10 years, probably because of an increase in abundance of food plants, the hare becomes so abundant that the demand for food exceeds the supply. The populations of lynx, which prey on the hares, also increase as their food supply grows. But when the hare population de-

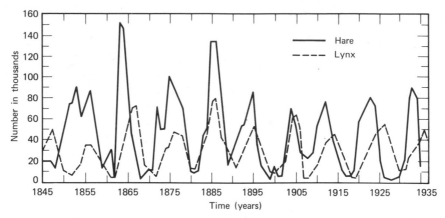

Figure 8.4 Changes in the abundance of the lynx and the snowshoe hare as indicated by the number of pelts received by the Hudson Bay Company. From Odum (1971), Fundamentals of ecology, 3rd ed., p. 191, copyright W. B. Saunders, Philadelphia.

creases dramatically (or crashes), the predators in turn suffer high mortality. The lemmings, with a 3- or 4-year cycle, show a similar variation in numbers in the North American Arctic. The migratory locust of Africa and Eurasia is also well known for its tremendous variation in population size.

Among the better documented marine examples are the pismo clam (*Tivela stultorum*) from the west coast of North America, which has shown large year-to-year variations in abundance. Because *T. stultorum* is an important clam for human consumption, careful records of population abundance have been kept for many years by the California State Fisheries Laboratory. The abundance of young individuals especially fluctuates irregularly from year to year. The abundance of older individuals does not vary so much because they include individuals spawned over a period of many years (Fig. 8.5; Coe, 1957). Even larger variation is found in species of the genus *Donax*, the butterfly or bean clam, which may have an abundance of thousands of specimens per square meter in the intertidal zone of exposed, sandy beaches and then may almost disappear for several years. Populations of *Mytilus edulis*, the common edible mussel, often fluctuate abruptly (Coe, 1957). Phytoplankton species show marked short-term fluctuations that are largely a function of nutrient concentration. The infamous red tides are the result of such variation in population size of certain dinoflagellate species. One of the best known recent examples of a great fluctuation in population size has been the tremendous increase in abundance of the Crown of Thorns starfish (*Acanthaster planci*) on coral reefs in the Pacific (Chesher, 1969; Moran et al., 1986; Colgan, 1987). In the early 1970s this species had become so abundant as to cause concern about survival of the reefs. Human activities may have been involved in the case of *A. planci* as well as in some other examples of population explosions. People have

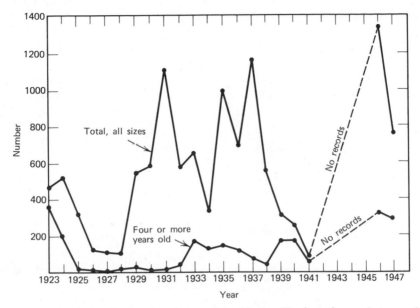

Figure 8.5 Variation in the abundance of the bivalve *Tivela stultorum* between 1923 and 1947 at Pismo Beach, California. From Coe (1957).

certainly had an effect in the many cases where they have introduced species into new areas (house sparrows, starlings, water hyacinths, etc.).

The cause for these fluctuations in population size can sometimes be determined, but often it is not clear. The cause of the decline in population size is often much clearer than the cause of the increase. Odum (1971) has suggested that four categories of factors are likely to be involved as follows:

1. The most obvious category consists of density-independent environmental factors. Certain physical (temperature) and chemical (salinity) environmental conditions favor the growth of the population, and when those conditions change, the population abruptly declines. Density-independent biological factors may also be involved. Disease and the production of poisonous waste products by other organisms (red tides) are often cited as causes for the catastrophic decrease in population size [see Brongersma-Sanders (1957) and Noe-Nygaard and Surlyk (1988) for fossil examples]. A change in larval-bearing currents is another example of a density-independent factor.

2. Chance or random fluctuation in a combination of physical and biological environmental factors occasionally results in conditions favoring rapid population growth. To an extent, chance factors are probably involved in population fluctuations, but often we invoke the chance aspect to cover up our ignorance of what is really involved. The extent to which biological and

geological occurrences of this sort are probabilistic or deterministic is a matter of considerable debate.

3. Density-dependent population interactions will certainly affect population growth and density. Clearly, the lynx in the Arctic fluctuates in abundance because its most important prey, the snowshoe hare, fluctuates. The predator may simply increase in abundance until it uses up too much of its prey and then crashes in abundance to remain at a low level until the prey has a chance to recover.

4. Complex interactions between all trophic levels of the ecosystem may have large effects on population sizes within the ecosystem. Variations in both the abundance and type of the primary producers particularly may have effects on the abundance of populations throughout the ecosystem.

Opportunistic and Equilibrium Species

All populations vary in size to a certain extent, but some, such as the examples cited above, vary much more than others. Species characterized by marked variations in population size have been called *opportunistic species* because they have the capacity to increase in abundance rapidly when the opportunity arises (Fig. 8.6). MacArthur and Wilson (1967) hypothesize that

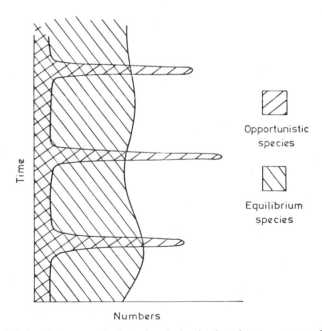

Figure 8.6   Schematic representation of variation in abundance patterns of opportunistic and equilibrium species in a stratigraphic sequence. From Hallam (1972), Models in paleobiology, T. J. M. Schopf (ed.), Freeman, Cooper & Company.

this is a characteristic evolved by these species to adapt to unstable, variable environments which provide periods of favorable conditions (with opportunities) separated by periods of unfavorable conditions (without opportunities). They have called the evolution of opportunistic species *r-selection*. The *r* refers to the logistic equation (8.2), in which *r* is the intrinsic rate of natural increase. Species populations with a large *r* value have the greatest capacity to increase rapidly to utilize trophic resources that have suddenly become available. They are characterized by J-shaped population growth curves and by high juvenile mortality (although the highest mortality record may not be preserved because it is in the larval stage).

Opportunistic species are generalists that are broadly adapted to a range of environmental conditions; that is, they thus occupy large ecological niches. They are likely to be common in environments dominated by physical stress, as in the case of the snowshoe hare, lynx, and lemming in the harsh climate of the arctic, or the locust in the desert or semiarid regions. *Tivela stultorum* and *Donax gouldi* are marine examples that live in shifting sands in the turbulent, intertidal zone. Levinton (1970) concluded that opportunistic species are most common in young habitats with high environmental stress and with environmental factors that fluctuate with a period close to the reproductive age of the species. Examples of the least favorable situation for opportunistic species would be the stable environments of the subtidal tropics and abyssal deeps. Occasionally, opportunistic species appear in what otherwise seems to be a stable environment, such as *Acanthaster planci* on coral reefs and *Mulinia lateralis* on the stable mud bottoms. The explanation for this is not clear, but even in stable environments conditions vary enough so that food resources may suddenly increase or a predator may disappear for some reason so that a species with high population growth potential suddenly undergoes a population explosion.

Populations that show little variation through time have been called *equilibrium species*. They are in equilibrium with their environment and apparently are maintaining a population size near the carrying capacity *K* for the environment (Fig. 8.6). These species have evolved through *K-selection* (MacArthur and Wilson, 1967) and are characterized by the ability to maintain their population size near the carrying capacity.

Opportunistic species are inefficient in their use of large amounts of energy for reproduction. Neither is their feeding efficiency maximized. They are generalists that can use a broad range of food but are not adapted to maximum efficiency in utilizing any one type of food. Equilibrium species minimize these inefficiencies by producing fewer offspring and thus using less energy for reproduction. They have small ecological niches, being highly specialized for a specific kind of food which they secure with great efficiency. They have S-shaped population growth curves, and their juvenile mortality rate is lower than in opportunistic species and ideally should be constant or highest in gerontic stages. Because of the narrowness of their adaptation, they are morphologically less variable than opportunistic species. Examples of equilibrium species are more common than opportunistic because they tend to occur in diverse communities. The gastropod *Conus* with its poison dart method of food capture, the giant clam *Tridacna* with algal gardens in its siphonal tissues, and deep-sea angler fish with luminescent lures are three modern examples.

An equilibrium species requires a stable, dependable environment with a constant food source. The major environmental stresses are biological, from competing populations, rather than physical. They occur in the opposite type of environment from opportunistic species, being especially common in the subtidal tropics and the deep sea.

Fossil Opportunistic and Equilibrium Species

Differentiating between opportunistic and equilibrium species in the fossil record could be a useful tool in distinguishing between physically stable and unstable environments. The great abundance of a given species in a single bed may suggest that it was an opportunist, but the concentration could be due to taphonomic factors such as current sorting. The following criteria are useful for recognizing opportunistic species (Levinton, 1970): (1) The species is very abundant in the assemblage (85 to 100%); (2) it is widespread in a given isochronous horizon, although it may be patchy in its distribution within that horizon; (3) it occurs in several horizons with barren intervals between; (4) it appears in a variety of facies; (5) it occurs abundantly with faunal associations that are otherwise easily distinguishable; and (6) its morphology is variable and suggests a generalized feeding mechanism. These criteria should be combined with evidence for taphonomic concentration.

Equilibrium species should have many characteristics that are essentially the opposite of opportunistic forms: (1) The species is restricted to a definite facies; (2) it occurs relatively continuously throughout the fossiliferous section; (3) it may be relatively abundant, but not more than a few percent in a well-preserved fauna; (4) it should occur in diverse faunas; (5) its morphology shows little variation between specimens; and (6) it is clearly specialized for a given mode of life.

Although some species are opportunistic and some are equilibrium, many have intermediate strategies. These are merely the extremes in a continuum. Going through a faunal list and designating each species as either opportunistic or equilibrium is not justified. Similarly, physically unstable environments do not necessarily include 100% opportunistic forms and stable environments 100% equilibrium. Even unstable environments contain a certain portion of dependable trophic resources that can be used by nonopportunists and trophic resources in stable environments fluctuate so that there is room for some opportunistic or at least semiopportunistic species.

Waage (1968) describes at least two bivalve species from the Cretaceous Fox Hills Formation of South Dakota which appear to be opportunistic. The

Timber Lake member of the Fox Hills Formation was largely deposited in an environment that could not be tolerated by readily fossilizable organisms. Occasionally, environmental conditions apparently improved, allowing the invasion of the area by populations of *Pteria* and *Cucullaea* which evidently increased very rapidly in abundance but were periodically killed. The nature of the fauna suggests an unstable, fluctuating environment.

Alexander (1977) describes six brachiopod species ranging in age from Cambrian to Triassic which he considers to be opportunists. He uses the criteria given by Levinton (1970) as his major basis for this conclusion. All of the species are small and thin shelled and have minimal ornamentation. Growth-line analysis suggests that they grow rapidly during their first year of life. Five of the species apparently lived on a fluid substrate in a generally unstable environment, which accounts for their opportunistic characteristics. They appear to be adapted to maintain themselves on the soft substrate in an environment with low oxygen concentration. One of the species apparently lived in an area of mobile (shifting) substrate that eliminated most other species.

Noe-Nygaard et al. (1987) describe molluskan assemblages from the lower Cretaceous of Denmark which they interpret as being composed of opportunistic species. One group of assemblages consists almost entirely of the bivalve *Neomiodon angulata*. Various lines of evidence, including size-frequency distribution (see below) and excellent preservation in a mud matrix, suggest that the bivalves were not transported but lived in a restricted lagoonal environment. The essentially monospecific nature of the assemblages, their great abundance at certain horizons and absence in intervening strata, the morphological variability of the species, and the broad facies distribution of *Neomiodon* in other areas are all typical features of an opportunistic, *r*-selected species. The environmental cause of the variation in abundance cannot be determined with certainty. Instability in a whole range of physical or chemical properties (such as temperature, salinity, dissolved oxygen) are possibilities. Noe-Nygaard et al. suspect that dinoflagellate blooms (red tides) are the most likely candidate. The great abundance in some strata in the section of cysts of one dinoflagellate species supports this interpretation. Toxins produced by modern dinoflagellates in such red tide blooms are known to cause mass mortality in many invertebrate species.

The strata also contain monospecific assemblages of the gastropod *Viviparus cariniferus*, which also apparently lived in a low-energy lagoonal setting. These assemblages consist entirely of adults of about the same size. The monospecific nature of the assemblages and their occurrence at limited horizons and scarcity in intervening horizons suggests that this was an opportunistic species. Noe-Nygaard et al. interpret the crashes in population size as being due to periodic drying of lagoonal ponds. The vagile gastropods probably migrated to the center of the pond as it dried and formed large concentrations there.

POPULATION STRUCTURE

The distribution of ages of individuals within the population, or the population structure, is determined by the intrinsic properties of population growth and the extrinsic properties of the environment. A number of papers and books on population structure discuss this topic. The classical work of Deevey (1947) on life tables and survivorship curves is especially useful, and Odum (1971) and Hedrick (1984) have good reviews of population dynamics. The paper by Hallam (1972) is brief but gives a good introduction from the paleontologic point of view, as do earlier papers by Craig and Hallam (1963), Kurtén (1964), and Craig and Oertel (1966). More recently, Thayer (1975c, 1977), Richards and Bambach (1975), and Noe-Nygaard et al. (1987) have made important contributions to the application of population dynamics to fossil populations.

Data on population structure can be presented in two basic forms: age-frequency or size-frequency plots and survivorship curves. Age-frequency or more commonly size-frequency plots are the most obvious form in which to present these data and are best adapted for study of paleoecological data. Survivorship curves are also very useful and have been especially used in biological studies.

Age- and Size-Frequency Plots

Age-frequency plots are commonly prepared in histogram form showing the number of individuals (or percentage of the total) belonging to each class. In the case of fossils (and indeed, in many living invertebrates) age is difficult or impossible to determine. Some organisms produce growth rings on their skeletons or growth bands within their skeletons which may be annual (see Chapter 4). This has been rather convincingly shown for some mollusks, corals, fish, and echinoids, but the technique must be used with caution. The usual assumption is that slowing of growth during the winter produced the growth ring, but storms or other events that disturb the animal may also produce a growth ring. In some cases, by careful study, such disturbance rings can be distinguished from seasonal rings, but care is required in using this technique. In vertebrates, growth rings and evidence of wear in teeth and various other skeletal features have been used to determine age (Kurtén, 1964). The age of arthropods, especially ostracods, have been determined from molt stages or *instars* (Fig. 8.7).

The more usual approach is to assume a relationship between size and age. This is at least approximately true for most invertebrates which continue to grow throughout life but at a decreasing rate in old age. Levinton and Bambach (1970) concluded that the relationship between growth rate and age for bivalves can be approximated by the equation

$$D = S \ln(T + 1) \tag{8.3}$$

in which D is the size, T the time, and S a constant characteristic of the particular population. For purposes of constructing generalized survivorship curves, this equation is probably as good as any. A simpler approach is to plot size on a logarithmic scale as was done by Thayer (1977). This takes into account the exponential decrease in growth rate, which is included in the equation above, so that the scale should be approximately arithmetically equivalent to age. The relationship between size and age may be more complex than this in detail. Many species seem to show a sigmoidal shape with a somewhat slower growth rate very early in life, a higher but gradually decreasing rate later, and a slow rate late in life. Perhaps an even greater problem is that growth rate varies between individuals so that specimens of equivalent size are not necessarily of equal age. This especially makes difficult the interpretation of polymodal curves (Thayer, 1977).

A size-frequency plot is an approximation of an age-frequency plot. Many studies of population structure have used size-frequency plots directly and not attempted the refinement of converting size into actual or relative ages. In general terms the population structure of a population can be interpreted directly from size-frequency plots (Fig. 8.8).

Biologists normally study the structure of populations of living organisms;

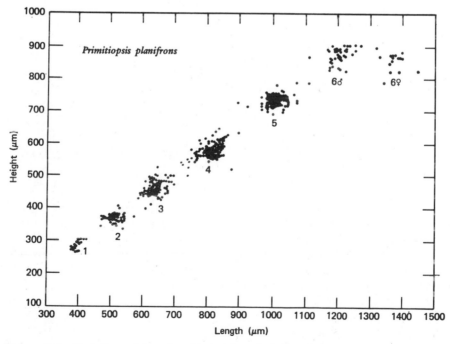

Figure 8.7 Variation of length with height of shells of the ostracod *Primitiopsis planifrons*. The numbers below the point clusters are molt stages. From Kurtén (1964).

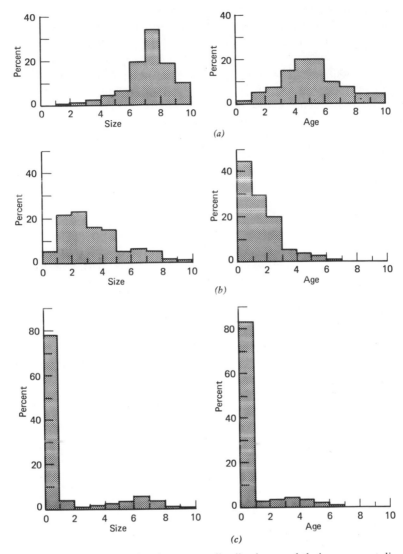

Figure 8.8   Representative size-frequency distributions and their corresponding age-frequency distributions. Growth rate (and thus size) is presumed to decrease logarithmically with age. (*a*) Population with increasing mortality with age; (*b*) population with approximately constant mortality; (*c*) population with decreasing mortality with age. From Richards and Bambach (1975), Journal of Paleontology, by permission of Society of Economic Paleontologists and Mineralogists.

that is, they determine the age of live organisms all living at the same time in a population. Rarely does a fossil assemblage consist entirely of organisms that were all living at the same time. Populations of organisms all living at the same time may occasionally occur when some catastrophic event causes the sudden killing of the entire living population and allows its burial separate from preexisting specimens. Hallam (1972) has called such fossil assemblages *census populations* to distinguish them from *normal populations* that result from the gradual accumulation of shells over a period of time.

Ager (1963) has given a colorful and insightful comparison of the two types of populations. He compares the census population to the human population of the city of Pompeii near Rome, which was abruptly covered by an ash fall from Mt. Vesuvius in 79 B.C. The living population of the city was suddenly killed and preserved at the spot where they were at the time of the ash fall. This would be a census population. The normal population is like the bodies in the cemetery outside the city walls which accumulated over a period of many years as the population died of various causes.

A number of examples of census populations have been described in the paleontological literature. For example, Hallam (1961) describes assemblages of brachiopods from lower Jurassic rocks in England, and Waage (1964) describes molluskan assemblages from upper Cretaceous strata in South Dakota which appear to consist of census populations. There are few definite rules for distinguishing census and normal populations, although Hallam (1972) gives some characteristics of census populations. A census population would be expected to occur in a very narrow stratigraphic interval, perhaps on a single bedding plane, and usually consists of only a few species. It would most likely be found in a quiet-water environment where mixing with fossils immediately above or below is minimized. The fossils might occur in a very thin, fossiliferous unit with relatively fewer fossils above and below. A study of growth rings or skeletal structure should reveal that all specimens died in the same season. Some sedimentologic or stratigraphic evidence of a catastrophic event would further support the census population interpretation. An example might be a turbidite bed immediately above the fossiliferous bed. All of these criteria are, at best, suggestive and do not prove that the fossils in question constitute a census population.

Actually, fossil populations probably cannot always be clearly separated into census and normal. Some populations may be in part census, resulting from a catastrophic event, but they may also include a normal population which was accumulating previously at the site of the catastrophic event. The event may also kill only a portion of the population (such as either smaller or larger individuals), but the remainder of the population may be able to survive (Peterson, 1985).

The most important factors affecting population structure are recruitment and mortality patterns. For the moment let us consider that recruitment is constant and stable so that we can discuss the effects of mortality pattern on population structure. We can divide mortality patterns into three intergrading categories:

1. Populations characterized by high juvenile mortality with lower mortality later in life. Such a population will have many young individuals and few older ones. This pattern is quite common among invertebrates that usually produce large numbers of larvae, most of which die young, with few individuals living to maturity. In many, if not most invertebrates, this high mortality occurs in the larval stage, which leaves no fossil record. The mortality pattern may change drastically after settling of the larvae and development of a mineralized shell. Thus the preservable portion of the record of the life of the species may not show the high-juvenile-mortality pattern. This has the effect of eliminating a portion of the record of the population structure. An age- or size-frequency plot for either a census or normal population having high juvenile mortality will have a strongly positively skewed curve with a large number of young (small) individuals and few old (large) individuals.

2. At the other extreme are populations with low mortality early in life but an accelerating rate as maximum longevity is approached. This pattern applies to some invertebrates and many vertebrates, most notably to humans. Improved medical treatment and public health have resulted in reduced pregerontic mortality, but maximum longevity has been changed little, if any. An age-frequency or size-frequency plot for a census population with this mortality pattern will show a slowly declining number of individuals until maximum longevity is approached, at which point numbers will decline abruptly. The plot of a normal population will have a quite different shape. There will be few young or small individuals because few "dead bodies" or shells are produced. A large number of skeletons near the maximum age or size are produced, yielding a negatively skewed curve with the mode to the right at the older or larger end of the spectrum.

3. In the intermediate case mortality is constant throughout life (i.e., that chance of death is the same regardless of age). Few populations probably show this pattern perfectly, but many approximate it with some irregularities. Postlarval stages of some invertebrates, some mammals, and some birds are examples. A census population for such a population would show an exponentially decreasing number of individuals with increasing age or size. This is because the number of individuals left in any age category is a function of the mortality rate and the number of individuals in the preceding age category. The exponential decay equation applies to this situation, which is perfectly analogous to radioactive decay:

$$\frac{dN}{dA} = mN \tag{8.4}$$

in which N is the number of individuals, A the age (time), and m the mortality rate. A normal population would show a positively skewed curve but less so than in the case of the high-juvenile-mortality population.

All intergradations between these patterns are possible. Mortality patterns between constant and high gerontic mortality will yield age- or size-frequency curves that are nearly symmetrical. The closer the mortality pattern approaches the high-gerontic case, the more different the age- or size-frequency curve will be between census and normal populations.

Patterns can be further complicated by having a more complex mortality pattern, such as high juvenile mortality followed by a period of low mortality with a final period of high gerontic mortality. This pattern was once prevalent for human populations and still occurs today in areas with inadequate medical care. A normal population produced from such a mortality pattern would have modes at young and old ages with a minimum between. An interesting exercise is to look for such a pattern from data collected from headstones in an old cemetery! Even more complex patterns are possible, such as seasonally increased mortality. If recruitment (i.e., spawning) is annual rather than constant, these patterns can lead to polymodal age- or size-frequency curves for normal populations, with the modes corresponding to the season of high mortality.

To this point we have considered recruitment to be constant in order that we might concentrate on the effects of mortality. In fact, recruitment, especially in invertebrates, is seldom constant. Many species spawn or give birth seasonally; thus there is a yearly pattern to recruitment. The success of the recruitment may vary from year to year depending on environmental conditions, especially in species living in unstable environments. This has been demonstrated for many groups of organisms but has been especially extensively studied in commercially important fish (Hutchinson, 1978).

Especially on age–frequency curves of census populations, this recruitment pattern produces polymodal distributions (Fig. 8.9). The modes correspond to time of recruitment. In living populations these modes will change position with time as the individuals grow. A given mode can be tracked from month to month or year to year in order to monitor changes in the population with time. The mode will gradually decrease in height as the individuals responsible for that mode die through time. The approximate age of individuals and maximum age in populations can be determined from size-frequency plots by counting the number of peaks. Spacing of peaks can also be used to determine growth rate. In size-frequency plots the modes will gradually merge to the right, corresponding to decreasing growth rate with age and compounding with age of differential growth rate between individuals.

If recruitment is not constant, the shape of age- or size-frequency curves for census populations tend to be dominated by the recruitment pattern. For this reason they must be used with care to interpret mortality patterns (Cadée, 1982). Because normal populations are time averaged, effects of differential recruitment tend to be eliminated. Modes corresponding to seasonal recruitment or good recruiting years are usually lost in the mixing process. Thus mortality pattern is the dominant control on the shape of the age- or size-frequency curve. A seasonal recruitment pattern in combination with seasonal mortality pattern may result in polymodal age- or size-frequency curves even in normal populations.

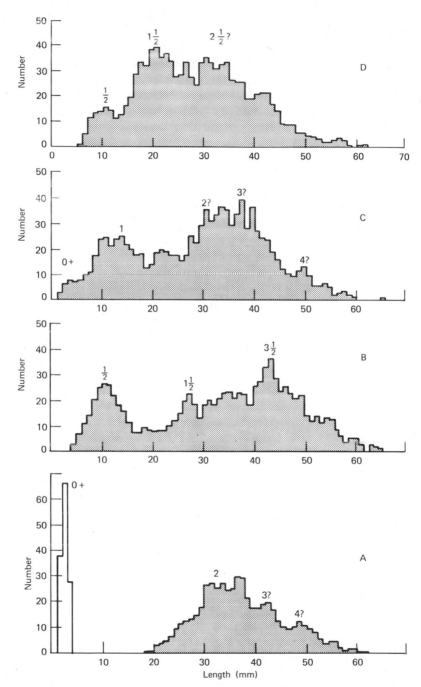

Figure 8.9 Polymodal size-frequency histograms of *Mytilus edulis* from Ferny Ness, Scotland. The numbers above the histograms are the presumed age of specimens in that mode in years. Collection dates: (*A*) April 1961; (*B*) November 1961; (*C*) April 1962; (*D*) November 1962. From Craig and Hallam (1963).

Life Tables and Survivorship Curves

Many features of the structure of populations can be visualized most readily from life tables and survivorship curves. Life tables (Deevey, 1947) contain the number of survivors from a starting population, mortality rate, and mean life expectancy at regular intervals during the life of the species (Table 8.1). They have been used extensively by population ecologists but not as much used directly in paleoecology. Life tables include, in tabular form, the information graphically shown in survivorship curves, which are constructed in the following way: Consider the individuals in a population that are all born or spawned at the same time (a cohort in ecological terminology). With time, individuals of the cohort gradually die, leaving a certain number of survivors. A graph of the number of survivors versus time since birth (or age of the cohort) is a survivorship curve (Fig. 8.10). The abscissa of the survivorship curve may be the age in some convenient time unit, but for ease of comparison of species with widely different longevity the percentage of the mean or maximum life span is sometimes used. Commonly, the ordinate of the survivorship curve is the number of individuals out of a starting 1000 (or some other arbitrary number), or it could be the percentage surviving. The ordinate is almost always given on a log scale. Such a scale results in a straight-line plot for the case in which the mortality rate is constant at all ages (Fig. 8.10).

The shape of the survivorship curve differs with the structure of the population. Three general types can be recognized, which correspond to the three types of age- or size-frequency curves discussed above (Figs. 8.10 and 8.11):

TABLE 8.1 Life Table for a Population of *Tivela stultorum* **from California**[a]

Age (years) x	Age as % Deviation from Mean Length of Life x'	Number Dying in Age Interval out of 1000 Born d_x	Number Surviving at Beginning of Age Interval out of 1000 Born l_x	Mortality Rate per Thousand Alive at Beginning of Interval $1000\,q_x$	Expectation of Life, or Mean Lifetime Remaining to Those Attaining Age Intervals (years) e_x
0–1	−100	550	1000	550	2.07
1–2	−52	202	450	449	1.87
2–3	−3	72	248	290	1.99
3–4	+45	60	176	341	1.60
4–5	+93	60	116	517	1.18
5–6	+142	38	56	678	0.90
6–7	+192	13	18	722	0.75
7–8	+237	5	5	1000	0.50

[a]From Hallam (1967).

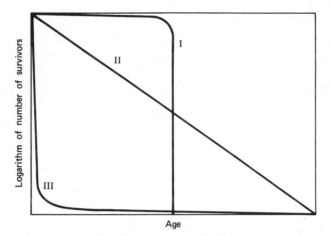

Figure 8.10 Schematic representation of three basic types of simple survivorship curves. I, mortality increasing with age: II, constant mortality rate: III, mortality decreasing with age. From MacArthur and Connell (1966).

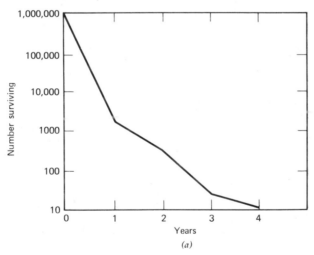

Figure 8.11 Three basic types of survivorship curves in natural populations. (*a*) Decreasing mortality in the prawn *Leander squilla*. From Kurtén (1964). (*b*) Increasing mortality in the recent brachiopod *Terebratalia transversa*. Upper curves (triangles) are plotted as a function of valve length on the arithmetic horizontal axis. The lower curves (circles) are plotted as a function of valve length on the logarithmic scale, which should be approximately equivalent to age (see text for discussion). The curves with open symbols represent one sample; the curves with solid symbols represent a second sample. From Thayer (1977), copyright Paleobiology. (*c*) Constant mortality in the bird *Vanellus vanellus*. From Hutchinson (1978), An introduction to population ecology, copyright Yale University.

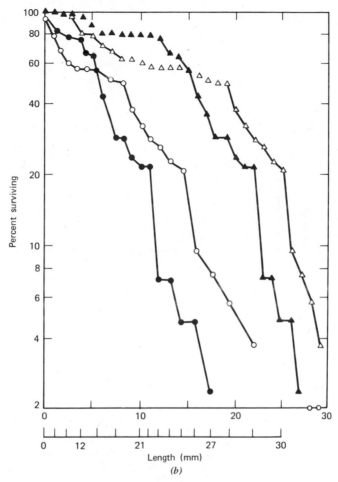

Figure 8.11 (Continued)

(1) Populations characterized by high juvenile mortality and lower mortality later in life have a curve that is concave upward (Fig. 8.11*a*); (2) at the other extreme are convex-upward curves characteristic of species with a low mortality rate early in life but an accelerating rate as maximum longevity is approached (Fig. 8.11*b*); and (3) the intermediate case is a descending straight line representing constant mortality throughout life (Fig. 8.11*c*). Few, if any, species would plot as a perfectly straight line but many approximate it with some irregularities. Cases 1 and 3 may be difficult to differentiate in age- or size-frequency plots, but they are more distinct on survivorship curves.

As can be seen from the examples shown here, survivorship curves contain irregularities and are not the perfect hypothetical curves shown in Fig. 8.10. A

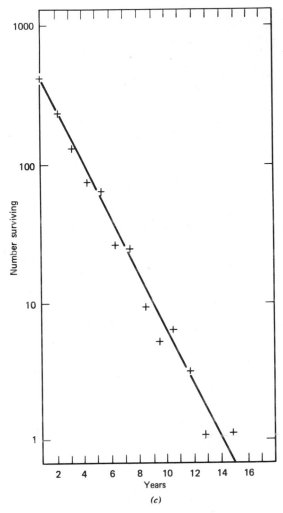

Figure 8.11   (Continued)

common example is a crude sigmoidal shape which indicates a higher mortality rate at the juvenile stage with a lower intermediate rate followed by a
higher rate later in life. A stair-step pattern is found in some species, such as
the arthropods, which have certain stages during their life history when they
are more vulnerable to mortality.

How can survivorship curves be constructed from fossil populations? Survivorship curves are often difficult to construct, even for modern populations. A
true survivorship curve, sometimes called an *age-specific* survivorship curve,
traces the survivorship of a single cohort and can be constructed only from
records kept over the life span of the entire cohort. This is difficult and time

consuming to do for modern populations and impossible for fossil populations. The best approximation of an age-specific survivorship curve is a *time-specific* survivorship curve. This curve is constructed by determining the number of living individuals in each age class at a given time (i.e., from an *age-frequency* curve). In a stable population that is adding new individuals at a constant rate while mortality continues at a constant rate within each age class, the time-specific survivorship curve will be identical to the age-specific curve. In fact, populations are seldom stable with a constant recruitment rate. This makes construction of a survivorship curve from a census population difficult (Cadée, 1982). Normal populations have the positive effect of smoothing out irregularities caused by nonconstant recruitment (see above) and are thus more useful than census populations for constructing survivorship curves.

Constructing a Survivorship Curve from a Normal Population

An approximation of a survivorship curve can be constructed from a size-frequency distribution for a normal fossil population by the following procedure:

1. Establish a series of age or size (linear dimension) classes for the population. The age classes should be of equal size. The number of age classes established will probably be determined by the precision with which these age classes can be recognized from growth lines or skeletal structures in the fossil. Size classes can be set at the convenience of the researcher, but for most cases 8 to 20 classes should be sufficient. Numerous finely subdivided classes require more specimens in the sample. Age classes are necessary to construct a true time-specific survivorship curve, and they obviously should be used if the age can be determined.

2. Determine the age or measure a linear dimension for all specimens in the population and assign them to the classes established in step 1.

3. If size rather than age is used for the x-axis of the survivorship plot, it may be converted to age or a time factor by use of an equation such as (8.3). The x-axis may also be a logarithmic scale to more closely approximate the age–size relationship. If an arithmetic scale is used, the probable nonlinear relationship between size and age should be considered in interpreting the results.

4. The ordinate should be logarithmic, with the scale most conveniently expressed as percent of specimens surviving.

5. At age zero or size zero, 100% of the specimens are surviving. At the end of the age or size of the first class, subtract all the members of that class from the total number of specimens. These are the specimens that did not survive beyond that class. Determine the percent remaining (or surviving) after the subtraction and plot the percentage at the boundary between classes 1 and 2. Repeat this procedure for each class, adding each time the members of that class to all younger or smaller fossils which did not survive to that age

or size. A logarithmic scale has no zero, so the plot must stop at the lower end of the last age or size class. The equation for calculating points on the survivorship curve is

$$l_x = \frac{N_t - \Sigma_x^1 \times N_i}{N_t} \, 100 \qquad (8.5)$$

in which l_x is the percentage of the population surviving at the end of class x, N_t the total number of fossils in the population, and $\Sigma_x^1 \times N_i$ the sum of the number of fossils in all classes between 1 and x. Figure 8.12 shows a hypothetical example of a survivorship curve constructed from an age frequency histogram by this procedure.

Modifying Factors of Fossil Survivorship Curves

Differential preservation and modification of the size distribution by current winnowing will modify the survivorship curve. Small, thin shells are more subject to solution and other diagenetic processes than are larger, more robust shells, and currents are more likely to winnow small shells from the assemblage. Some of the earliest interest in size-frequency plots was as a method of investigating whether or not the fossils had been transported (e.g., Menard and Boucot, 1951; Boucot, 1953; Fagerstrom, 1964). The hypothesis was that current transport or sorting would produce normal bell-shaped curves whereas *in situ* populations would be positively skewed. This may often be correct, but clearly *in situ* populations may produce normal or even negatively skewed size-frequency curves (Craig and Hallam, 1963).

Small (juvenile) individuals frequently are underrepresented in fossil populations (Cadée, 1982). In many cases this is probably because of physical and chemical destruction. Other factors may be involved as well. Cadée (1968, 1982) gives examples of populations in which the adults migrate away from the juvenile habitat. Shimoyama (1985) discusses an example of gastropod populations which appear to be modified by size-selective use of the shells by hermit crabs. Differential destruction of juvenile skeletons results in the common bell-shaped distributions in fossil populations. Several researchers (e.g., Kurtén, 1964; Dodd et al., 1985; and Cummins et al., 1986c) have suggested that although such distributions make impossible the study of juvenile mortality, the adult mortality can still be studied if the larger individuals are preserved. The effects of transportation and mixing may be less important than once thought (Walker and Bambach, 1971; Fürsich and Flessa, 1987). Accordingly, Hallam (1972) believes that the evidence suggests that size-frequency distributions are largely accounted for by growth and mortality rates. Noble and Logan (1981) found that taphonomic processes did not appear to be important in determining the shape of size-frequency distributions in modern brachiopod populations from the Bay of Fundy.

Powell et al. (1986) and Cummins et al. (1986b) studied living and dead

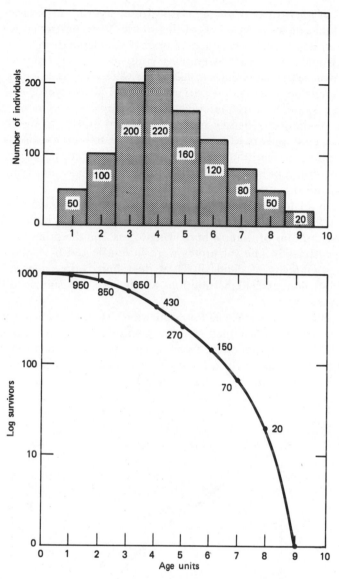

Figure 8.12 Hypothetical age-frequency histogram and survivorship curve derived from it. The numbers on the histogram refer to the number of individuals in each age class. The numbers on the survivorship curve refer to the number of individuals surviving to the indicated age. From Hallam (1972), in Models in paleobiology, T. J. M. Schopf (ed.), Freeman, Cooper & Company.

specimens of the bivalve *Mulinia lateralis* from bays on the Texas coast. They concluded that most of the shells of this species were destroyed within a few months after death of the specimens. Loss was size dependent, being higher in small individuals. The death assemblage consisted of two components: specimens near or at the sediment surface which were constantly being produced and destroyed and more deeply buried specimens, which were preserved for a longer time span. Storms periodically stirred up the sediment and added specimens to the more deeply buried, preserved group. In the Texas bays, taphonomic loss appears to be in large part due to solution of the shells by brackish water undersaturated with respect to $CaCO_3$. The pattern of preservation may be different in a setting with $CaCO_3$-saturated water (which is more common under open marine conditions).

In a study of four modern infaunal bivalve species in Japan, Tanabe and Arimura (1987) concluded that taphonomic processes modified size-frequency distributions of shallow burrowing forms. On the other hand, deeper-burrowing species in fine substrates were little affected by taphonomic processes. Assemblages of dead shells of deep-burrowing species, especially articulated specimens, had distributions which were very similar to the associated living specimens.

Arthropods are very suitable for population structure studies because their relative age can be determined from molt stages. However, because arthropods grow by molting, not all the fossil specimens may actually represent dead bodies. Thus the normal rules for survivorship curves do not apply without modification. If the arthropod fossils were all derived from dead bodies, the

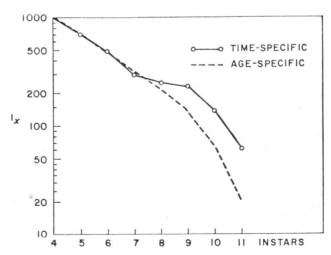

Figure 8.13 Survivorship curves for the Silurian ostracod, *Beyrichia jonesi* calculated by the age-specific and time-specific methods (see text for details). From Kurtén (1964).

procedure for normal populations described above could be used. If the fossils are all molts, their abundance is a measure of the number of living organisms that lived successfully through that molt stage. Thus, in a sense, the molts represent "live bodies" rather than dead. In this case the fossils are equivalent to a census population and can be used directly to construct a time-specific survivorship curve (or an age-frequency curve). In most cases the actual fossil arthropod assemblage is a mixture of molts and actual bodies, so neither approach will yield a true survivorship curve. Kurtén (1954) constructed survivorship curves for a fossil ostracod population based on each of these assumptions (Fig. 8.13). In the example he used the curves do not differ much, so he concludes that either method can be used without much error. In the more general case, however, the results could be quite different; hence the problem should not simply be ignored.

Information from Population Structure

In summary, potentially, six categories of information can be obtained from studying the structure of fossil populations:

1. *Survivorship Patterns.* These patterns contain useful information in terms of understanding the biology of the species and can also yield useful environmental information. Variation in the survivorship curves between populations of the same species may reflect differences in environmental conditions, such as substrate hardness.

2. *Seasonal Spawning Patterns.* These can be detected from census populations and perhaps less reliably from normal populations. This would suggest that environmental parameters vary seasonally.

3. *Seasonal Mortality Patterns.* These can be detected from polymodal normal fossil populations. They also suggest seasonally variable environments.

4. *Growth Rate.* Spacing of modes in polymodal distributions can, under ideal conditions, yield this information. As indicated in Chapter 4, growth rate may be a function of environmental parameters, especially temperature.

5. *Maximum Age.* The number of modes in polymodal distributions can give an estimate of the maximum age of individuals in the population.

6. *Current Sorting.* Normal, bell-shaped distributions may suggest current sorting, particularly if other current indicators are present. Care must be taken in making this interpretation because populations with low juvenile mortality may produce bell-shaped distributions.

Examples of Population Dynamics Studies

A number of studies of the population dynamics of living and fossil populations have been conducted. We discuss only a small sampling here. Shimoyama

(1985) studied living populations of the gastropod *Umbonium (Suchium)* *moniliferum* in a bay in Japan. He monitored the size-frequency distribution of the species for almost 3 years. The distribution varied from month to month as new specimens were recruited (spawned), grew, and died. Modes corresponding to annual spawning classes could be traced for about a year. Shimoyama used the data from the gradual decrease in size of the peaks from month to month as a method of determining mortality rate in the species during the previous month. From these data he calculated the size-frequency curve of dead shells that should have resulted from this mortality pattern. This would be the predicted normal population generated from the living population.

Shimoyama (1985) then compared survivorship curves based on actual samples of dead shells with that predicted from the mortality pattern (Fig. 8.14). Major differences were interpreted as resulting from postmortem processes. He concluded that many samples were little modified by these processes, but some modification resulted from winnowing by tidal currents. He concluded that other samples of snails resulted from transport by hermit crabs, which preferentially selected a limited range of shell sizes. He also studied Holocene fossil populations of *Umbonium* and concluded that they also had been modi-

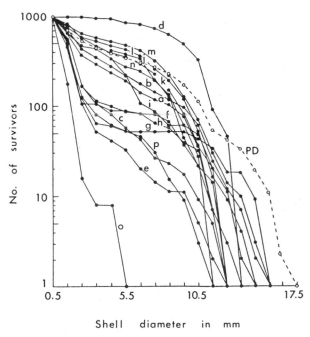

Figure 8.14 Survivorship curves of samples of dead specimens of the snail *Umbonium (Suchium) moniliferum* from Japan. The dashed line (PD) is survivorship curve for a living population. From Shimoyama (1985), by permission of Elsevier Science Publishers, Amsterdam, Netherlands, and the author.

fied by hermit crabs. Modification of fossil distribution by hermit crabs is being increasingly recognized as an important taphonomic process (Frey, 1987).

Sørensen (1984) studied population dynamics in two bivalve species from Pleistocene sediments which had been deposited at bathyl depths in the Mediterranean Sea. This is probably an especially good example of a population dynamics study in terms of minimal effect by taphonomic processes. The aragonite shells are well preserved in an essentially current-free environment. Some shell breakage (biomechanical?) is the only apparent taphonomic process. Sørensen supplemented his study of size-frequency distribution with a study of presumed annual banding within the shells.

Kelliella miliaris, a suspension-feeding endobyssate form, showed a pattern of logarithmically decreasing growth rate and near constant mortality rate (Fig. 8.15). *Phaseolus ovatus,* an infaunal deposit feeder, was interpreted as having a slow–fast–slow growth pattern and increasing mortality with age. The growth pattern was in part determined on the basis of the bimodal size distribution. The modes correspond to periods of slow growth and the intervening "trough" to a period of rapid growth (not many shells of this size were produced because they quickly "grew through" that size range).

Sørensen interprets the differences in growth and survivorship as being largely due to the mode of life of the two species. *Phaseolus* has a relatively low but constant food source and thus a lower juvenile (or young) mortality. Limited food results in slowing of growth during periods of reproduction, but growth rate increases between these periods, as shown by the slow–fast–slow growth pattern. *Kelliella* depends on suspended food, which varies seasonally, and must live at the surface on a muddy, unstable substrate. Thus they have a higher rate of juvenile mortality than *Phaseolus,* but the rate does not change with age.

Dodd et al. (1985) completed a detailed study of population dynamics in populations of the echinoids *Dendraster* and *Merriamaster* and the bivalve *Anadara* from Neogene strata exposed in the Kettleman Hills of California (Fig. 8.16). *Dendraster* and *Anadara* populations showed the typical pattern of few or no small specimens. *Merriamaster* populations had a more complete range of specimen size. This lack of small individuals suggests that young *Dendraster* specimens never die! Almost certainly, small *Dendraster* specimens occasionally died, so there must be some other explanation of the lack of small specimens. Perhaps the young lived in a different environment than the adults, but this explanation seems unlikely as small specimens were never found, even separated from adults. Rapid juvenile growth rate and low juvenile mortality were probably the most important factors in explaining the lack of specimens, but taphonomic factors may contribute to the lack of small specimens. Larger (and older) specimens should not be differentially affected by these factors. Thus survivorship curves for these specimens should show mortality patterns for this stage of life.

Survivorship curves for most *Dendraster* and *Anadara* populations are con-

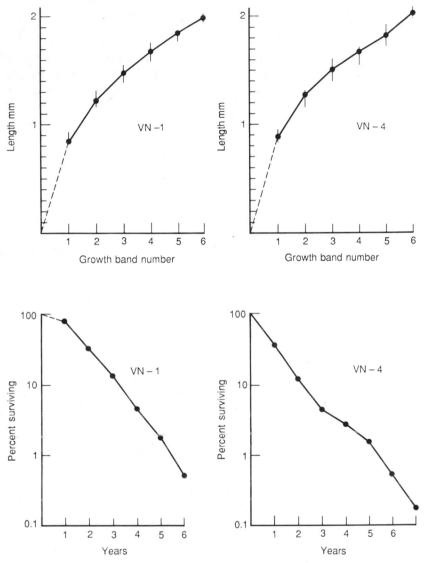

Figure 8.15 Growth curves for two samples of *Kelliella miliaris* determined from growth bands (upper graphs). Survivorship curves for same samples of *K. miliaris* (lower graphs). From Sørensen (1984), Lethaia, by permission of Norwegian University Press (Universitetsforlaget AS), Oslo, Norway.

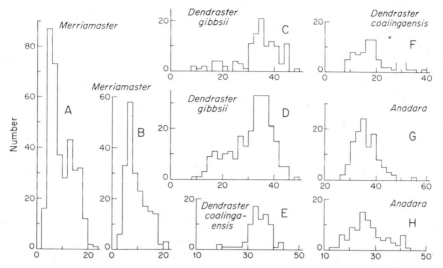

Figure 8.16 Size-frequency histograms for samples of the sand dollars *Merriamaster* and *Dendraster* and the bivalve *Anadara* from the Neogene strata of the Kettleman Hills, California. From Dodd et al. (1985), by permission of Elsevier Science Publishers.

vex, indicating increasing mortality rate with age. These populations apparently lived on firm substrates in turbulent water. Other species living in these environments display a similar survivorship pattern (Surlyk, 1972; Richards and Bambach, 1975). *Merriamaster* populations and a few *Anadara* populations have linear to slightly convex survivorship curves, indicating constant mortality. These populations occur in somewhat muddier sediment, which was presumably deposited under somewhat less turbulent conditions. Environmental interpretations for the Kettleman Hills Neogene based on population structure agrees with that based on various other techniques.

Noe-Nygaard et al. (1987) studied size–frequency distributions for two molluskan species from lower Cretaceous strata in Denmark. The occurrence of the bivalve *Neomiodon angulata* meets the various criteria listed above for census populations. As indicated above, the authors speculate that catastrophic death was caused by toxins produced by a dinoflagellate species (red tide). In most cases the size–frequency plots are bimodal, suggesting seasonal recruitment (Fig. 8.17). The first mode includes specimens in their first year of growth and the second includes 2-year-old and some older specimens. Their combined study of size–frequency distribution and growth ring spacing suggests a maximum age of 4 years with gradually decreasing growth rate. Mortality seems to occur preferentially in the summer (or 3 to 4 months after resumption of growth following a period of cessation).

Richards and Bambach (1975) studied 16 brachiopod assemblages of Ordovician, Silurian, and Pennsylvanian age in one of the most extensive studies of

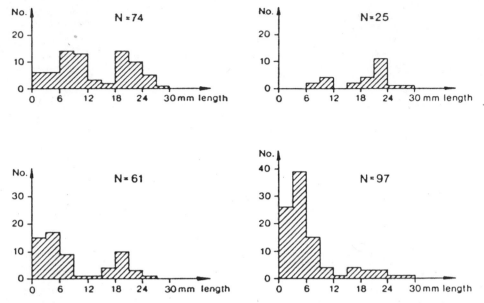

Figure 8.17 Size-frequency histograms of four samples of *Neomiodon angulata* from Cretaceous strata of Denmark. From Noe-Nygaard et al. (1987), by permission of Society of Economic Paleontologists and Mineralogists.

population dynamics in fossils. They found two types of populations in their study: one with a strongly positively skewed size-frequency distribution (with an abundance of small specimens) and another with a negatively skewed distribution (with an excess of large individuals and few small ones). The survivorship curves for the first group showed a high juvenile mortality rate and a decreasing mortality rate with age (Fig. 8.18). The survivorship curves for the second group show a mortality rate that is low in juveniles and increases with age (Fig. 8.19). All the populations with high juvenile mortality are from assemblages that the authors interpret to have lived on a soft, muddy substrate. The juveniles had difficulty coping with the turbidity and resuspended sediment as well as the unstable substrate; consequently, the mortality rate was high. The second group lived on firmer substrates, and consequently, the mortality rate among juveniles was low. The higher mortality rate in larger individuals simply reflects the gerontic mortality.

DISPERSION PATTERNS

Individuals within a population may be spatially distributed in three basic patterns: (1) *random,* (2) *regular* (*overdispersed*), and (3) *clumped* (*underdispersed* or *contagious*) (Fig. 8.20). Dispersion can be viewed on many

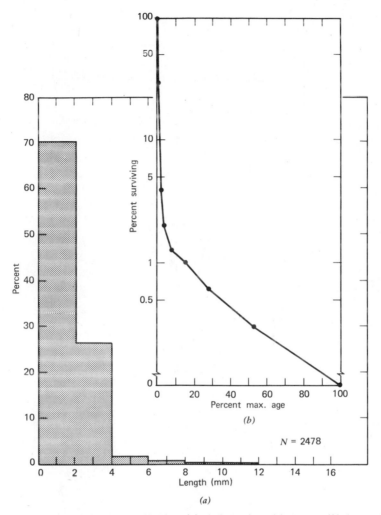

Figure 8.18 Size-frequency distribution (*a*) and survivorship curve (*b*) for a population of the brachiopod *Composita* cf. *subtilita* from Pennsylvanian strata at LaSalle, Illinois. From Richards and Bambach (1975), Journal of Paleontology *49:* 790, by permission of Society of Economic Paleontologists and Mineralogists.

different scales, and the type of dispersion pattern may vary depending on the scale. On the global scale all populations show clumping. For example, populations of species living in the shallow sea will be found only on the shelf areas and will be absent from the broad expanses of open ocean and from terrestrial areas. Thus they will be clumped in the shelf areas. Concentrations or clumps of species might be found on a given segment of the shelf, but these clumps may be randomly distributed. Within an individual clump the distribution may

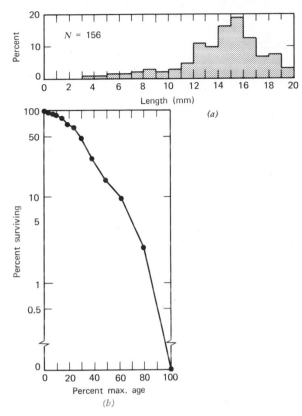

Figure 8.19 Size-frequency distribution (*a*) and survivorship curve (*b*) for a population of the brachiopod *Lepidocyclus capax* from Ordovician strata at Abington, Indiana. From Richards and Bambach (1975), Journal of Paleontology, *49:* 795, by permission of Society of Economic Paleontologists and Mineralogists.

be evenly spaced or regular. Thus a single species may show all the basic patterns of distribution, depending on the scale of the observation. The type of dispersion pattern may be obvious in some cases but not in others. In the latter case a statistical test must be used to determine the type of distribution. Standard statistics texts, such as Sokal and Rohlf (1969), discuss the procedure for making such a test. Pemberton and Frey (1984) briefly discuss the nearest-neighbor and coefficient-of-dispersion methods as applied to a paleontological example.

Random Distribution

In a random distribution pattern, the location of any individual is independent of the location of all other individuals. Because of the method of distribution

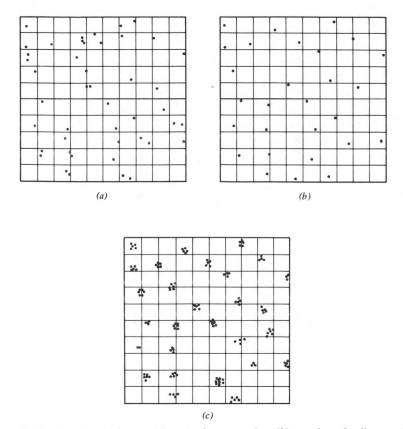

Figure 8.20 Random (*a*), overdispersed or regular (*b*), and underdispersed or clumped (*c*) distributions of points. Each of these distribution patterns can be recognized in organisms. From MacArthur and Connell (1966).

of many species by the apparently random settling of larvae, this distribution might be expected to be very common among invertebrates. Numerous studies have shown that larval settling is not really random but that the larvae of many species can, at least to some extent, select the site where they settle. Based largely on distributions of terrestrial organisms, Odum (1971) states that random dispersal patterns are uncommon. Jackson (1968) observed a random distribution on mud flats of the suspension feeding bivalve *Mulinia lateralis.* Two-year-old individuals of the bivalve *Gemma gemma* from the same area also were randomly distributed, although younger individuals showed clumping. This is probably because *G. gemma* is ovoviviparous; that is, the eggs hatch and the larvae develop within the shell of the mother. Thus the very young are concentrated around their mothers. With time they disperse enough to assume a random distribution. The opposite pattern may be expected to occur in other cases; that is, the very young may have a random distribution because of larval settling, but with time those individuals that settled in less favorable localities will die, leaving the older individuals in a clumped distribution.

A random distribution may indicate one or a combination of several factors: (1) The environment in the area is uniform (or the species is able to tolerate a wide range of environmental variability); (2) individuals of the population do not interact strongly with one another (negative interactions tend to result in regular distribution and positive interactions result in clumping); (3) currents may move the individuals about, giving them a random distribution (although more commonly currents will produce clumps of shells); and (4) reproduction (especially in sessile forms) is by random settling of larvae.

Regular Distribution

Regular distributions have been more commonly described in terrestrial environments. This distribution is the result of either an active or a passive effort to minimize competition between individuals in the population. Regular distributions may result from the dying of young individuals or larvae that have settled too close to previously established individuals. In vagile forms (or by larval selectivity) the spacing may result from individuals moving to unoccupied sites where there is a maximum of distance from other individuals. Regular distributions are most obvious and have been studied most actively in vertebrates such as birds, which often exhibit strong territorial behavior. Plants also commonly have regular distributions because of the need to minimize competition for light or moisture. Invertebrates sometimes show regular distributions (usually within clumps) because of competition for limited space. The crowded but regular distribution of barnacles or mussels on rocks in the intertidal zone is an example.

Clumped Distribution

All species are clumped on one scale or another. Indeed, the concept of the population almost implies clumping. An individual population is a clump or aggregation that is separate from other aggregations. Clumping may simply result from inhomogeneities in the distribution of suitable habitat for the species. The clumping of organisms requiring a hard substrate on isolated rocks would be a small-scale example of this. On the other hand, clumping may result from a positive interaction between individuals of the species. Extreme examples of this would be in the social insects such as bees and in colonial organisms such as corals. Perhaps the ultimate modern example of the positive interaction of individuals within a species (and the ultimate clump?!) is the coelenterate *Physalia,* the Portuguese man-of-war. In *Physalia* the individual almost loses its identity and becomes an organ for the superorganism. Some individuals specialize for feeding, others for defense (much to the dismay of the swimmer who is stung by them), and others for flotation (Barnes, 1974), but all are clumped into a single jellyfish-like colony.

Clumping may be achieved by several processes: (1) The larvae may postpone metamorphosis until they contact a suitable substrate; (2) the larvae may settle randomly but die out between clumps; (3) the food supply may be concentrated so that the individuals move to those concentrations; (4) the young may be born live, have an extremely short larval stage, or develop from eggs deposited near a parent so that the young concentrate in the area of the parent; and (5) currents may physically concentrate the living specimens.

Clumping may offer a number of advantages to the individual:

1. The baffling effect of aggregation of individuals such as crinoids or alcyonarians reduces the current velocity sufficiently to allow suspended food in the water to be captured.
2. Clumped distributions are also an advantage in sexual reproduction in bringing potential mates together.
3. The clumping or schooling of individuals provides a defensive advantage to many swimming fish and invertebrates as well as to many terrestrial vertebrates, such as the caribou and bison. Predators that will attack isolated individuals often hesitate to attack a school or herd.
4. Many additional advantages accrue to the individual if clumping results in social and colonial animals. In this case, individuals may become highly specialized, so that the colony can operate more efficiently than separate individuals. The massiveness of a colonial skeleton such as a coral allows existence in a very turbulent environment. Coordinated beating of cilia allow the production of much stronger food-bearing currents than would be possible by the individual. The defense of a colony such as a stinging coral or Portuguese man-of-war is much more effective than that of a single, small individual. Many additional advantages of the colonial and social existence could be cited (Wilson, 1975a).

The individuals within a colony or other clump of individuals are often very evenly distributed. In some cases this may simply be a result of close packing of individuals. But at least in some cases it appears to have developed to minimize competition on the fine scale while taking advantage of clumping on the slightly larger scale.

Dispersion in Fossils

Is the distribution pattern found in living populations preserved in fossil populations or assemblages of dead shells? Several studies (e.g., Johnson, 1965; Peterson, 1977; Cummins et al., 1986a; Tanabe and Arimura, 1987) have compared the distribution of dead shells (potential fossils) and living organisms. The degree of correspondence depends on the species involved, the type of environment, and various diagenetic or taphonomic processes; but a few generalities emerge. Two processes that are especially important in changing distributions are transport and time-averaging.

One cannot assume that all fossils that are found together actually lived together at the same time (Fürsich, 1978). The diversity in an accumulation of fossils is commonly greater than that of the living organisms that actually occupied the spot at any one time. Similarly, the distribution of organisms found together in a single bed is the averaging of the distribution over a period of time and not at any one instant. Random distributions should be unaffected by time averaging because the averaging process simply adds more randomly distributed individuals and can do nothing to either clump or distribute individuals more evenly. Cummins et al. (1986a) used computer simulation to consider the question of the effect of time averaging on clumped distributions. They concluded that clumping would be retained despite time averaging. The geologic record includes many examples of clumps that have maintained their identity through time. Reef and bank assemblages are perhaps the most obvious examples. Reefs owe their existence to the fact that the organisms form a clump and do not migrate with time. Cummins et al. did not consider the effect of time averaging on regular distributions, but intuition suggests that these distribution patterns would be randomized.

Transportation usually has the effect of concentrating skeletons on the basis of their hydrodynamic properties and thus produces more clumped distributions. Jackson (1968) notes a probable example of this for dead *Gemma gemma* shells on modern mud flats in Connecticut. Can clumped distributions of biological origin be distinguished from transportation-produced clumps? Cummins et al. (1986a) concluded that they can be distinguished on the basis of covariation in abundance of populations of different species in the clumps. In most cases populations in biologically produced clumps should not covary in abundance. Exceptions would be species with strong interactions, such as commensalism, parasitism, symbiosis, and so on. Concentrations produced by transport would frequently show covariance because many species would behave similarly in hydrodynamic terms. Cummins et al. (1986a) tested this model for modern samples of dead shells from locations in the Texas coastal region. They found the greatest amount of covariance in the location with the strongest currents and in smaller species that are more readily moved by currents. Not only does covariance allow distinguishing between transport-produced and biologically produced clumps, it also allows recognition of assemblages produced by transport and thus of more limited value in paleoenvironmental interpretation.

Trace fossils offer some advantages over body fossils for study of dispersion. The main advantage is that in most cases transportation can be eliminated as a factor affecting distributions. Pemberton and Frey (1984) studied the distribution of trace fossils at two locations in Canada. They observed at

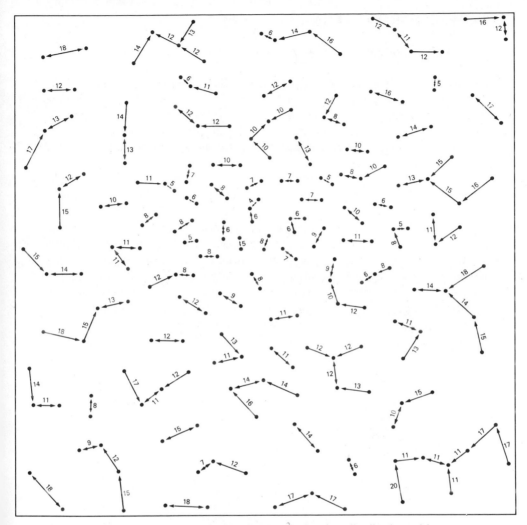

Figure 8.21 Map of typical quadrat (0.25 m^2) showing distribution of burrows on bedding surface. Arrows join nearest neighbors. Distance shown in millimeters. From Pemberton and Frey (1984), Lethaia.

one location that concentrations of *Skolithos* occurred as clumps. Within the clumps the distribution was random when the population density was low but regular where it was high (Fig. 8.21). They interpreted this regular distribution as resulting from suspension-feeding organisms which were regularly partitioning limited space. A random-to-regular distribution in the U-shaped burrows of *Diplocraterion* at another locality also suggested a suspension-feeding mode of life. Their analysis thus led them to a conclusion about the biology of the trace-making organism.

MORPHOLOGIC VARIATION

No two individuals of a population are exactly alike, and an important characteristic of a population is the morphological variability among individuals of
the population. The amount and nature of this variation are in part determined by environmental parameters. Some species, particularly those that are
highly specialized, have a small range of variability. Other species, especially
those that are more generalized, tend to be morphologically more variable
within a given population. The morphological variability may either be the
result of *genotypic* (under direct genetic control) variability or of *phenotypic*
(not directly controlled by genetic differences) variability.

All sexually produced organisms have some genetic variability between
individuals. This is, of course, the basis for evolution through natural selection. Some of the morphological variability that results from the genetic variation probably has no functional significance, but some variants will be slightly
better suited than others for a given habitat. Thus the genetic variability
would itself have a selective value for organisms living in variable environments. Using an electrophoresis technique to determine the relative number
of genetic loci that are polymorphic, Ayala et al. (1973, 1975) and others have
measured the amount of genetic variability within populations. Surprisingly,
the available data indicate more genetic variability in forms from stable envi-

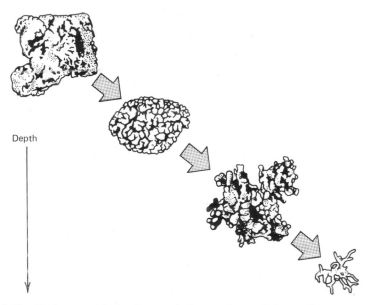

Depth

Figure 8.22   Variation with depth or turbulence of growth form of crustose red algae
(a hypothetical example). From Wray (1971).

ronments than unstable. The explanation of this still is not clear. Apparently, this indicates, however, that morphological variability within populations of invertebrates is due more to phenotypic than to genotypic causes. Ayala et al. (1975) suggest as one explanation that genetic variability among specialists allows them to become even more highly specialized to their particular habitat on an individual basis.

Bonner (1965) has suggested two types of phenotypic effects on morphology. One, which he calls *multiple choice variability*, results in two or more morphologies being possible for the species. Each type can be called an *ecophenotype* of the species. The particular morphology that develops is determined by the environmental conditions under which the species grows. An example of multiple-choice variability can be found in the coralline alga *Goniolithon* (Bosence, 1976), which may develop a crustose form under high-turbulence conditions, a short branching form at intermediate conditions, and delicately branching form in quiet conditions (Fig. 8.22). This ability to develop a number of morphologies is due to the genetic constitution of the species, but each morphologic type has the same basic genetic makeup.

The other type of phenotypic variation is called *range variation*. During development of the species, a certain range of morphologies is possible because of the multitude of biochemical processes involved and the effect of environment on these processes. In some cases a certain variant suits the organism to its particular habitat, and in other cases there is no obvious advantage. The variation in eccentricity of modern specimens of *Dendraster excentricus* (Raup, 1956) and in Pliocene species of that genus (Stanton et al., 1979) may be examples of range variation (see Chapter 5).

The morphology of the genus *Anadara* has been analyzed in the Pliocene strata of the Kettleman Hills (Alexander, 1974). A number of morphologic features in *Anadara* populations are correlated with environmental conditions. The distribution of mean specimen sizes in populations in the *Pecten* zone indicates that specimen size decreases with salinity of the depositional environment (Fig. 8.23). The largest specimens are found on the east flank of North Dome, an area of normal marine salinity, as indicated by various lines of evidence, such as taxonomic uniformitarianism, oxygen and carbon isotopic composition, and diversity. Specimens from lower-salinity areas such as the west flank of North Dome and the east flank of Middle Dome are smaller. This is probably an ecophenotypic effect. From comparison of mean size with paleotemperature, as determined by faunal (Chapters 2 and 9) and geochemical (Chapter 3) evidence, size is positively correlated with temperature.

The mean value of the number of ribs on *Anadara* shells is positively correlated with salinity. Specimens collected on the east flank of North Dome from the *Pecten* zone have as many as 28 ribs on the shell (Fig. 8.24). Specimens from lower-salinity areas usually have fewer ribs, and one sample from the west flank of North Dome has a mean of only 24. The positive correlation

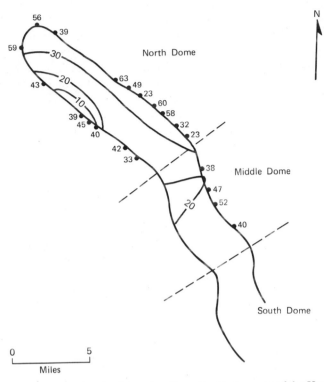

Figure 8.23 Mean length (mm) of *Anadara* from the *Pecten* zone of the Kettleman Hills Neogene. Numbered lines are salinity isopleths in per mille as determined from oxygen isotope analysis (see Chapter 3). From Alexander (1974), Journal of Paleontology, *48:* 636, by permission of the Society of Economic Paleontologists and Mineralogists.

between rib number and salinity is not strong but is similar to the results of other studies. Eisma (1965) noted a similar reduction of rib frequency with salinity in populations of modern *Cerastoderma edulus* from the Netherlands. He interpreted this relationship as being due to the increasing undersaturation of the low-salinity seawater with respect to $CaCO_3$.

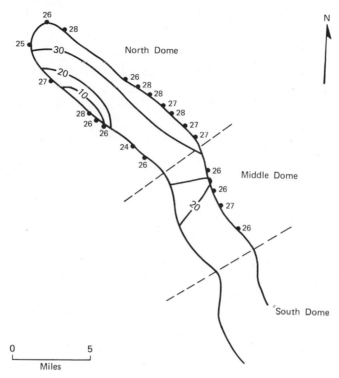

Figure 8.24 Mean rib frequency (ribs per valve) in *Anadara* from the *Pecten* zone of the Kettleman Hills Neogene. Numbered lines are salinity isopleths in per mille as determined from oxygen isotope analysis (see Chapter 3). From Alexander (1974), Journal of Paleontology, *48:* 641, by permission of Society of Economic Paleontologists and Mineralogists.

9

Ecosystems and Communities in Paleoecology

Much of the material of paleoecology and its interpretation is concerned with single taxa and individuals. Examples of this comprise the bulk of the preceding part of this book. One example would be the analysis of the chemistry, microstructure, or morphology of a fossil as these attributes were determined by the particular environment in which the fossilized organism lived. Other examples would be the study of population characteristics of a species as they reflect the life history and post-mortem preservation, or the method of taxonomic uniformitarianism, in which the tolerances and preferences of living organisms are transferred to fossil representatives of the same taxon in order to determine paleoenvironmental conditions.

These techniques exemplify *autecology,* in which the analysis focuses on the characteristics of the individual organism or population. The contrasting approach, of *synecology,* focuses on describing, understanding, and interpreting organisms in the context of the other organisms with which they coexisted. Synecology, can take place at two levels: (1) study of the organism in relation to associated organisms, emphasizing interactions, interdependencies, and coevolution within the framework of a common environment; and (2) analysis of the community—the assemblage of co-occurring organisms. Community analysis, the topic of this chapter, depends on recognizing and interpreting community parameters as they are understood by ecologists, focusing specifically on those parameters as they are preserved in the fossil record and thus are potentially useful in the reconstruction of ancient environments.

The motivation for community paleoecology has been the early and continuing recognition that the world is subdivided geographically into biologically defined units that range in size from broad regions encompassing several continents or large parts of oceans to areas so small that they can be viewed at a single glance. Regardless of the scale, each unit is characterized (1) by compositional and structural attributes resulting from the nonrandom aggregation of taxa, (2) by an internal homogeneity such that the biotic differences between samples or localities within the unit are less than between samples from different units, (3) by boundaries that can readily be defined, although they may be gradational and difficult to locate in detail, and (4) by temporal persistence and geographical recurrence. These characteristics are meaningful, although expressed qualitatively; they can be developed more precisely and quantitatively.

Ecologic units of very different sizes have been described and can be arranged in a nested hierarchy within an area, with those at each level encompassing several of the next lower level. The broader, regional units, such as *realms, biomes,* and *provinces,* determined by historical events and gross features of the environment are reviewed in Chapter 10. In the community, at the other end of the scale, interactions between species also appear to play an important role in establishing the unit.

The community has been an exceedingly popular topic of study in ecology, as reflected in a comprehensive literature, including several recent symposium volumes (Cody and Diamond, 1975; Strong et al., 1984). However, the biological foundation necessary for the paleoecologic analysis of communities is not well established. As an example, a consensus among ecologists themselves is lacking as to the fundamental phenomena that determine the nature of a community—for instance, the relative importance of biological parameters such as competition versus the physical environment in determining community composition and structure. In addition, many of the ecosystem concepts and models developed by the ecologist appear to be of limited value in paleoenvironmental reconstruction because the necessary community characteristics are generally not preserved in the fossil record. However, the incorporation of ecologic concepts and methods into paleoecology is a slow and continuing process, and community concepts that presently appear to be of little use in the reconstruction of ancient environments may become useful as our knowledge increases. In this chapter we focus first on the nature of the community, and then examine how it can be interpreted to tell about ancient environments and ecosystem characteristics.

THE COMMUNITY CONCEPT

The description by Möbius (1877) of the biocoenosis of a North Sea oyster bank provides a valuable starting point in the discussion of community concepts. Möbius was trying to understand the fluctuations in productivity of the oyster banks. His significant contribution was to recognize that it was necessary to consider the total biota and the physical environment as a unit (the *ecosystem* in present terminology) because the diverse elements of the ecosystem interact strongly and in complex ways. As any one element of the environment changes, other elements, such as the abundance of a certain species, will change in response, causing still other aspects of the ecosystem, both physical and biological, to be modified. Despite such short-term fluctuations, however, the gross aspect of the ecosystem remains constant because of the long-term stability of the physical environment and because of interactions of the taxa that are present.

Inherent in this community concept is the notion that organism interactions are critical in determining the composition and structure of the community at any instant, as well as the changes that take place during community succession. The extreme expression of this is the notion of a community as a superorganism, with the component species being analogous to the different organs that all interact for the benefit of the body (Clements and Shelford, 1939). The alternative extreme notion is that communities do not exist as real entities but are only statistical constructs devised by human beings to subdivide the biological continuum—that organisms that characteristically co-occur are associated primarily by common physical-environmental tolerances and preferences rather than by biological interdependencies, and lateral biotic changes are gradational rather than sharp discontinuities bounding distinct and internally homogeneous communities (Muller, 1958). Both of these points of view have been championed in numerous papers during the past 50 years. As in most controversies, examples of either extreme point of view may be cited, but intermediate examples are even more common, and the modern consensus tends to avoid either extreme (Fager, 1963).

A community, as defined by the characteristics enumerated at the beginning of the chapter, consists of a group of organisms occurring together, with distinctive taxonomic composition and structure, with definite geographic distribution, and recurring in space and time. This is a very general and pragmatic definition. It avoids the notion of Clements and Shelford (1939) of the community as a superorganism, and it recognizes the reality of biotic patterns that seem to be more than just the random association of organisms due to overlapping ranges.

Several specific terms that have been used in paleoecology are useful for more precise community definitions. The toal biotic component of the ecosystem is the *biocoenosis*. The biocoenosis is seldom studied in ecology because of the difficulties inherent in so comprehensive an analysis; it cannot be described or analyzed in paleoecology because many, if not most of the original taxa are not preserved in the fossil assemblage. Community analysis in paleoecology is based on only one or a few taxonomic groups, such as foraminifers, brachiopods, shelled benthic macroinvertebrates, or terrestrial vertebrates, because these are all that are preserved. This data base, although only a small part of the biocoenosis, is no more incomplete than that in ecology, in which community analysis is also generally based on only one or a few taxonomic groups. If the paleocommunity analysis of a marine fauna is based on the total preservable part of the biocoenosis, comprising, for example, the shelled benthic macroinvertebrates, such a partial representation of the biocoenosis has been named an *organism community* (Newell et al., 1959). The organism community is defined on the basis of its biotic characteristics. The distinction between biocoenosis and organism community is important because it emphasizes the difference in information content and therefore the potential difference in completeness and precision of interpretation. By either method of definition, the community is based on the organisms and their distributions.

Communities may also be defined on the basis of the physical environment—*habitat communities*. Without knowing about the biota, we may recognize habitats such as bay or ocean, lake or stream, shelf or abyss, and then describe as habitat communities the biota of each (Newell et al., 1959). Paleontologic examples would also be the "redbed community" or the "black-shale community." These are valuable as generalized descriptive entities. They are conceptually deficient, however, because they are based on the assumption that observers themselves can recognize the significant environmental gradients and tolerances that control the distributions of the organisms and communities. Because the biota reflects the complex and multidimensional environment as it is rather then how we imagine it might be, community recognition based on the organisms themselves results in more real and precise community boundaries and characteristics than if based on postulated habitats.

In paleoenvironmental studies, the theoretical question of environment versus biologic interaction is emphasized if we express it in terms of the equation

$$\text{community characteristic} = f(\text{environment, biologic interaction})$$

It is evident that we need to know the interaction term in order to determine the ancient environment from community characteristics. This is largely impossible, however, because most of the biocoenosis, and therefore most of the interactions, are not preserved in the fossil record. In fact, the interaction term is generally not discussed and is assumed not to be significant in most paleoenvironmental reconstructions based on community analysis. In contrast, however, much of the literature on the potential environmental impact of the extinction of endangered species has considered that interactions are a major term in the equation—that extinction of only a few species might so upset community structure that the ecosystem would become unstable and collapse. As another example of the potential significance of the interaction term, the roles of predation and competition relative to that of the physical environment in determining community are hotly debated in ecology. By and large, the application of the community concept in paleoenvironmental reconstruction has been on a very meager theoretical foundation. The use of the community has been very pragmatic without being concerned about whether community boundaries are abrupt or gradational and whether a community is a real, integrated structure or only the congregation of overlapping ranges of individual taxa, perhaps bounded by locally somewhat steeper environmental gradients (Hoffman, 1979).

When the term *community* is used in paleoecology it should be clearly defined because of the wide range of possible meanings that have been applied to it. The basic paleontologic unit, the *fossil assemblage*, clearly is quite different from the original biocoenosis because of the normally meager preservation of the original biota. Taphonomic analysis of the fossil assemblage may improve our knowledge of the biocoenosis but generally cannot go far toward reconstructing it in any detail (Stanton and Nelson, 1980). If the assemblage is analyzed as an organism community consisting of the skeleton-bearing and readily preserved organisms in the original biocoenosis, it is comparable to the community studied by the ecologist that encompasses only one taxonomic group. The ecological study of the skeleton-bearing organisms in modern coral-reef communities is really very similar to the paleoecological study of reef communities in the fossil record. As the community concept has been carried over from ecology to paleoecology, the term *community* has become fashionable, but may be nothing more than a synonym for *assemblage*. Ideally, community analysis in paleoecology stimulates visualization of the fossils as part of an ecosystem and spells out the relations among fossil assemblage, organism community, and biocoenosis.

The community has been used in paleoecology to study two types of problems. One is paleoenvironmental reconstruction, the major concern of this book. The second is the evolution during earth history of biological organization or ecosystem structure—the interactions among the organisms in the community, of community structure, and how these phenomena guide the evolution of individual species.

Two distinct community attributes have been most studied and used in paleoecology. One is the taxonomic composition, which has generally been in terms of presence or absence of taxa, but ideally also considers relative abundances and population dynamics of the component taxa. The other attribute is community structure as expressed by diversity and dominance on the one hand and trophic relationships on the other. Analysis of the first aspect is analogous to the taxonomic uniformitarian method in autecology. Analysis of structural characteristics independent of the taxonomic composition is comparable to the analysis in autecology of characteristics such as functional morphology and population dynamics of individual taxa in that it is potentially independent of the taxonomy of the organisms being analyzed, and thus, it is hoped, less time dependent.

The term *community* has been used for entities of a range of magnitudes. In addition to the usage here, it has also been applied to geographically broader units such as depth zones (Natland, 1933), which are determined essentially by uniform environmental conditions. We believe that these two entities are different in concept as well as scale and should be distinguished in nomenclature. We would confine the term *community* to the more limited grouping, in which organism interactions may likely play an important role, and refer to the broader units as *life zones* (Watkins et al., 1973).

COMMUNITY RECOGNITION

The distribution of species is complex; the area in which each species is found may have irregular boundaries and be discontinuous. The abundance of a species may differ widely and irregularly in time and space. If all the species are considered concurrently, few will have coincident boundaries and abundances. The problem, then, in community recognition is to identify natural groupings of organisms that reflect both the environmental parameters and the biological structure of the ecosystem. Because the number of organisms may be large, the tendency, particularly in the days before large-capacity computers, was to use only the most abundant, *dominant,* species in the analysis. Now, however, even the rarest species can be included if desired in the analysis. Two basic approaches, *Q-mode* and *R-mode* analysis, have been used in community recognition. Both are valid but lead to different results and interpretations.

Q-Mode Analysis

Samples are compared on the basis of their taxonomic compositions, and those that are relatively similar are grouped together so that the average similarity value for samples within a group—homogeneity—is greater than the average similarity value between samples in different groups. *Q*-mode analysis has been the most commonly used approach in ecology and paleoecology. Petersen (1915) used it in a qualitative way in his classic study of the living biota of the Kategat. The subsequent restudy of his original data by computer-based *Q*-mode analysis (Stephenson et al., 1972) provided an excellent comparison of the results of community recognition before and after the advent of computers. *Q*-mode analysis is the approach used in quadrat sampling in plant geography, and it is the approach used by the marine biologist in observing successive grab or dredge samples brought on board ship and perceiving similarities and differences in the biotic parameters of the samples. Boundaries drawn between adjacent communities form a map of the distribution of the communities. In geologic terms, the rock unit containing a community is a *biofacies.* Any time horizon through the biofacies represents the mapped distribution of the community on the earth's surface at that time and is equivalent to the present-day distribution pattern of a community. Comparison of samples in *Q*-mode analysis is largely on the basis of taxonomic composition, but it may be based on other attributes, such as morphologic characteristics, as has been the case in the description of plant associations.

R-Mode Analysis

In R-mode analysis, taxa are grouped on the basis of their similarities in distribution among samples. Those taxa that co-occur are grouped together, whereas taxa with distributions that are mutually exclusive or that are not strongly correlated are placed in different communities.

The strength of R-mode analysis is that it emphasizes patterns of co-occurrence and of mutual exclusion among species, and thus identifies possible biological interactions within the community. Its weakness is that R-mode communities are more difficult to map because individual samples are commonly difficult to relate uniquely to the communities that have been established. A sample may contain species diagnostic of more than one R-mode group and at the same time lack some of the species considered diagnostic of any one group. Because Q-mode analysis establishes mappable biofacies that can readily be compared to sedimentary facies, it has been the more common method in paleoecology.

The communities resulting from either Q-mode or R-mode analysis are best characterized by the species that are common and have a high degree of *fidelity* (are largely restricted to a single community). Also diagnostic are the species that are less common, and may not occur in many of the samples, but still have a high degree of fidelity for a particular community. Common but ubiquitous species are nondiagnostic, although they may be functionally important within the community.

An essential criterion used in mapping biocoenoses and communities is that sample-to-sample differences (*heterogeneity*) within a community are less than those between communities. Heterogeneity within a community is the result of patchy, uneven distributions of species. The extent to which patchiness in a community is confused with between-community heterogeneity depends on sample size and density relative to patch dimension. General guidelines for designing the sampling program for community paleoecology are given by Stanton and Evans (1971) and by Harper (1977).

The preceding discussion focuses on subdividing reality into discrete units. The fact that such units have been widely recognized by botanists, ecologists, and paleontologists indicates that this is the general nature of things. Two contrary arguments can be made to this, however. The first is that it is human nature to subdivide and pigeonhole—that perhaps many of the boundaries are forced. The second is that commonly used statistical techniques such as cluster analysis compel samples to be segregated into discrete groups. In contrast, if an environmental gradient was smooth, with no sharp fluctuations, and if the distribution of organisms along that gradient was largely determined by physical aspects of the environment rather than by biological interactions that would cause species distributions to coincide, the biota at the ends of the gradient might be quite different, but without any significant boundaries subdividing the biotic transition into discrete units. Gradient analysis is the appropriate approach in analyzing data of this sort. Hennebert and Lees (1985) provide an introduction to this in the context of a paleontologic problem. Underwood (1978) discusses the problem with more mathematical rigor.

TAXONOMIC-UNIFORMITARIAN ANALYSIS

The taxonomic-uniformitarian approach in paleoecology has historically been the major method of determining ancient environmental conditions. The fundamental steps in the method, although not always laid out in this way, are: (1) list the fossil assemblage; (2) list the comparable living taxon for each of the fossil taxa; (3) list for each environmental parameter, such as depth or salinity, the tolerances of each of the modern taxa; and (4) assume that the overlap in tolerances of the assemblage of modern taxa defines the probable range of values for that parameter for the fossil assemblage (Fig. 9.1). The method relies on the substantive uniformitarian assumption that an environmental tolerance for a fossil taxon is like that of the closest living relative. This becomes increasingly dubious with geologic age and with decreasing degree of relationship between the fossil community and the modern analog. In addition, the method

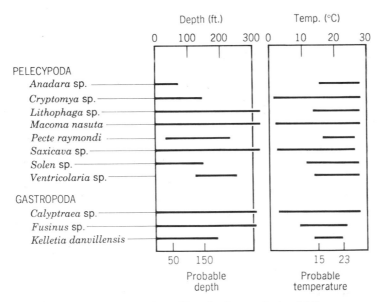

Figure 9.1   Example of taxonomic uniformitarian analysis of Miocene assemblage, using known depth and temperature ranges of living analogues to estimate depth and temperature of Miocene habitat.

depends on the abundance and precision of environmental data available for the modern analogs—data that are generally sparse. For example, temperature, salinity, and water depth are the most basic parameters of the marine environment, but the limiting values of these for most modern species are not known. Instead of actual established data, tolerances are generally estimated from limits of geographical distributions. Other potentially significant parameters, such as the ranges of variability of temperature, salinity, and depth, are seldom available, and very little is known as well about the effect of one parameter on the tolerance of an organism to another parameter.

The use of the community rather than the species as the entity with which to transfer modern environmental information into the geologic record results in a more integrated and comprehensive interpretation. Instead of separating the fossil assemblage into component taxa and estimating the value of each individual environmental parameter for the assemblage from the overlap of the values for each of the modern related taxa, the assemblage as a whole is compared with modern communities, and the modern environmental conditions of the most comparable community are applied to the fossil assemblage. The uniformitarian assumption imposes the same limitations as when the individual taxa are used as entities for transferring environmental information. However, the advantage of the community approach is that when a modern community has been identified as most similar to the fossil assemblage, it may be possible to determine more readily the parameters of the modern environment than of the individual taxa. The significant environmental parameters that are causing the differentiation of the modern biota into communities can be determined and this knowledge can then be applied to the fossil assemblage to provide a broader and more comprehensive reconstruction of the paleoenvironment.

Using the community, as defined and compared by its taxonomic composition, as an information transfer unit is most effective in the Neogene. With increasing age difference between modern and fossil communities, taxonomic similarity decreases because of evolution, extinction, and changes in range. Thus the information that can be transferred by this method becomes increasingly generalized, potentially to the point of being trivial. Consequently, a taxonomy-based analysis of ancient communities becomes largely internalized, comparing the adjacent communities with one another, with sedimentologic characteristics, and with inferred paleogeographic differences.

COMMUNITY STRUCTURE ANALYSIS

The alternative approach, which avoids somewhat the problem of decreasing taxonomic similarity with increasing geologic age, is to analyze structural characteristics of the community that are determined by the environment, that are independent of the taxonomic composition of the community and that are thus relatively independent of age. The structural characteristics that have been most commonly considered are number of taxa, or richness; their relative abundances; and their functional and trophic roles in the community. This approach is based on the observation that structural characteristics in different communities from similar modern environments are similar. This is true even though the communities may be from different parts of the world and may be taxonomically very different, at least at the species or genus level. This leads to the working hypothesis that each environment requires certain physiological attributes and life histories at the level of the individual taxa, and consequently, certain structural characteristics at the community level. In other words, a certain play requires a definite cast of characters, but the actors in those interacting roles are not limited to specific people. The structural approach is readily carried out in the study of Cenozoic communities, in which direct comparison with the structure of similar modern communities is possible. In the study of more ancient communities, comparison of the fossil community with modern communities is less precise because of the possibility of evolution of the community structure itself. In this case, gradients and relative differences in structural measures are most useful because absolute values of the measures may have changed as individual taxa and the community evolved. Three principal community characteristics have been used in paleoenvironmental reconstruction. One is *diversity* and the related measures, *dominance* and *equitability*. the second is *trophic structure*, and the energy flow through the community. The third is the functional approach of guild analysis.

DIVERSITY TERMINOLOGY

Patterns of diversity among living organisms have been intensively studied by ecologists, and numerous diversity measures and their physical and biological determinants have been proposed. Thus a strong ecological background is available for the paleoecologist in measuring and interpreting diversity in the fossil record. The compilation by Magurran (1988) is a useful reference to diversity. In addition to descriptive and interpretive material, it contains worked examples of different diversity indices. Because this book provides a comprehensive bibliography, many of the references important to the development of the concept of diversity are omitted in the following discussion.

The concept of diversity encompasses both the number and proportions of taxa. Both *diversity* and *species diversity* have been used to denote the number of taxa or a combination of the number of taxa and their proportions. Because of this lack of uniformity in usage, the term *diversity* is suitable for general discussion, but the specific concept intended must be stated when a more precise meaning is desired. The term *richness* is to be used when referring simply to the number of taxa.

The relative abundance or importance of taxa within a sample has been described by terms such as *dominance, relative importance, evenness,* and *equitability.* Evenness and *equitability* describe the degree to which the taxa in the community are equally abundant. Lloyd and Ghelardi (1964) suggested that evenness be expressed in absolute numerical terms, whereas equitability should be in relation to some hypothetical abundance distribution. This distinction has generally not been made, and, following Magurran (1988), evenness will be used here. *Dominance* is the inverse of evenness. Diversity indices that combine richness and evenness, that is, that are based on the proportional abundances of species, have been referred to simply as *diversity* or as *dominance diversity*. These diversity indices are nonparametric because their calculation does not depend on assumptions about an underlying distribution model to which the species abundance distributions might conform. To avoid the confusion resulting from the use of diversity in this way, Peet (1974) terms these *heterogeneity indices*. The term *dominance diversity* causes even less confusion, however, and is recommended.

Calculations of evenness and dominance diversity are normally based on the number of individuals per taxon. This implies that all individuals of a taxon are ecologically uniform and that all taxa are equally distinct from one another and occupy niches of equal size. Because these assumptions are clearly not valid in detail, units based on other attributes, such as productivity or biomass, have been proposed as alternatives to numerical abundance of taxa. These are more difficult to calculate, but they will be discussed later in this chapter because they have several distinct advantages.

DIVERSITY MEASURES

Publications by Sanders (1968), Whittaker (1972), Hill (1973), Peet (1974), and Magurran (1988) provide an introduction to the extensive literature on the measurement of diversity and on the indices that have been proposed, their properties, and their relative merits. One major distinction in diversity indices is between those that are, or are not, based on some frequency distribution pattern that is assumed to be generally valid in nature. For example, α, proposed by Fisher et al., (1943), is based on a hypergeometic frequency distribution of the number of individuals of species occurring together. They have shown that this distribution is true in many instances (the relative abundance of moths caught in a light trap, the frequency distribution of fleas on the heads of prison inmates, etc.). This may not be a universal relationship, however, and in fact the interesting and general question when comparing species abundance distributions of fossil data with the various models that have been proposed is whether the frequency distribution function is different in different environments, or under different taphonomic situations, or whether it has changed through earth history.

Diversity indices can be grouped into those that measure richness, those that measure a combination of richness and evenness (dominance diversity), and those that measure evenness. In measuring diversity, the sample size and the taxonomic level to be used (whether species, genus, etc.) depend on the

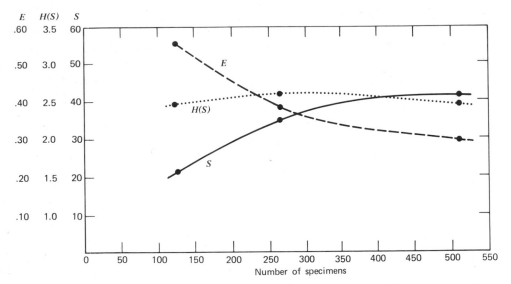

Figure 9.2 Relation between sample size and diversity and equitability measures. *S* richness; *H(S)*, entropy; *E*, equitability as measured by relative entropy. Curves based on a foraminiferal sample from the continental shelf off North Carolina. After Gibson and Buzas (1973), Geological Society of America Bulletin.

nature of the problem. Because sample size and taxonomic level strongly influence the calculated value, they must be taken into account when comparing values from different studies or from samples of different size within a study. Figure 9.2 illustrates the dependence of several diversity indices on sample size.

To yield significant interpretive results, a diversity index should be ecologically sound and statistically rigorous. However, these two requirements are difficult to satisfy simultaneously. It is particularly important to keep in mind the effect on diversity of the partial and biased preservation due to taphonomic processes. This is a problem that is largely peculiar to the paleontologic use of this ecologic approach. The taphonomic bias cannot be eliminated, but at least it should be as close as possible to uniform among samples being compared. Because of the generally poor but variable preservation of the original biocoenosis, differences in diversity between samples may be more the result of preservation differences than of original environmental differences (Lasker, 1976). Several of the diversity measures that have been used most commonly in paleoecology are described briefly below.

Richness

Richness, *S*, the number of species or taxa, is the simplest measure of diversity. However, it is strongly dependent on sample size. Several indices have been proposed that will minimize sample size bias. An example is *Margalef's diversity index,*

$$D_{MG} = (S\text{-}1)/\ln N$$

where S is the number of species and N is the number of individuals. Although there is a growing tendency to use natural logarithms in diversity calculations, any log base can be used, with of course the corresponding effect on the calculated value of the index. The possibility of using different log bases means that if diversity values are compared, they must all be calculated in the same way. Margalef's index is based on the commonly observed fact that if the cumulative number of species is plotted against cumulative number of individuals in the sample, the curve is approximately logarithmic. The relation between sample size and richness is determined by two considerations: (1) Common species will be present in small samples and as sample size increases, progressively rarer species tend to be recorded, but it requires an ever-larger increase in sample size to add each new species to the list; and (2) if the larger sample size means that a larger volume or area of the original environment is included in the sample, heterogeneities in environment and biota are more likely to be incorporated in the sample. The similar relation of richness to sample size that has been recognized in island biogeography is

$$S = CA^z,$$

in which A is sampling area, C depends on the taxon being counted, and the value of z has been determined empirically to lie within the range of 0.20 to 0.35 (MacArthur and Wilson, 1967).

The *rarefaction* method (Sanders, 1968) attempts to compensate for the effect of sample size by graphically determining for a sample what the richness would be if the sample size had been smaller. In the example of Fig. 9.3, the total number of individuals and taxa in each sample are indicated by the point at the right-hand end of each line. Because of differences in sample sizes, however, these S values are not easily compared. If the relative proportions of taxa in each sample are known, however, and assuming that these proportions would be constant in samples of smaller sizes for each locality, the rarefaction curve is easily constructed. Then the richness of samples of different size can be compared from values on the respective rarefaction curves at any common sample size. Hurlbert (1971) has pointed out that the method proposed by Sanders is not mathematically correct, so that richness is slightly but generally overestimated in previously published curves.

Dominance Diversity

The index most widely used in paleontology has been entropy, based on information theory. It is discussed here as an example of this class of index.

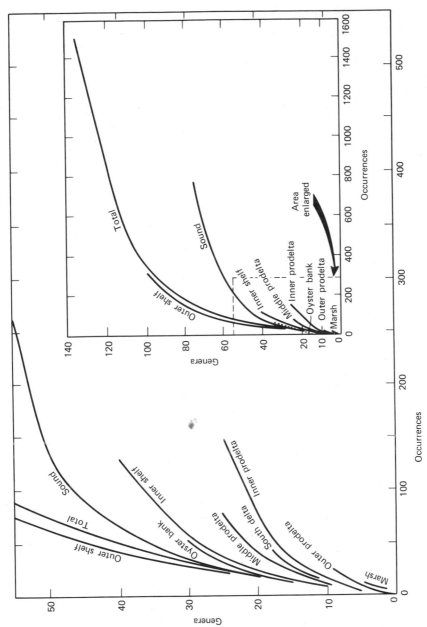

Figure 9.3 Rarefaction curves for total living fauna and modern communities on south and east of Mississippi Delta. Curves are based on shelled macroinvertebrates. Occurrences are the summation for a community of the number of samples from which each genus was collected. See Fig. 9.6 for location of communities. After Stanton and Evans (1972). Journal of Paleontology. Society of Economic Paleontologists and Mineralogists.

Entropy is a measure of the uncertainty in predicting the identity of a randomly selected individual from a community. The *Shannon index*

$$H = -\sum_{i=1}^{s} P_i \ln P_i$$

is most commonly used because it is easily calculated if n_i/N, the proportion of the ith species in the sample, is substituted for P_i, the proportion of that species in the community. The use of sample values for community values introduces a bias into the results, but this is small considering the inherent inaccuracies of the paleontologic data. Again, the natural logarithm is used, although $\log_2$ would perhaps formally be more correct in accordance with the information theory origin of the index. H is determined both by richness and by evenness, and consequently is more difficult to interpret than indices based on richness along. H is greatest when evenness is highest—for a given number of species the maximum value of H is ln S, when all the species in the sample are equally abundant.

  Simpson's index, based on the probability that two individuals drawn randomly from a sample are of the same species, is

$$\sum_{i=1}^{s} P_i^2$$

Because the value of this index varies inversely with dominance diversity, other forms, such as its reciprocal, have been proposed. Neither the variants nor the basic index have been widely used in paleontology.

Evenness

The most commonly used *index of evenness* is

$$E = \frac{H}{H_{max}}$$

the ratio of the actual entropy to the entropy of the community if all the species were equally abundant (ln S). The value of E ranges from 1 when all species are equally abundant to zero as dominance increases and evenness decreases. Other similar ratios based on entropy indices have been described and evaluated by Sheldon (1969).

DIVERSITY CAUSES AND PATTERNS

A valuable approach to the understanding of diversity is through consideration of the environment as a multidimensional volume (the *environmental*

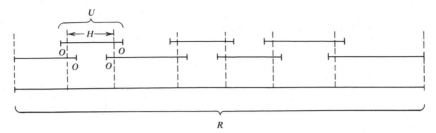

Figure 9.4 Relation of resource and niche characteristics. See text for description. After MacArthur (1972, figure 7-1, p. 171), from Geographical ecology: Patterns in the distribution of species, copyright Harper & Row, Publishers, Inc.; reprinted by permission of the publisher.

hyperspace) containing the total set of habitable conditions. The axes of the hyperspace are environmental parameters; those for physical parameters such as temperature or temperature variability are readily visualized, but those for biologic parameters such as food resources are also necessary to describe the ecosystem comprehensively in hypervolume terms. The environmental hyperspace, then, contains as many compartments as there are species present, and each compartment represents one species' niche. The shape of each compartment and its dimensions on the hypervolume axes are determined by the environmental tolerances for each species. Richness is equivalent to the number of niches (or hypervolumes) into which the hyperspace is divided. The number of niches is determined by (1) the size of the hyperspace as represented by the maximum dimensions of all the axes, (2) the size of each of the niches or hypervolumes, and (3) the extent to which the niches overlap. This can be visualized (Fig. 9.4; MacArthur, 1972) for a single axis (an environmental parameter or resource) as

$$S = \frac{R}{U} \left(1 + 2\frac{O}{H} \right)$$

where $S =$ richness or number of species
$ R =$ axis length or dimension of the resource or parameter (e.g., the total temperature range) within which life is possible
$ U =$ mean total niche width
$ O =$ mean overlap
$ H =$ approximately the mean niche width

In other words, richness is proportional to resource breadth and niche overlap and is inversely proportional to average niche size. This equation helps us to focus on the essential causes of diversity differences and gradients, both geographically and during geologic time. Niche width and resource breadth are most easily taken into account; niche overlap is not readily determined for living organisms and has been little studied in paleoecology. Evenness in the context of the equation is a measure of how the different species subdivide the energy flux through the ecosystem.

Diversity varies in both time and space. It may vary on a relatively short time scale during the annual seasonal cycle, or on a successively longer time scale of community succession, transgressive–regressive cycles, eustatic cycles, and to the even-longer-term consequences of evolution. In this chapter we concentrate on spatial variations in diversity.

In the modern world, richness of the total biota and richness within individual higher taxonomic units decreases from the equator toward the poles (Stehli, 1968). This latitudinal gradient is evident within each of the environments that might be studied, whether marine or terrestrial, or a subdivision of these such as marine shallow shelf or rocky shore (Fig. 9.5). At regional and smaller scales, diversity gradients are ubiquitous and measurable at all taxonomic levels. Both richness and evenness of marine invertebrates decrease along a transect from the stable environmental conditions of the open shallow shelf to the more rigorous and variable conditions nearshore and in estuaries and lagoons. Typical examples of this trend are from the Baltic Sea for both vertebrates and invertebrates (Segerstråle, 1957), from the vicinity of the Mississippi Delta for benthic shelled macroinvertebrates (Figs. 9.3 and 9.6; Stanton and Evans, 1971), from numerous estuaries and lagoons for invertebrates (Emery and Stevenson, 1957), and from the Atlantic coast in the Cape Hatteras area for ostracodes (Hazel, 1975). Richness also increases from the shelf into the deeper water of the ocean basins for benthic macroinvertebrates (Hessler and Sanders, 1967) and foraminifers (Fig. 9.7; Gibson and Buzas, 1973).

The understanding of modern diversity gradients can be approached theoretically with the preceding equation, in which richness is shown to be proportional to resource dimension and inversely proportional to average niche width. In less abstract terms, the possible causes of modern diversity patterns have been comprehensively reviewed by Sanders (1968), MacArthur (1972), Valentine (1973a), Pianka (1978), and Magurran (1988). The wide range of explanations indicates that there is no simple and universal explanation, but that diversity at any site is determined by a complex interplay of historical, biological, and physical factors within the ecosystem. The factors that appear to be of prime importance in paleontologic studies are time, stability, and resource.

Time

Diversity should increase as organisms evolve, accommodating themselves within the ecosystem to the physical environment and to other organisms. Evolution should have two effects: Biological interactions, in providing new opportunities (new ways to make a living), add axes to the hypervolume or extend existing axes, thus increasing the resource dimension. The other effect

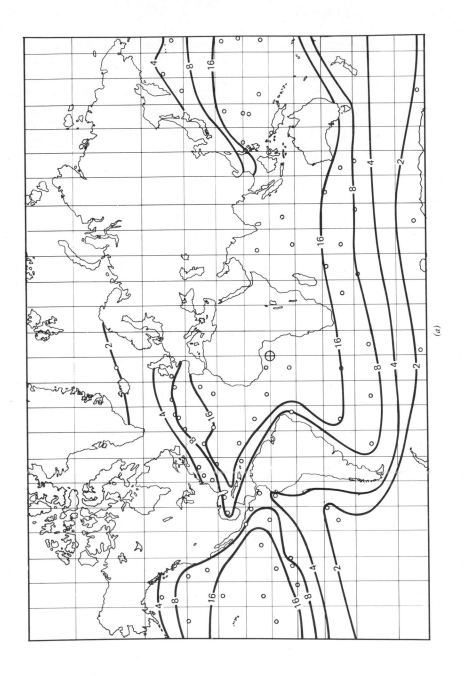

(a)

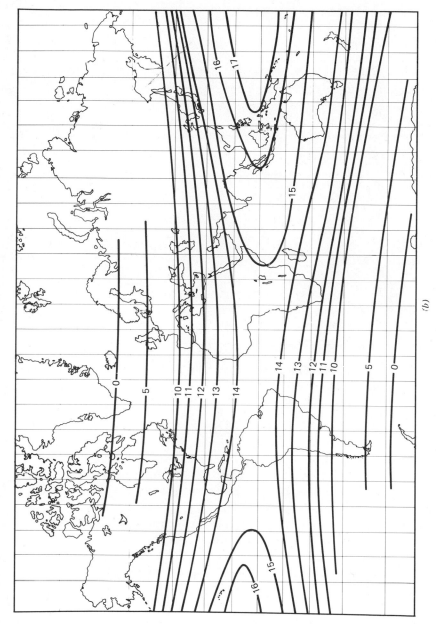

Figure 9.5 Latitudinal pattern of deversity (number of species) of recent planktonic foraminifers; (*a*) contoured data points; (*b*) quadratic surface fitted to data points in (*a*). After Stehli and Helsley (1963), copyright 1963 by the American Association for the Advancement of Science. Science *142*:1057–1059, figures 1 and 2.

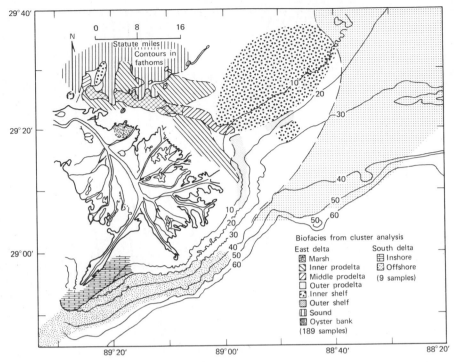

Figure 9.6 Biofacies adjacent to the Mississippi Delta, determined by Q-mode cluster analysis of 189 samples east of the delta (Parker, 1956) and 9 samples south of the delta. After Stanton and Evans (1972). Journal of Paleontology, Society of Economic Paleontologists and Mineralogists.

of evolution is increased specialization (i.e., niche subdivision or decreased niche size) (Hutchinson, 1959). During earth history, for example, hyperspace size has increased when biological advances opened up first terrestrial and later aerial habitats to be colonized.

Diversity change within specific habitats is not as easily established. Tabulations of taxa occurring on marine soft bottoms may have been fairly stable through geologic time. On the other hand, diversity has clearly increased in the reef ecosystem since scleractinian corals first appeared as reef dominants in the Triassic.

Stability

Stability plays an important role in determining diversity in three ways:

1. It determines the amount of geologic time available during which evolution can proceed in any particular environment toward the attainment of the optimum degree of specialization and thus of niche subdivision and diversifica-

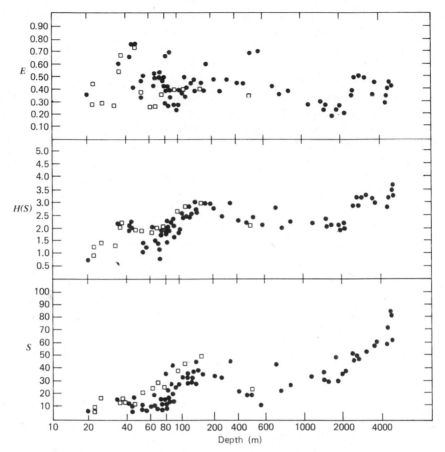

Figure 9.7 Graph of S, $H(S)$, and E versus depth for foraminifera from stations in the Cape Cod to Maryland area of the continental shelf off the eastern United States. Circles are stations north of Maryland, open squares are stations off Maryland. After Gibson and Buzas (1973), Geological Society of America Bulletin.

tion. A short-lived, unstable environment will not allow such specialization. Because specialized organisms cannot survive conditions outside their narrow tolerance ranges, in an unstable environment they tend to be generalists, with large niches. This is commonly referred to as r-selection—that is, the best-adapted organisms are those with large reproductive potential and rapid growth and maturation. In contrast, the adaptation in stable environments of specialized organisms with lower reproductive potential and slower growth rates is referred to as K-selection, in reference to the control on the population size by the carrying capacity of the environment.

2. Stability determines the extent to which succession progresses, with generally increasing diversity, toward the climax stage. In these cases of evolu-

tion and succession, diversity is determined by a long-term or a short-term process, and the time span during which each process can continue is controlled by the stability of environmental conditions. In the first case, we are dealing with the evolutionary process during perhaps millions of years; in the second case, the successional process may last only a few months or even less.

3. Stability functions as a resource or dimension of the hyperspace. Although the maximum or average value of an environmental parameter, such as temperature, is commonly thought of when specifying conditions for life, the fluctuations in that parameter diurnally, seasonally, or during longer periods of time and whether periodic or erratic may actually be much more important as a limiting factor for the presence of a species or for the level of biotic activity in the ecosystem. In general, equable, stable conditions are optimal for most organisms, whereas few organisms are adapted for highly variable conditions. This is true if the environmental fluctuations are periodic. It is even more true if the fluctuations are unpredictable.

Resource

Any characteristic of the environment, such as space, salinity, or food, for which organisms compete, or that beyond a certain value, limit their existence, represent an axis of the hyperspace and are considered as a resource. Resources determine diversity because they establish the total volume of the hyperspace that potentially may be utilized. A narrow range for an environmental parameter at a locality is equivalent to little heterogeneity in the environment and thus to low diversity. This is most readily apparent to us from our experience of living in terrestrial environments. A grassy plain will contain a certain limited complement of organisms. The addition of a clump of trees greatly expands the resources along several axes of the hypervolume—most evidently in the ranges of temperature and sunlight and in types of shelter and food for animals—and correspondingly increases both plant and animal diversity.

Resource and stability are of primary importance in understanding and applying diversity in paleoecology. The effect of these two parameters on diversity is summarized in Fig. 9.8 and analyzed in more detail by Valentine (1971b). As a summary example, the combined effects of time, resource, and stability can be illustrated in the low diversity of a desert. Deserts are typically characterized by low and sporadic precipitation and by temperature that is typically high during the summer daytime, but that fluctuates over a wide range diurnally and seasonally. After an infrequent but perhaps torrential rain, dormant seeds will germinate and plant life will flourish, but only for a short time until arid conditions return. The low environmental stability results in fluctuations in primary productivity and thus limits time for succession to lead to a complex and diverse community structure.

Diversity is also low in deserts because of limited resources. Critical environmental conditions such as low and fluctuating moisture, high and fluctuat-

Resources	Stable	Unstable
Poor	**Box 1** K—selection Small stable population, highly specialized Highest diversity	**Box 2** Compromise selection Moderate population size and specialization minimizing fluctuations Low diversity
Rich	**Box 3** r—selection and K—selection Mixed population size and specialization High diversity	**Box 4** r—selection Fluctuating population, often large, unspecialized Lowest diversity

Figure 9.8 Evolutionary process, population characteristics, and diversity (richness) in communities formed under different conditions of resource stability and quantity. After Valentine (1971b), Lethaia, *4:* 57.

ing temperature, and low and fluctuating primary productivity at the base of the food pyramid are each represented by values on an axis of the hyperspace that fall largely outside the tolerable range for most organisms. Thus the resource values in a desert environment define a small hyperspace. Because the number of niches is generally correlated with hyperspace volume, diversity is low. In addition, because of the variable environment, the organisms that can live in the desert must either be generalists adapted to a wide range of conditions, or specialists that migrate or are dormant during intolerable conditions. Consequently, individual niches are large, subdividing the hyperspace coarsely and resulting in low diversity.

TROPHIC STRUCTURES

The way in which energy flows through a community is a unique and distinctive structural attribute of the community. It is determined by the physical parameters of the ecosystem that control the relative abundances of the organisms present, as well as the rate and time distribution of productivity and the spatial distribution of nutrients and food for the different organisms. It is determined by the biological parameters of the ecosystem: niche differences and overlaps, food requirements and preferences, and trophic interactions between organisms.

The trophic structure of a community may be described in two ways. The organisms may be categorized according to *trophic level* and arranged in an ecologic pyramid based on number of individuals, on biomass, or on energy flow. Alternatively, the food web within the community may be constructed so that the pathways of energy flow are charted. The first approach is a more general one; the second is more detailed and contains much more information but is more difficult to carry out in a paleoecologic study. Both approaches are well surveyed in the ecological literature (e.g., Pianka, 1978), and in a paleoecologic context (Scott, 1978).

The most fundamental way to describe the trophic structure of a community is in terms of *energy flow.* Energy input to the ecosystem is primarily as sunlight. This results in plant growth *(primary production)* that supports herbivores, and these, carnivores, scavengers, and parasites. Ultimately, the organic material is buried in the sediment or decomposed to the original inorganic elements. The efficiency of converting organic material at one level to organic material at the next level is on the order of 10 to 20%. This is evident as we individually try to balance our caloric intake with the weight we wish to gain or lose. It is commonly portrayed in the *ecologic pyramid* (Fig. 9.9). The width at each level reflects the decreasing energy available at successive levels and the corresponding decreasing number or biomass of herbivores and successive levels of carnivores.

If the ecosystem were thoroughly understood, energy flow could be described in detail. Such an attempt is illustrated in Fig. 9.10. A comprehensive *food web* would indicate the amounts of energy resulting in the primary production of each type of plant, the extent to which each plant was utilized by the different herbivores or was decomposed back to the original elements, and the food utilizations at the successive carnivore levels. Because such a chart would require complete knowledge of the ecosystem, however, none exists and even generalized ones of the type illustrated in Fig. 9.10 are uncommon. The food web is difficult to establish for two basic reasons. The first is that the food utilized by the different animals is difficult to determine because an organism may feed on several levels at the same time, it may feed on different levels through time, shifting its preferences with age, size, and changing food availability, or it may have very narrow food preferences that differ from locality to locality. In addition, detailed analysis is necessary to be sure that an organism is actually utilizing what it eats. Amphipods, for example, feed on plant detritus but derive nutritive value only from the microorganisms attached to and breaking down the detritus (Fenchel, 1970), and mollusks may generally feed on detritus in the same way, as indicated, for example, by the gastropod *Hydrobia* and the bivalve *Macoma* (Newell, 1965). The second reason is that it is difficult to measure energy and material flux into, within, and out of the ecosystem.

Analysis of community structure by reconstruction of the food web is diffi-

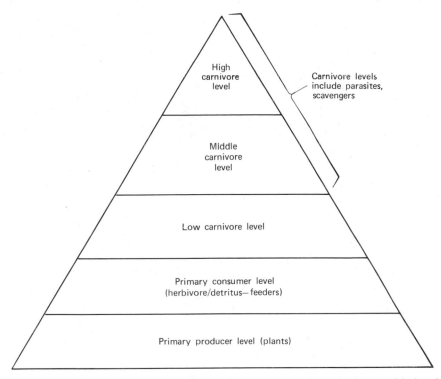

Figure 9.9 Ecologic pyramid. Decreasing width at successively higher trophic levels reflects 10 to 20% efficiency of flow of energy and corresponding decrease in number of individuals.

cult for modern communities and quickly leads to complex flowcharts with multiple pathways and feedback loops. It is even more difficult for fossil communities because much of the necessary information is not preserved. Consequently, analysis of the trophic structure of fossil communities is best approached through attributes that are more easily described and measured. Analysis in terms of levels of the trophic pyramid is one approach that has several advantages: (1) It permits grouping the fossil taxa into a few trophic-level categories and thus provides a general view of the trophic structure even though the more specific food chains are not reconstructable; and (2) construction of the ecologic pyramid from the proportions of organisms at each trophic level provides a simple means of comparison of the fossil community with other fossil and present-day marine communities. Using the proportions of organisms based on biomass rather than numbers is more informative because biomass is much less altered by the taphonomic processes, and because biomass provides more information about the energy flow and the functioning of the community as a unit (Staff et al., 1985, 1986).

An initial approach to the trophic structure of marine communities in

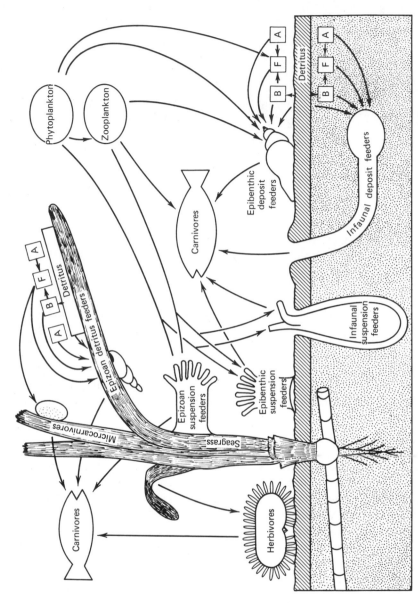

Figure 9.10 Generalized diagram of the energy flow in a seagrass community. B, bacteria; F, foraminifera and other microherbivores; A, microscopic algae. After Brasier (1975).

paleoecology was based on the proportion of deposit-feeding and suspension-feeding organisms at the primary consumer level (Walker, 1972). Turpaeva (1957) demonstrated that the primary-consumer level of a benthic community is generally dominated by organisms of one feeding type (trophic group), that a community is usually dominated by a few abundant species, and that if these dominant species are ranked by abundance, successive species are of different feeding types. Futhermore, Savilov (1957) showed that these trophic community characteristics were determined by environmental conditions. The explanation for these relationships is that the two major sources of food for primary consumers in marine communities are organic matter that is either suspended in the water or is on or in the sediment, and the relative proportions of these food resources depend primarily on water turbulence. Because particulate organic matter behaves as detrital particles, under very quiet conditions it will settle out of water onto the sea floor, whereas under more turbulent conditions it will remain in suspension or will be swept into suspension if lying on the seafloor. The texture of the substrate is determined by the same energy conditions—higher-energy conditions producing coarser sediment, and lower-energy conditions, finer sediment. The correlation between deposit feeders and finer sediment is further reinforced by the fact that dissolved organic compounds are preferentially absorbed onto clay particles, from which they can be removed by deposit-feeding organisms (Sanders, 1956; Driscoll, 1969).

The proportion of organisms feeding on food in suspension versus food on or in the substrate has been correlated with water energy in modern environments by Sanders (1956, 1968), Savilov (1957), and Driscoll and Brandon (1973). As a first approximation, water energy is correlated inversely with depth, but shallow protected and thus quiet settings also have a high proportion of deposit-feeding organisms. A preponderance of suspension-feeding organisms has also been found concentrated locally in deeper water, where internal waves impinge on the shelf edge (Savilov, 1957).

Other factors of secondary importance in affecting the relative abundances of deposit- and suspension-feeding organisms include substrate stability, which may be affected in part by bioturbation; oxygen content, which may be controlled in part by the amount of organic material settling onto the substrate; and the amount and diversity of the particulate food resources (Rhoads and Young, 1970; Rhoads et al., 1972; Aller and Dodge, 1974).

Although trophic proportions would seem to be strongly correlated with water energy, this method of analysis has proven to be of limited value for a number of reasons. The foremost is that the proportion in the living community is not well preserved in the fossil assemblage.

Reconstructing the trophic web of a fossil community is potentially useful in defining the preservational completeness of the biocoenosis by indicating levels or components of the ecologic pyramid or food chain that are apparently inadequately represented. It is also valuable in exploring the possible evolution of community structure. Much of the research in structural analysis of communities has focused on identifying and using characteristics that are relatively time independent. One characteristic that may be particularly useful is the *guild*. A guild is a group of organisms within a community that utilize a common resource. The concept was originally applied to birds (Root, 1967) and is valuable because it focuses on potential competitive interactions in the community. The concept has been applied to paleoecology in two ways. Bambach (1983) used it to categorize organisms in terms not only of food source, but also of morphology and life habits. Fagerstrom (1987) has used it to categorize reef organisms by their morphology and functional role. Although these usages of the term are different from the original definition, the guild is an interesting way to categorize community members because it provides a means of analyzing changes in communities independently of the large-scale taxonomic changes that have occurred through geologic time.

APPLICATION OF THE COMMUNITY IN PALEOENVIRONMENTAL RECONSTRUCTION

The full range of compositional and structural characteristics described above have been used in the analysis of fossil assemblages, and numerous examples are in the current literature. A small representative sampling of approaches in community paleoecology is described in this section.

Taxonomic Uniformitarian Analysis

Ancient communities have been recognized and compared with one another and with modern communities on the basis of taxonomic composition in numerous studies. The degree to which the communities are taxonomically similar is generally considered diagnostic of the degree of similarity of the paleoenvironments. The way paleocommunities are interpreted has varied widely. In many studies, biofacies are recognized and the diagnostic biota characteristic of each biofacies is referred to as a community. In some cases, the composition of the community and their spatial distribution within the stratigraphic framework, in conjunction with associated lithologic criteria, are interpreted without making any comparison with modern communities.

An example of a study of this type is the description of late Ordovician communities of the central Appalachian region by Bretsky (1969). Within more or less contemporaneous strata, Bretsky was able to define the depositional environment in general terms of bathymetry, distance from shore and deltaic areas of sediment influx, and substrate texture and composition. The three communities recognized in these strata, the *Sowerbyella—Oniella*, *Orthorhynchula—Ambonychia*, and *Lygospira—Hebertella* communities, then, define distinct tracts on the seafloor (Fig. 9.11) and, by inference, bathymetric zones (Fig. 9.11A).

Similar studies in the Lower Paleozoic (e.g., Ziegler et al., 1968; Walker, 1972; Tipper, 1975) have led to the awareness that the communities can be

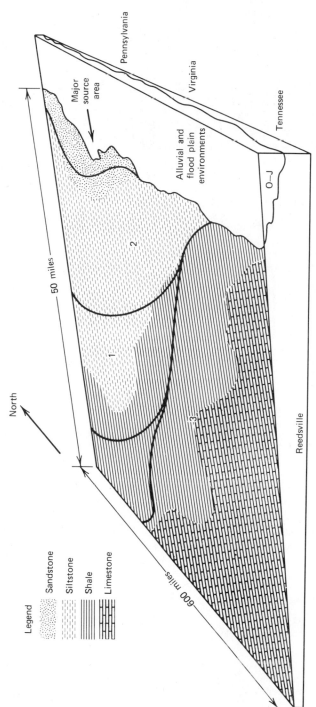

Figure 9.11 Distributions of sediments and biofacies during deposition of the Late Ordovician Oswego barrier-lagoonal deposits along the northeastern portions of the shoreline in the central Appalachians. The communities characteristic of each biofacies are (1) *Sowerbyella—Onniella* community; (2) *Orthorhynchula—Ambonychia* community; (3) *Zygospira—Herbertella* community. See Fig. 9.11A for the composition and inferred environment of each community. After Bretsky (1969), Geological Society of America Bulletin.

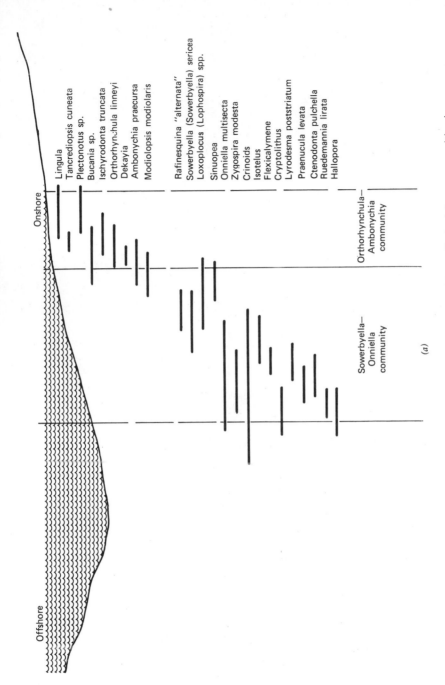

Figure 9.11A Composition and inferred environment of Late Ordovician communities in the central Appalachians. Geographic distribution of the communities is illustrated in Fig. 9.11. (*a*) Northern communities; (*b*) southern community. After Bretsky (1969). Geological Society of America Bulletin.

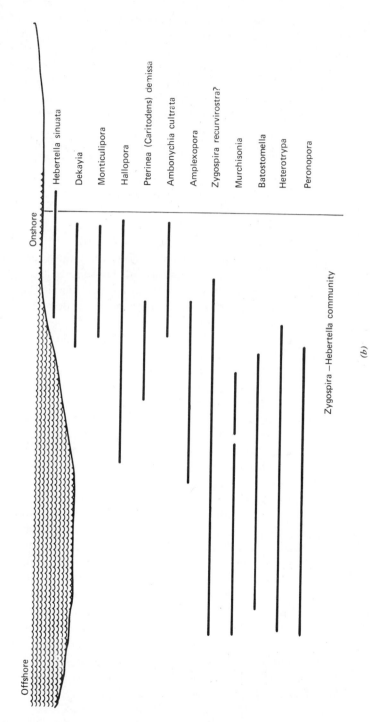

Onshore

Offshore

Hebertella sinuata

Dekayia

Monticulipora

Hallopora

Pterinea (Caritodens) demissa

Ambonychia cultrata

Amplexopora

Zygospira recurvirostra?

Murchisonia

Batostomella

Heterotrypa

Peronopora

Zygospira–Hebertella community

(b)

Figure 9.11A Continued

compared and related (1) to environmental similarities and (2) to evolution of community composition and perhaps structure (Bretsky, 1968). A comparison of Ordovician and Devonian communities of New York has shown that the communities of different age but from presumably the same environment are very similar, reflecting the ecologic control of community composition. The similarities, as indicated by the subtidal communities (Fig. 9.12), are at a high taxonomic level but reflect the presence of equivalent niches (Walker and Laporte, 1970).

Pictorial representations of communities as in Fig. 9.12 are instructive because in showing the form and life habit of the organism they provide much more information than a list of names. For this reason, they are common in community paleoecology papers, and have been gathered together, as well, for a range of environments throughout the Phanerozoic (McKerrow, 1978). They are misleading, however, because probably only a small part of a fossil assemblage was alive at any one time. Numerous comparison of present-day living communities and dead assemblages have shown that the preservable component of the community present at any particular time is less diverse than the fossil assemblage accumulating in the underlying sediment. This may be largely due to seasonal or short-term variability, and not reflect any major environmental changes. Thus it is safe to assume that a fossil assemblage contains elements of several different communities living at the site successively, and such sketches must be considered as composite and not as representations of the community at any particular time.

A Pliocene flora from the Mount Eden beds of southern California (Axelrod, 1938) is an early example in which a plant community is used as an entity to transfer information from the Recent to the past. The flora consists of 30 species representing 21 genera and 16 families (Table 9.1). From a knowledge of modern plant communities occurring in the diverse habitats of southern California, Axelrod concluded that the Pliocene flora was composed of elements derived from several different plant communities. By analogy with modern communities, the fossil communities were also diagnostic of very specific habitats determined by differences in elevation and slope exposure, which in turn controlled temperature, soil moisture, and humidity (Table 9.2). The recognition of modern plant communities analogous to components of the Pliocene flora has two significant payoffs. The first is that the mixed nature of the flora can be established. The second is that the environments of the modern analogous communities can be described in much greater detail and in terms of more parameters than would be possible from considering the individual taxa one by one. With this data base, Axelrod was able to describe the depositional setting as a low basin of shallow lakes and marshes with adjacent alluvial fans, rolling hills and plains, and nearby higher hills. Further, he was able to describe such subtle parameters of the climate as summer and winter rainfall and temperature ranges and seasonality.

The reconstruction of the Pliocene environment in the Kettleman Hills region of central California relied heavily on a comparison of Pliocene commu-

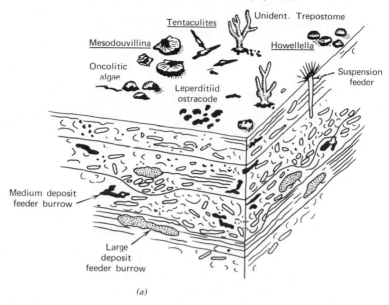

(a)

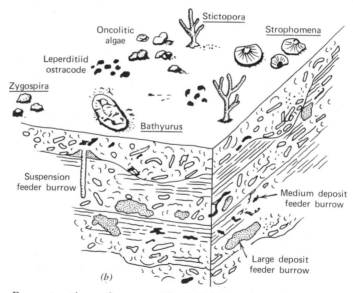

(b)

Figure 9.12 Reconstructions of communities in living position and as they accumulate as fossil assemblages; (*a*) Devonian Manlius Formation; (*b*) Ordovician Black River group. Both communities lived in a low-intertidal environment, which presumably accounts for the high similarity in composition and form, although they are of very different age. After Walker and Laporte (1970). *Journal of Paleontology.* Society of Economic Paleontologists and Mineralogists.

TABLE 9.1   Mount Eden Flora: Systematic List of Species[a]

Pteridophyta
  Equisetales
    Equisetaceae
      Equisetum sp.
Spermatophyta
  Gymnospermae
    Coniferales
      Pinaceae
        Pinus hazeni
        Pinus pieperi
        Pinus pretuberculata
        Pseudotsuga premacrocarpa
      Cupressaceae
        Cupressus preforbesii
    Gnetales
      Gnetaceae
        Ephedra sp.
  Angiospermae
    Monocotyledonae
      Pandanales
        Typhaceae
          Typha lesquereuxi
    Dicotyledonae
      Salicales
        Salicaceae
          Populus pliotremuloides
          Salix coalingensis
          Salix sp.
      Juglandales
        Juglandaceae
          Juglans beaumontii
      Fagales
        Fagaceae
          Quercus hannibali
          Quercus lakevillensis
          Quercus orindensis
          Quercus pliopalmeri

Dicotyledonae—*continued*
  Rosales
    Platanaceae
      Platanus paucidentata
    Rosaceae
      Cercocarpus cuneatus
      Prunus preandersonii
      Prunus prefremontii
    Leguminosae
      Prosopis pliocenica
  Sapindales
    Anacardiaceae
      Rhus prelaurina
    Sapindaceae
      Sapindus lamottei
  Rhamnales
    Rhamnaceae
      Ceanothus edensis
      Ceanothus sp.
  Ericales
    Ericaceae
      Arbutus sp.
      Arctostaphylos preglauca
      Arctostaphylos prepungens
  Gentianales
    Oleaceae
      Fraxinus edensis
  Asterales
    Compositae
      Lepidospartum sp.

[a]The Mount Eden flora contains 30 species, respresenting 21 genera and 16 families. Twenty-two of the species are dicotyledons, of which 7 are arborescent, 3 are normally small trees, and 12 definitely shrubby. Of the remainder 6 are conifers, and the monocotyledons and pteridophytes are both represented by single species.

TABLE 9.2   Mount Eden Flora Subdivided Into Habitat Communities[a]

Desert-border element	Savanna and woodland
Ephedra sp.	*Arbutus* sp.
Lepidospartum sp.[b]	*Juglans beaumontii*[b]
Prosopis pliocenica	*Pinus pieperi*
Prunus preandersonii	*Quercus hannibali*
Prunus prefremontii	*Quercus lakevillensis*
Quercus pliopalmeri[b]	*Quercus orindensis*
Sapindus lamottei[b]	Chaparral
Lake-border or marsh[c]	*Arctostaphylos preglauca*
Equisetum sp.	*Arctostaphylos prepungens*
Typha lesquereuxi	*Ceanothus edensis*
Riparian	*Ceanothus* sp.
Fraxinus edensis[b]	*Cercocarpus cuneatus*
Juglans beaumontii[b]	*Fraxinus edensis*[b]
Lepidospartum sp.[b]	*Quercus pliopalmeri*[b]
Platanus paucidentata	*Rhus prelaurina*
Populus pliotremuloides	Coniferous associations
Salix coalingensis	*Cupressus preforbesii*
Salix sp.	*Pinus hazeni*
Sapindus lamottei[b]	*Pinus pretuberculata*
	Pseudotsuga premacrocarpa

[a]After Axelrod (1938).

[b]May normally occur as a dominant in more than one habitat.

[c]Most of the riparian genera may also be present about the borders of lakes or marshes.

nities with modern communities (Stantion and Dodd, 1970). The molluskan communities of the San Francisco Bay area, California, were established by Q-mode cluster analysis (Fig. 9.13). Six of the eight communities derived from the cluster analysis are within the bay and are of particular value in the comparison with the Pliocene communities. The physical conditions in which each community occurred are illustrated in Fig. 9.14, and the faunal composition of each community, in Fig. 9.15. The distribution pattern of the communities is symmetrical with the bay geography. Communities labeled C,D, and E in Fig. 9.15 are concentrically arranged around the bay entrance; F, the oyster bank community, is largely restricted to the south arm of the bay; G and H are restricted to the inner part of the north arm.

Going from the bay entrance into either of the arms, the environment changes from equable and normal marine to progressively less stable because of increasing isolation from the buffering effects of the open ocean, and increasing responsiveness to the climate and adjacent land. Community diversity decreases along this environmental gradient from the outer bay through the middle and inner bay, primarily by the loss of marine taxa as the environment exceeds their tolerance limits; the total number of genera increases only slightly as species from more restricted settings are added. Thus diversity is

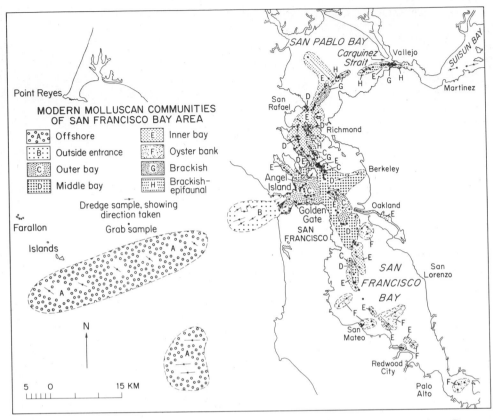

Figure 9.13 Distribution of modern molluskan communities in San Francisco Bay and the adjacent Pacific Ocean. After Stanton and Dodd (1976). Lethaia, *9*.

proportional to "marineness," not to average environmental conditions within the bay or to environmental heterogeneity. The diversity gradient is similar from the entrance going into both arms, as indicated by the distribution of community E.

Considering only the soft-bottom, infaunal component of the fauna, community C represents the basic marine open outer-bay fauna. The other communities are derived from community C and consist of those genera that are tolerant to the various stresses present in different parts of the middle and inner bay. Most of the abundant and diagnostic taxa of the inner-bay infaunal community are also present and even common near the bay entrance. Exceptions are *Mya*, which is rare in the outer bay but dominant in the inner bay, and *Solen*, which is confined to the midbay.

The gradual elimination of taxa without replacement from the bay entrance back into the inner bay is not as obvious within the epifaunal component of

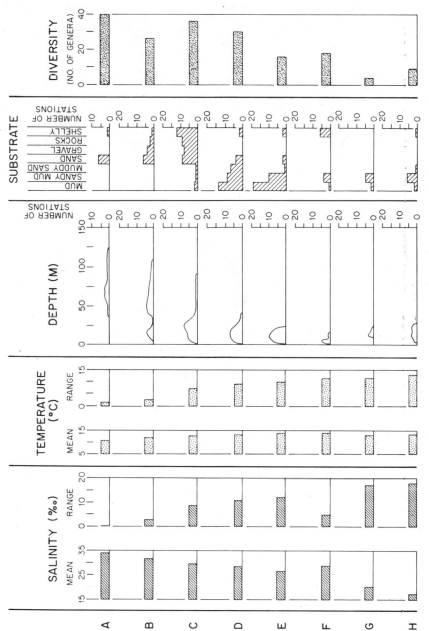

Figure 9.14 Environmental and diversity characteristics of modern molluskan communities in the San Francisco Bay area. Communities: A, offshore; B, outside entrance; C, outer bay; D, middle bay; E, inner bay; F, oyster bank; G, brackish; H, brackish-epifaunal. After Stanton and Dodd 1976). Lethaia, 9.

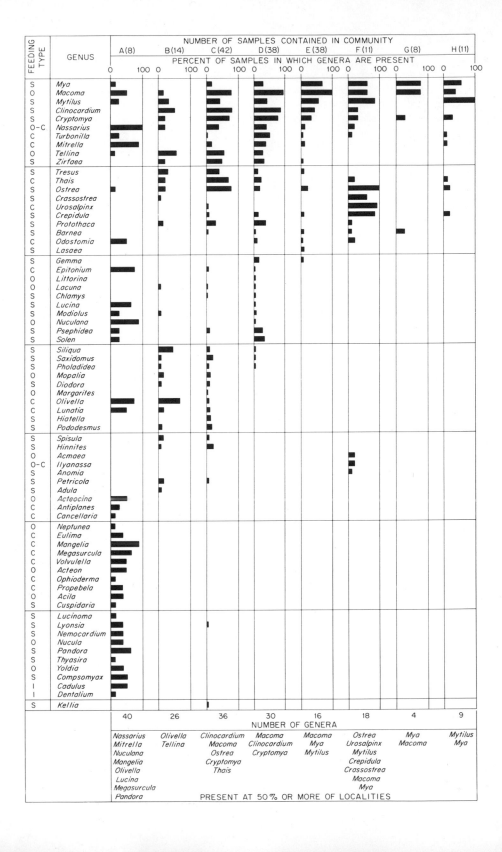

the fauna. The oyster bank community in the inner bay has been sampled at only a few localities near the bay entrance. Consequently, the occurrence of a number of epifaunal genera only in the south arm of the bay (Fig. 9.13) may be due to insufficient sampling and actual scarcity of sublittoral shell bottoms elsewhere in the bay.

The environmental parameters listed in Fig. 9.14 seem to be those of major importance in determining the faunal distribution patterns and thus are most applicable to paleoecology. Water depth is a major independent variable, but the whole of San Francisco Bay is shallow enough that probably few of the sublittoral species have particular adaptations that would limit them to narrower depth ranges. Depth exerts a strong influence on current and wave action and consequently on substrate texture; therefore, apparent fauna-depth correlations probably reflect the substrate-depth correlation. The plot of bottom temperature shows that average temperature is fairly constant in the bay but that the temperature range increases greatly with distance from the bay entrance into either arm. Salinity range is small near the bay entrance and throughout most of the bay except in the inner part of the north arm (communities G and H), where freshwater inflow is large.

The environmental homogeneity for an individual community, particularly evident in terms of substrate, confirms that the communities reflect discrete environments. Although the substrate may be uniform for a community, localities with the same substrate texture but grouped into other communities on the basis of the fauna indicate that the fauna reflects more than substrate or any other sigle environmental parameter.

Comparison of the Pliocene communities with these modern communities is possible because the fossil fauna is similar to that of the model at the generic level, with few extinctions or new arrivals. Beyond taxonomic similarity, however, community structure may change as the role of individual species within the community changes with evolution, but significant changes appear to have taken place in only a few genera. The generalized distribution of depositional environments during deposition of the Pliocene sediments is illustrated in Fig. 11.13. The cyclic stratigraphic framework provides an independent means of cross-checking the conclusions based on the community analysis.

A more detailed study, of the *Pecten* zone, has recognized eight communities by *Q*-mode cluster analysis of 93 samples from 54 measured sections. The distribution of the biofacies that they define in the east and west flanks of the

---

Figure 9.15 Composition of the modern molluskan communities of the San Francisco Bay area. The bar length represents the frequency of occurrence of the genus within each community. Communities: A, offshore; B, outside entrance; C, outer bay; D, middle bay; E, inner bay; F, oyster bank; G, brackish; H, brackish-epifaunal. Feeding types: *S*, suspension feeder,; *O*, deposit feeder utilizing resources on the substrate; *I*, deposit feeder utilizing resources in the substrate; *C*, carnivore, parasite, or macrophagous scavenger. From Stanton and Dodd (1976). Lethaia, *9*.

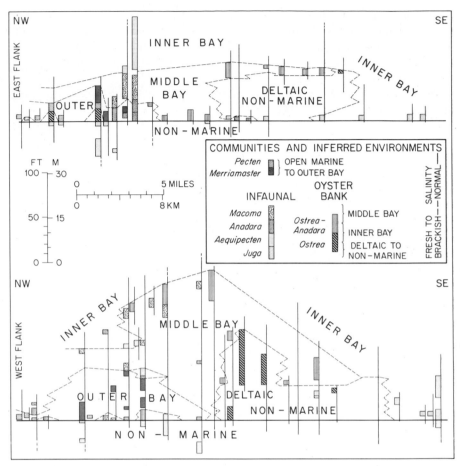

Figure 9.16 Distribution of communities and depositional environments in the *Pecten* zone along the east and west flanks of the Kettleman Hills. The datum line is the base of the *Pecten* zone. Vertical lines indicate the extent of stratigraphic section examined. The patterned bars indicate the distribution of communities. After Dodd and Stanton (1975). Geological Society of America Bulletin.

Kettleman Hills is illustrated in Fig. 9.16, and the taxonomic composition of the communities, in Fig. 9.17. The inferred paleoenvironment is again based in large part on comparison of these Pliocene communities with those in the San Francisco Bay area. In both the general study and that of the *Pecten* zone, the initial environment description is in terms of geographical position within the bay setting. This provides the basis for detailed environmental description in terms of specific parameters as they can be discriminated in the various bay settings.

Figure 9.17 Composition of the communities in the *Pecten* zone. The bar length indicates the percentage of samples of the community which contains the taxon. Feeding types: *S*, suspension feeder; *D*, deposit feeder; *C*, carnivore, parasite, or macrophagous scavenger. After Stanton and Dodd (1976). Lethaia, 9.

Community Structural Analysis

Diversity

Diversity is an integral aspect of community structure and is thus useful in the description and analysis of paleocommunities, as it is of temporal and paleo-biogeographic patterns and trends.

The Pliocene strata of the Kettleman Hills described above provide an example of the use of diversity in a local paleoenvironmental study. In the total Pliocene section (Fig. 11.13) diversity is high in the Jacalitos and Pancho Rico Formations (74 and 80 genera, respectively), intermediate in the lower part of the Etchegoin Formation (33 to 47 genera in the *Siphonalia* zone and underlying units), and low in the upper part of the Etchegoin and throughout the San Joaquin and Tulare formations (4 to 32 genera per unit), except for the *Pecten* zone with 59 genera (Table 9.3). Within this overall gradient of decreasing diversity in progressively younger strata, diversity fluctuates widely from unit to unit in the Etchegoin Formation above the *Siphonalia* zone and in the

TABLE 9.3   Diversity in the Pliocene Strata of the Kettleman Hills, California[a]

Formation	Zone of Woodring et al. (1940)		Number of Genera/Unit
Tulare			2
	Upper	*Amnicola*	11
	Lower	*Amnicola*	17
			6
	Upper	*Mya*	14
San Joaquin			7
	Acila		29
			4
	Pecten		59
			11
	Neverita		21
			9
	Cascajo Cgl.		32
Etchegoin			7
	Littorina		10
			6
	Upper	*Pseudocardium*	28
	Siphonalia		47
	Macoma		33
	Patinopecten		38
Jacalitos Pancho Rico			74/80

[a]After Stanton and Dodd (1970).

San Joaquin Formation. San Francisco Bay again provides a relevant modern analog for these diversity patterns. Diversity in the Recent communities there (Fig. 9.13) is plotted in Fig. 9.14 and discussed in the preceding section. The overall gradient in the Pliocene section parallels that from the open ocean to bay head in San Francisco Bay, as does the compositional trends, also discussed previously in this chapter. Consequently, the interpretation of the overall diversity gradient is that it reflects changes in environmental stability caused by increasing isolation from equable open-marine conditions.

The vertical and lateral gradients within individual zones are of diversity decreasing vertically and from north to south. The lateral gradient corresponds to an environmental transition from relatively open marine to brackish and nonmarine. The vertical gradient is interpreted similarly, as discussed more fully in Chapter 11: The base of a zone marks the influx of marine conditions, either as a result of sea-level fluctuations or spasmodic tectonic activity, but then the embayment became increasingly isolated from the open ocean and graded to fresh water and terrestrial conditions.

An interesting variant on this interpretation is provided by the *Pecten* zone. Community diversity is portrayed by the rarefaction curves of Fig. 9.18. In general, the diversity values agree well with the interpretation based on taxonomic composition but an exception to this generality is the *Merriamaster* community, which is indicative of an open marine-to-outer bay environment

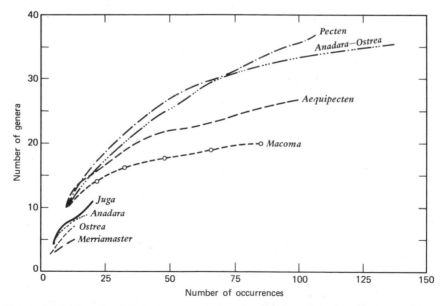

Figure 9.18 Rarefaction curves portraying the relative diversity of communities in the *Pecten* zone. After Dodd and Stanton (1975). Geological Society of America Bulletin.

on the basis of the composition, for it contains abundant individuals of the sand dollar *Merriamaster*. The community is found in well-sorted and commonly cross-bedded sandstone, indicating a current-swept habitat comparable to that in which modern sand dollars are found. Lithology and bedding geometry suggest sand patches and bars on the seafloor comparable to those outside San Francisco Bay (Yancey and Wilde, 1970) but on a smaller scale, rather than to those in tidal channel or beach settings. The *Merriamaster* community is closely associated with the *Pecten* community, always overlying it where the two occur in the same section. The *Merriamaster* community probably existed in essentially the same conditions of normal salinity, good communication with the open ocean, and environmental stability as did the *Pecten* community, but its low diversity reflects stress conditions of the mobile sand substrate. This specific example illustrates that localized conditions must be taken into account in interpreting patterns of diversity as well as the more common interpretation in terms of geographical gradients in environmental stability and in restriction from the uniform, well-mixed, open-marine environment.

Trophic Proportions

The study by Walker (1972) of Ordovician to Mississippian communities in the eastern United States provides a convenient starting point for our description of the application of trophic structure in paleoecology. Walker showed that the fossil communities were like the communities described by Turpaeva in that each was dominated by one trophic group, each trophic group was dominated by one species, and the feeding types of the ranked dominants within a community alternated. Subsequently, Rhoads et al. (1972) and Wright (1974) have shown that Cretaceous and Jurassic communities of Wyoming and South Dakota are also characterized by apparently distinctive trophic structure and are each restricted to particular depositional environments. These Mesozoic communities differ, however, from the modern communities described by Turpaeva in two respects: (1) Suspension feeders are much more common as dominants, and (2) alternation of feeding types among the dominants is much less common. In fact, many of these fossil communities are homogeneous, containing only suspension-feeding or only deposit-feeding organisms, suggesting at first glance unusual environments in which trophic resources were limited to either the water column or the substrate.

Application of trophic structure in paleoecology is difficult because of the incomplete preservation of the original biocoenosis. Environmental conditions determine the trophic proportions of the biocoenosis, but the biocoenosis consists of many more species and individuals than are likely to be preserved in the fossil assemblage. Thus to interpret the trophic structure of a paleocommunity, it is necessary to assume that (1) each fossil community is a sample drawn from a corresponding original biocoenosis, and boundaries and distribution pattern of the fossil community coincide with those of the biocoenosis, and that (2) the trophic structure of the biocoenosis is preserved in the paleocommunity. The living macrobenthic invertebrate communities of the southern California shelf were investigated to evaluate these assumptions (Stanton, 1976). Communities based on the total fauna were compared with those based on the shelled potentially fossilizable component of the fauna. A high degree of correspondence between the total and the "fossil" communities indicates that the first assumption is valid. Strong divergence between the trophic proportions of the total and "fossil" communities, however, indicates that the second assumption is not valid (Fig. 9.19). The latter conclusion has been further tested and confirmed by analyzing the trophic proportions of the Pliocene communities described previously for the Kettleman Hills (Stanton and Dodd, 1976), and independently by Bosence (19779a, b). Although trophic proportions seem to work as a paleoenvironmental tool in some cases, the basic assumptions appear to be generally invalid to the point that the technique has little predictive value.

Trophic Structure

Reconstruction of the trophic web of a paleocommunity requires a census of the community and a knowledge of the feeding habits of the organisms present. These, unfortunately, are generally not known for modern communities, let alone for paleocommunities. Although the trophic web that can be developed for a paleocommunity is incomplete, it is valuable because it illucidates the taphonomic processes that created the biased record as preserved in the fossil assemblage, because it highlights organism interactions within the community, and because it provides information about both the unpreserved component of the original community and the original environment.

The notion that dinosaurs may have been warm-blooded is based in part on trophic evidence. Because warm-blooded carnivores have a higher metabolic rate than cold-blooded carnivores, the shape of the ecologic pyramid in terms of relative abundance of carnivores versus herbivores should differ in the two cases. The shape of the pyramid can potentially be determined for Cretaceous vertebrates and thus the warm bloodedness of the carnivorous dinosaurs estimated (Bakker, 1975).

A systems analysis of shallow-water benthic ecosystems results in the flowchart in Fig. 9.20. This flowchart provides a framework for community analysis by presenting the interactions of biologic and physical materials, their quantities, and their rates. To reconstruct the trophic web of a paleocommunity, the fossil assemblage must be thoroughly sampled and identified and the relative proportions of the different taxa determined. Proportions are most easily based on numbers of individuals, but are more precise if based on biomass, taking into account the growth rates and longevity of the different species, or on actual energy flow. The taphonomic processes that have formed the fossil assemblage from the original biocoenosis must be analyzed to determine, as far as possible, what the original community was like. Finally, food preferences of the members of the community are determined from three lines of evidence: (1) indications

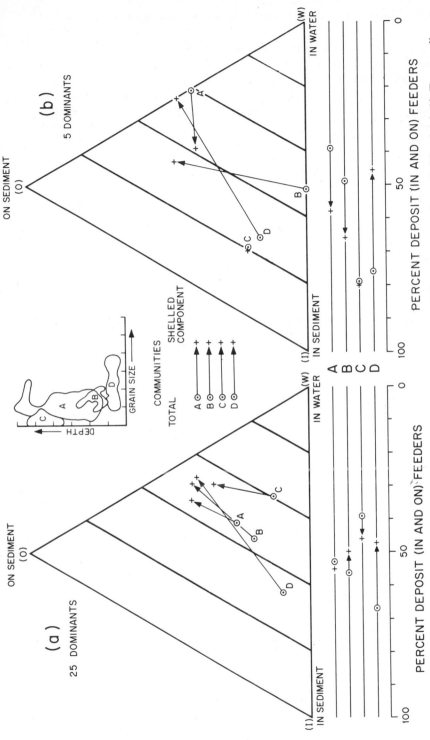

Figure 9.19 Trophic proportions of communities of benthic macroinvertebrates on the southern California shelf. Ternary diagram (*a*) is based on the 25 most abundant species in each community; ternary diagram (*b*), on the 5 most abundant species utilizing trophic resources in suspension, in the sediment, or on the sediment. Values circled are for total communities; values indicated by crosses are for shelled communities derived from total communities. Depth–texture characteristics of communities are generalized. The bar scales below the ternary diagrams represent the data in terms of proportions of deposit versus suspension feeders. After Stanton and Dodd (1976). Lethaia, 9.

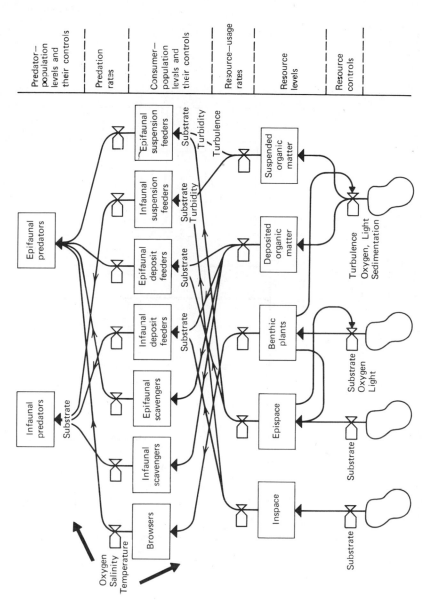

Figure 9.20 Diagrammatic structure of a shallow water benthic ecosystem. *Clouds*, system-independent states; *rectangles*, levels; *arrows*, rates (or flows); *faucets*, rate controls. After Hoffman et al. (1978).

369

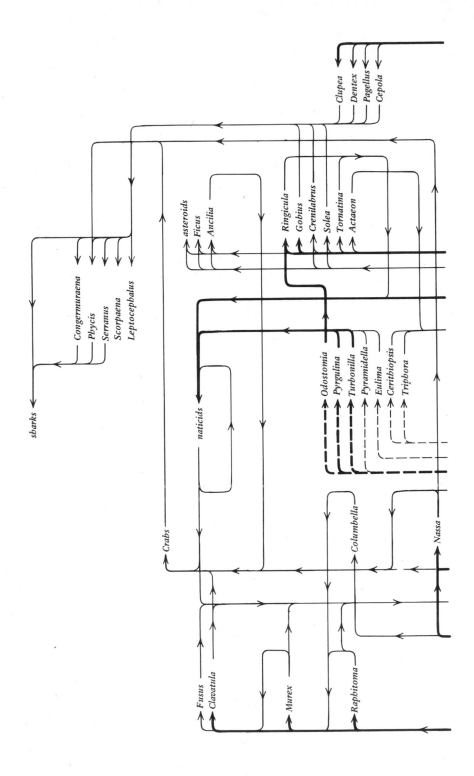

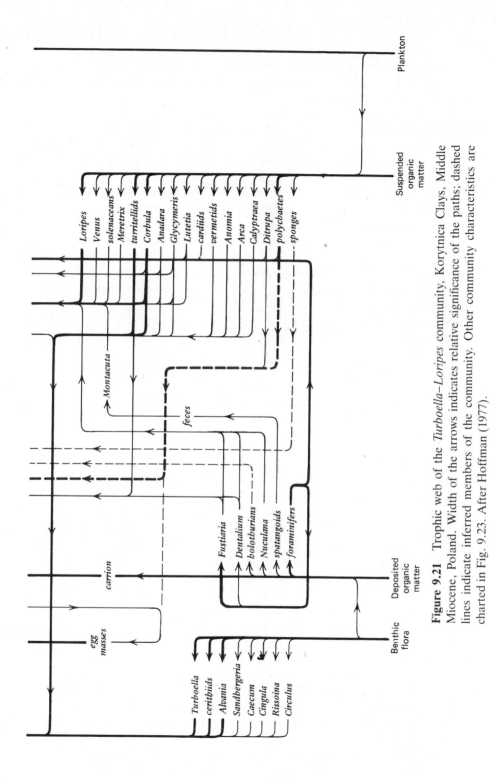

Figure 9.21 Trophic web of the *Turboella–Loripes* community, Korytnica Clays, Middle Miocene, Poland. Width of the arrows indicates relative significance of the paths; dashed lines indicate inferred members of the community. Other community characteristics are charted in Fig. 9.23. After Hoffman (1977).

of predation, such as gastropod borings; (2) food preferences of the living relatives of the fossils; (3) functional-morphologic indications of the mode of life of the fossil.

Two examples of the reconstruction of the tropic web component of the chart are illustrated in Figs. 9.21 and 9.22. The trophic web in Fig. 9.21 is for a Middle Miocene community in the Korytnica Clays of Poland. The composition of the community is illustrated by list and relative proportions in Fig. 9.23. The trophic web in Fig. 9.22 is for the community of the Main Glauconite bed in the Middle Eocene Stone City Formation, Texas. The composition of the community and relative proportions of taxa are illustrated in Table 9.4, together with the feeding characteristics of modern relatives of the fossils. The two trophic webs are similar in construction and are based on the same assumptions. They differ only in the extent to which all the components of the Miocene community are plotted individually, and in the extent to which the different pathways are quantified by numerical abundance in the Eocene trophic web.

In the Miocene example, the role of each taxon is indicated, as determined from specific predator–prey evidence or as inferred from present-day closely related taxa. Because many of these inferences are not strong, the trophic web presents a large amount of detail, but the accuracy of much of it may be relatively low.

The Eocene trophic web (Fig. 9.22), on the other hand, has been kept simple by grouping together taxa with similar trophic positions. Specific information such as the predation by naticid gastropods and crustaceans on each molluskan taxon (Table 9.4 and Fig. 9.22), could be incorporated in the web by separating out each individual genus or species. This has not been done, however, because most of these interrelationships are speculative for the Eocene community, being based on modern feeding information, which is itself incomplete. For example, the inferred predators on bryozoans and the prey of carnivorous gastropods could each add an entirely new dimension to the web. In the process, however, we would only be tracing out the modern story, more and more tenuously in the Eocene community, and by doing so we would be denying evolution of trophic structure during the intervening period.

The Eocene trophic web also simplifies the multiple and changing roles of many taxa within the community. For example, the crustaceans are placed at a single trophic level in Fig. 9.22, but from several lines of evidence it is inferred that they fit into the food web at a number of positions. Modern representatives of the Eocene fish have varied diets, consisting in large part of soft-bodied benthic invertebrates (Fig. 9.24). Worms and crustaceans, identified above from several lines of evidence as major components of the Main Glauconite bed community, thus also fit into the food web as the major food source for the fish; mollusks were possibly also utilized as food by the scianids. As in the case of the crustaceans, it is simplistic to group the fish and cephalopods into one position in the trophic web; each had distinct requirements, and feeding interactions between them were common. The input cate-

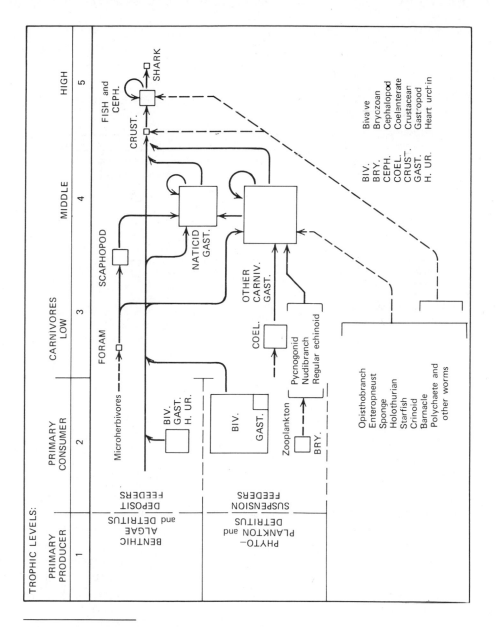

Figure 9.22 Trophic web of the community in the Main Glauconite bed of the Stone City Formation, Middle Eocene, Texas. Box sizes are proportional to numbers of individuals at each position. Solid lines and capital lettering indicate components present in the fossil assemblage and feeding relationships documented in the fossil assemblage or based on modern relationships. Dashed lines and lowercase lettering indicate inferred components and relationships in the original assemblage, based on modern trophic data involving components not preserved. After Stanton and Nelson (1980). Journal of Paleontology. Society of Economic Paleontologists and Mineralogists.

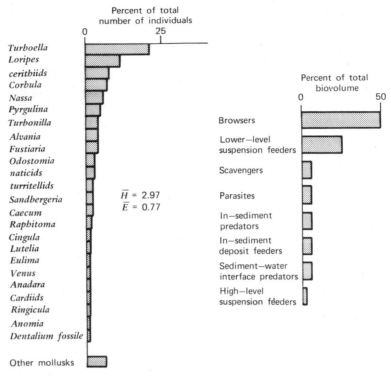

Figure 9.23 Taxonomic composition and relative abundance and trophic proportions of the *Turboella–Loripes* community, Korytnica Clays, Middle Miocene, Poland. The tropic web for the community is portrayed in Fig. 9.21. After Hoffman (1977).

gory of "Fish" includes these fish themselves and was used to avoid confusing feedback loops in the diagram. To complicate the picture further, many fish change their diets as they grow, decreasing, for example, the consumption of benthic invertebrates and increasing that of other fish.

Both of these food webs are based on incomplete data and include much speculation. Although only pale shadows of the trophic structure that must have characterized the original communities, they can provide valuable information about the original community and environment. They give information about the unpreserved components of the original community and thus about the extent to which the original community structure is evident in the fossil assemblage and about the taphonomic aspects of the environment. For example, the Eocene fossil assemblage is strongly weighted toward low and middle-level carnivores (levels 3 and 4), but even this part of the original community is poorly preserved, for crustaceans, which are rare in the assemblage but abundant in the original community, are at these levels of the ecologic pyramid. The primary consumers of level 2 (inferred to be deposit-

TABLE 9.4 Composition of the Main Glauconite Bed Community[a]

	1 G-S	2 %	3 T.L.	4 Bor.	5 Chip.	6 Feeding Habits
Coelenterata		6.5				Carnivorous
Scleractinia	2-2	6.2	3	—	—	
Alcyonaria	1-1	0.3	3	—	—	
Bryozoa	6-7	2.9	2			Phytoplankton
Scaphopoda	2-2	1.9	3	C	C	Carnivorous on forams, small bivalves
Gastropoda		49.3				
Fissurellidae	1-1	0.1	2-3	—	R	Detritus and carnivorous on small sponges
Turbinidae	1-1	0.1	2	—	R	Benthic diatoms, filamentous algae, eaten by Fasciolariidae
Vitrinellidae	2-2	0.1	2-3	A	P	Algae, parasitic on worms
Architectonidae	1-4	0.4	4	R	C	Carnivorous on anemones, corals
Turritellidae	2-6	2.2	2	A	A	Suspended detritus and phytoplankton
Caecidae	1-1	0.2	2	—	C	Interstitial diatoms
Scalaridae	2-3	0.1	4	—	C	Carnivorous and parasitic on coelenterates
Eulimidae	2-3	0.2	4	P	—	Parasitic on echinoderms (starfish, holothurians, regular echinoids), polychaetes
Pyramidellidae	1-1	0.1	4?	C	—	Parasitic/carnivorous on polychaetes, coelenterates, mollusks, starfish
Naticidae	3-3	16.1	3-4	A	A	Carnivorous on bivalves, gastropods. scaphopods
Ficidae	1-2	0.7	3	—	P	Carnivorous on urchins and other echinoderms
Cymatiidae	1-1	0.2	3-4	R	P	Carnivorous on mollusks, asteroids, echinoids
Muricidae	1-1	0.1	3	—	P	Carnivorous on bivalves, barnacles, gastropods
Pyramimitridae	1-1	0.1	?	—	P	???
Buccinidae	2-5	3.0	3	R	C	Scavengers; carnivorous on bivalves, crustaceans, worms
Nassariidae	1-1	0.6	2	A	A	Nonselective deposit feeder: diatoms, detritus; scavenger
Fasciolariidae	2-2	2.6	3-4	P	P	Carnivorous on gastropods, bivalves, polychaetes, barnacles

TABLE 9.4 (Continued)

	1 G-S	2 %	3 T.L.	4 Bor.	5 Chip.	6 Feeding Habits
Volutidae	1-2	1.1	3	R	P	Carnivorous on bivalves
Olividae	1-1	0.3	3	R	C	Carnivorous on small mollusks, foraminifers
Marginellidae	1-1	0.4	4?	—	P	Carnivorous on ???
Mitridae	1-1	0.1	4?	—	R	Carnivorous on crustaceans, sipunculid worms
Cancellariidae	2-3	2.0	3	R	C	Carnivorous on soft-bodied interstitial microorganisms
Conidae	1-1	0.1	4	—	P	Carnivorous on herbivorous polychaetes, fish, gastropods
Terebridae	2-2	2.5	4	C	P	Carnivorous on worms, enteropneusts
Turridae	12-14	12.9	3-4	P	C	Carnivorous on annelids, nemerteans
Retusidae	1-2	2.9	3-4	R	C	Carnivorous on other opisthobranchs, foraminifers
Mathildidae	2-3	0.2	?	—	C	???
Ringiculidae	1-1	0.2	3	C	—	Carnivorous on polychaetes, foraminifers
Bivalvia		36.6				Feed from suspension or sediment surface; dietary preferences generally not known, probably largely microflora and detritus, but nonselective, including bacteria, microfauna
Nuculidae	1-1	2.1	2	P	P	Deposit feeder
Nuculanidae	2-4	3.0	2	—	C	Deposit feeder
Arcidae	1-1	0.2	2	—	R	Suspension feeder
Noetidae	1-2	2.6	2	C	R	Suspension feeder
Ostreidae	2-2	1.1	2	R	P	Suspension feeder
Anomiidae	1-1	1.6	2	R	C	Suspension feeder
Cardititdae	1-1	2.4	2	R	C	Suspension feeder
Diplodontidae	1-1	0.5	2	—	C	Suspension feeder
Semelidae	1-1	0.3	2	C	R	Suspension feeder
Tellinidae	1-2	0.2	2	—	P	Deposit feeder
Mactridae	1-1	0.1	2	R	P	Suspension feeder
Veneridae	1-1	0.3	2	—	R	Suspension feeder
Corbulidae	3-3	23.3	2	C	C	Suspension feeder

Taxon						
Cephalopoda (*Aturia, Belosepia*)	2-2	0.1	3-5	—	—	Carnivorous on Crustaceans, fish, mollusks
Echinodermata (heart urchin)	1-1	0.1	2	—	—	Nonselective deposit feeder
Arthropoda (crustacean)	1-1	0.1	3-4	—	—	Carnivorous, scavenger
Foraminifera	—	—	3	—	—	Diatoms, bacteria
Chordata						
Elasmobranchii	3-3	0.1	5	—	—	Carnivorous on fish, cephalopods
Congridae	1-1	0.1	3-5	—	—	Carnivorous on bottom-living fish, crustaceans, cephalopods
Beryciformis	1-2	0.8	3-5	—	—	Carnivorous on benthic crustaceans
Serranidae	1-1	0.1	3-5	—	—	Carnivorous on benthic crustaceans and polychaetes as juveniles; on small fish as adult
Scianidae	3-3	0.8	3-5	—	—	Carnivorous on benthic polychaetes and crustaceans, mollusks?; planktonic crustaceans, fish, and squid
Ophididae	3-3	0.8	3-5	—	—	Carnivorous on benthic crustaceans (shrimp, crabs, stomatopods); juvenile fish, polychaetes
Soleidae	1-1	0.1	4-5	—	—	Carnivorous on benthic crustaceans

[a]Col. 1: number of genera and species within the taxon; col. 2: percent of individuals in macrofossil assemblage belonging to taxon; col. 3: trophic level of taxon in ecologic pyramid, as in Fig. 9.22—1: primary producer; 2: primary consumer; 3–5: low to high levels of carnivores; col. 4: abundance of bored individuals in taxon—A: greater than 20% of individuals; C: 10–20%; P: 5–10%: R: less than 5% of individuals are bored; col. 5: abundance of crustacean-chipped specimens in taxon—percentage ranges as in column 4; col. 6: feeding information for the taxon. After Stanton and Nelson (1980).

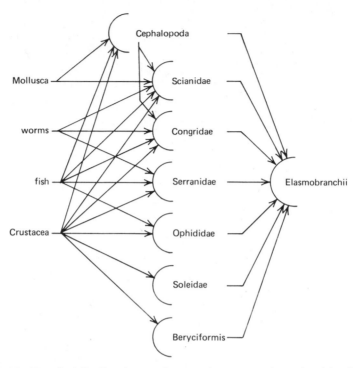

Figure 9.24   Detailed feeding interactions at the top-carnivore level in the trophic web of Fig. 9.22. Main Glauconite bed of the Stone City Formation, Middle Eocene, Texas. After Stanton and Nelson (1980), Journal of Paleontology, Society of Paleontologists and Mineralogists.

feeding worms) are even less well represented than the low- and middle-level carnivores.

The reconstruction of the trophic web can provide information about community evolution. In the Eocene community, for example, the food preferences of the crustaceans and the naticid gastropods appear to be different from those at present. Most important, reconstruction of the trophic web is a step toward reconstructing the paleoecosystem and so provides a basis for paleoenvironmental reconstruction.

Analysis of paleocommunities on the bassis of biomass rather than numerical abundance offers several very significant advantages. As discussed previously, if the composition and trophic characteristic of the macrobenthic communities of the southern California shelf are compared with the potentially fossilizable component of these communities, the similarities are poor because the numerical dominants are largely soft-bodied infaunal deposit feeders, whereas the fossilizable taxa are largely suspension feeders (Fig. 9.19).

The detailed and more quantitative study of the communities in the Texas bays has led to a similar result (Staff et al., 1986). As indicated in Fig. 9.25,

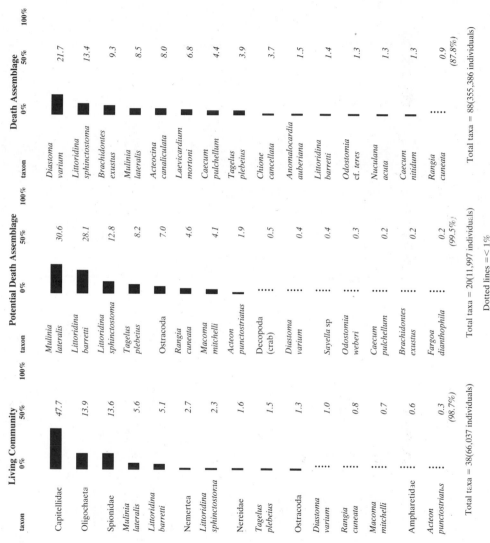

Figure 9.25 Numerical dominants in Copano Bay. Rankings are based on the percentage of the total number of individuals represented by each taxon. From Staff et al. (1986), Geological Society of American Bulletin 97: 436.

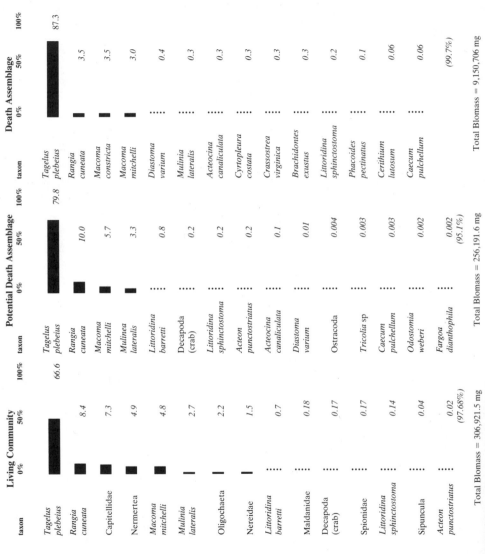

Figure 9.26 Biomass dominants in Copano Bay. Rankings are based on the percentage of the total biomass represented by each taxon. From Staff et al. (1986), Geological Society of America Bulletin *97:* 439.

the living community is numerically dominated by worms, and the potential death assemblage is very different from the living assemblage. In contrast, the living community and the potential death assemblage are quite similar if based on biomass (Fig. 9.26). The reason for this is that the numerical dominants are individually small as compared to individuals of the bivalves *Tagelus* or *Rangia*. Feeding types and habitat proportions are also better estimated from biomass data than from numerical proportions (Table 9.5).

A procedure for calculating biomass has been described by Stanton et al. (1981). The total biomass of the species incorporates estimates of the volume of each individual, and the size-frequency distribution for that species in the community. An important benefit of working with community biomass is that the dominants of the community, being the common and larger taxa, are recovered relatively early in the sampling program. The small organisms, even if abundant, are relatively unimportant in biomass terms, so large samples re-

TABLE 9.5   Feeding Types and Habitat Proportions for the Laguna Madre and Copano Bay Sites as Percents of Total Numerical Abundance (no.) or Biomass (B)

	Living Community		Potential Death Assemblage		Death Assemblage	
	No.	B	No.	B	No.	B
Copano Bay						
Feeding types						
Suspension feeders	10.9	73.9	18.3	96.3	36.2	96.2
Deposit feeders, herbivores	85.8	17.8	78.2	3.2	51.5	3.5
Predators, parasites	3.3	8.3	3.2	0.4	12.3	0.3
Habitat proportions						
Epifauna	10.7	16.1	93.3	21.0	91.9	5.8
Laguna Madre						
Feeding types						
Suspension feeders	15.6	50.8	51.0	96.5	77.3	95.5
Deposit feeders, herbivores	77.4	48.2	29.9	2.3	19.2	3.4
Predators/parasites	7.0	1.1	19.1	1.1	3.7	1.0
Habitat proportions						
Epifauna	10.9	1.5	47.0	2.3	73.4	11.2

Reprinted from Biomass: Is it a useful tool in paleocommunity reconstruction? by G. M. Staff et al. (1985), from Lethaia, by permission of Norwegian University Press (Universitätsforlaget AS), Olso.

quired to find the small and rare taxa are commonly not necessary. This procedure also highlights, of course, the overwhelming significance in the structure of the community of the large but perhaps rare organisms. The sampling program must be designed so that these can be included in the analysis.

Analysis of the paleocommunity as just described is based on produced biomass—that is, an estimate of the energy flow that is used for organism growth. The complete analysis of energy flow would include the energy expenditure not only for growth, but also for maintenance (or respiration) and for reproduction. Powell and Stanton (1985) have demonstrated the procedure for making these calculations (Fig. 9.27). It involves combining properties of

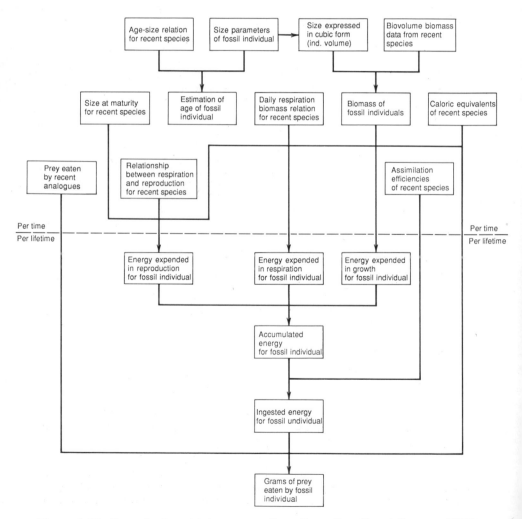

Figure 9.27   Steps in determining energy flow. From Powell and Stanton (1985). Palaeontology *28:2*.

TABLE 9.6 Numerical Abundances, Size and Age Data at 25% and 50% Maximum Age, Produced Biomass, and Cumulative Biomass for a Representative Sample of Species in the Fossil Assemblage

	A	B		C		D		E	F
	Total No. of Individuals	% Surviving to 25% Max. Age	No. Surviving to 25% Max. Age	% Surviving to 50% Max. Age	No. Surviving to 50% Max. Age	Size at 50% Max. Age (cm)	Relative Age at 50% Max. Age	Produced Biomass (as Biovolume)	Cumulative Biomass (as Biovolume)
Polinices aratus—(naticid)	881	4.3	38	0.5	4	.61	3.1	4.4	4.2
Retusa kellogii—(acteocinid)	191	80	153	21	40	.52	2.3	2.3	3.0
Michela trabeatoides—(turrid)	171	3.7	6	1.2	2	1.12	12.1	2.4	16.0
Latirus moorei—(fasciolariid)	152	20	30	8	12	1.56	34.9	7.0	117.6
Hesperiturris nodocarinatus—(turrid)	137	11	15	3.5	5	1.41	24.6	2.0	26.2
Bonelitia parilis—(cancellariid)	113	34	40	18	20	.52	2.3	0.8	0.9
Buccitriton sagenum—(buccinid)	111	22	24	2.8	3	.74	4.5	0.9	1.3
Notocorbula texana—(corbulid)	1,140	13	148	2	23				
Vokesula smithvillensis petropolitana—(corbulid)	226	47	106	6	13				

[a]Biomass is estimated from biovolume, calculated by treating *Retusa* as a cylinder and all other gastropods as cones, with shell length as height and maximum width as diameter.

[b]From Powell and Stanton (1985).

the fossils (biovolume and size frequency–age frequency distributions) with properties based on present-day analogous taxa (age or size at maturity, relation between biomass and respiration, relation between energy expended in respiration and in reproduction). From these, the energy assimilated by a fossil individual during its lifetime can be determined. Taking this one step further for a carnivore, if the caloric value of its prey species and the assimilation efficiency can be estimated, the amount of prey consumed by the predator during its lifetime can also be estimated. The greatest source of uncertainty in these calculations is the limited amount of data available for living organisms. The reasonably narrow range of values available for modern mollusks suggests that the resulting estimates for fossil mollusks can be quite useful. The necessary measurements for living organisms of other taxa are much less abundant, so the resulting calculations for the corresponding fossils must be interpreted more generally. The value of incorporating biomass and energy flow into the tropic analysis is demonstrated for the more abundant pelcypods and gastropods from the Stone City Formation, described previously (Table 9.6). The numerical abundances listed in column A are reflected in the relative sizes of the boxes in Fig. 9.22. This chart suggests that the naticid gastropod *Polinices aratus* was a dominant carnivore in the community. This is supported by the great abundance of naticid borings in a wide range of prey individuals. In columns B, C, and D, however, it is apparent that *Polinices* individuals are relatively small and therefore constitute only a small part of the community biomass (column E). Column E is a rough estimate of the total combined energy expended for produced biomass and for maintenance or respiration. Because the larger and longer-lived specimens spent more time as mature and reproductive individuals, when the energy expenditure for reproduction is also added (column F), the biomass dominance of the larger specimens is further enhanced.

10

Paleobiogeography: The Provincial Level

Paleobiogeography, the geographic distribution of ancient organisms, is of basic concern in paleoecology. Much work in applied paleoecology is concerned with describing distribution of fossils in space and time and interpreting what that distribution means in terms of environmental patterns. Although the terms *paleobiogeography* and *biogeography* do not carry any specification of scale, they are usually used in the context of relatively large-scale distribution of organisms, such as large regions of continents, oceans, or total worldwide distribution of a species or other taxonomic group. In other words, paleobiogeography is usually considered as an aspect of large-scale, "big-picture" paleoecology. Biogeography does not have to be studied in the context of ecology but simply in terms of the descriptive geographic distribution. Much of the early work in biogeography was of this nature, which Valentine (1973a) has termed *geographic biogeography.* Our primary interest in this chapter is the history of change of biogeographic distribution, *historical biogeography,* or of the environmental aspects and controls on biogeography, *ecological biogeography.*

The difference between biogeography and paleobiogeography is not sharp because the distribution of living organisms is so strongly influenced by the history of development of that distribution. In biogeography, certainly "the past is the key to the present." Thus understanding the modern distribution of organisms without considering the geologic history of how it got that way is impossible. However, one can study the distribution of organisms in the past for its own sake, without direct regard to the present. Paleobiogeography would seem to be the appropriate term for such studies.

Paleobiogeography is of interest to the paleoecologist for a number of reasons. Although much paleoecological work is of a local or regional nature, the ultimate goal of applied paleoecology is to determine worldwide environmental conditions for all of geologic time. A combined study of paleoecology and paleobiogeography may help to solve certain large-scale geologic problems in other geologic subdisciplines. This approach has been extensively used, for example, to help in our understanding plate tectonics history (e.g., Briggs, 1987; Stehli and Webb, 1985; Ross and Ross, 1985). An especially good example is paleobiogeographic studies of suspect terranes in western North America (see discussion below; Smith and Tipper, 1986). The geographic distribution of organisms is obviously strongly influenced by environmental factors, so that a study of geographic distribution helps in interpreting the geographic variation of environmental parameters. For example, one of the regions of greatest change in the biota is between the tropical (or subtropical) and temperate zones. A study of paleobiogeography may help to locate the position of that transition at various times in the geologic past. Finally, the geographic distribution of organisms is strongly influenced by geologic processes and geologic history, particularly tectonic history. The paleobiogeography of vertebrate fossils and land plants in particular have long been used in interpreting the development and breakdown of barriers between continents through time. The distribution of the *Glossopteris* flora is an especially well known example (Chaloner and Lacey, 1973).

In addition to the difficulties associated with any work in paleoecology, paleobiogeography has some special problems. Because of the large areas involved, usually one person cannot hope to do all the basic data gathering on distribution of the biota and so must depend in part on the work of others. However, each worker may have a somewhat different concept of what constitutes a particular species or genus, so that published lists of fossil biotas will not be directly comparable. Comparing illustrations will help, or, even better, examining the specimens used by earlier workers, but these are not always available. Long-distance correlation is also difficult, so that time equivalences of biotas may be imprecise. One of the purposes of paleobiogeographic studies is to look for areas with different biotas; thus by definition, correlation is difficult between areas with different biotas. Because of the irregular distribution of outcrops or even subsurface samples and the unevenness of preservation, fossils may not be available from critical areas for determination of biotic distributional boundaries. Actually, this may be a blessing: Boundaries can easily be drawn through an area of no data, whereas they are often difficult to place when data are gradational!

People have undoubtedly been aware of the geographic patterns of plant and animal distribution since ancient times. However, the earliest attempts to formalize these observations into biogeographic regions were made in the nineteenth century. Woodward (1856) first defined marine provinces based on molluskan species distribution. In 1859, Forbes and Godwin-Austen published *The Natural History of the European Seas*, which included a description of biogeographic provinces in Europe. Some of the most basic work in biogeography was done by Charles Darwin and especially A. R. Wallace (1876), both of whom worked with terrestrial biotas. Landmark publications in the field of biogeography include books of Hesse, Allee, and Schmidt (1951), Ekman (1953), and Darlington (1957), each of which contains a wealth of information of value to present-day workers in this field. Hedgpeth (1957a) published an especially useful review for paleoecologists.

Interest by biologists in biogeography seems to have expanded in recent years. This is reflected in the number of textbooks and reviews on both general biogeography (e.g., Pielou, 1979; Cox and Moore, 1985) and marine biogeography (Briggs, 1974). Interest has been especially high in the relatively new concepts of cladistic biogeography and vicariance. Several conferences have been held on these topics and books and review articles published (e.g., Cracraft, 1983; Sims et al., 1983; Humphries and Parenti, 1986). Cladistic biogeography uses the techniques of cladistic taxonomy, which attempts systematically to determine phylogeny of organisms based on degree of similarity, to determine the relationship between biogeographic units. Vicariance explains differences in organisms between areas as being due to the production of new species by the creation of barriers subdividing an area in which species ranges were originally continuous. This conflicts with the idea that species migrate to areas from centers of dispersal. Considerable controversy has developed between supporters of vicariance and centers of dispersal (McCoy and Heck, 1983; Briggs, 1987). Clearly, both processes occur. Only the relative importance of the two processes in particular cases is uncertain.

Until relatively recently few reviews have dealt specifically with paleobiogeography. Interest in the relationship between paleobiogeography and plate tectonics has led to a number of conferences on paleobiogeography, which have resulted in published proceedings volumes (e.g., Middlemiss et al., 1971; Hallam, 1973a; Ross, 1974; West, 1977; Gray and Boucot, 1979). A very good review of this subject, especially its theoretical aspects, is that of Valentine (1973a), who has long been active in research in this field. Part A of the *Treatise on Invertebrate Paleontology* (Robison and Teichert, 1979) includes discussions of the paleobiogeography of each of the Phanerozoic periods. Jablonski et al. (1985) discuss the implications of paleobiogeography to many paleobiological problems.

CONCEPTS

The basic data of biogeography are the distributions of species and other taxa. The basic unit of interpretation is the *biotic province*. Many definitions of the biotic province have been suggested, as have several alternative terms for the concept. Perhaps the most meaningful one is that of Valentine (1963), who defines a province as a region in which communities maintain characteristic compositions. This definition is a good expression of the concept of the province as it is now widely understood, but is difficult to use in practice for recognizing provinces and boundaries between them. The most important aspect of recognizing provinces is in recognizing their boundaries. These are clearly areas in which the composition of the biota of many communities change markedly and in which many species end their ranges (Jablonski et al., 1985). The areas between such zones of rapid transition constitute the provinces. The location of these transition zones changes with time so that provinces are ever changing in distribution.

One of the earliest definitions of a province was an area in which 50% of the species are *endemic,* that is, restricted to that area (Woodward, 1856). This definition has the advantage of being quantitative but is still difficult to apply because many boundaries could be chosen that would set off areas in which the biota would be 50% endemic. In the past, provinces were usually defined subjectively simply as areas with distinctive biotas. Despite the difficulty in finding an adequate definition of province, most workers in the field seem to recognize one when they see it. The original provinces recognized in Europe during the mid-nineteenth century by Forbes and Godwin-Austen (1859) and worldwide by Wallace (1876) have not been greatly modified by later workers. Methods of defining provincial boundaries are discussed in more detail later in this chapter.

Terms other than *province* have been used for biogeographic units. Some are equivalent in meaning to *province* and others are for larger or smaller units. *Realm* is frequently used for a large unit, on a continental or multi-continental scale. For example, four realms are commonly recognized for the late Paleozoic: Gondwanan, Euramerican, Angaran, and Cathaysian (Chaloner and Lacey, 1973). Wallace (1876) used the term *region* for units of the same general scale. Schmidt (1954) proposed a hierarchy of units which, arranged from largest to smallest, are realm, region, subregion, and province. Others have also recognized subprovinces. This may be more complexity than is needed, but realm and province imply a convenient distinction of scale. Sylvester-Bradley (1971) gives a thorough treatment of the problems of the definition of provinces.

Ideally, a province would be recognized on the basis of the entire biota. This is rarely done, and indeed is probably rarely practical, because a worker is likely to be familiar enough with only one phylum, class, or even smaller taxonomic unit to use it for recognizing provinces. Thus much of the work on recognizing modern shallow marine provinces has been based on the distribution of a single phylum, the Mollusca. Most of the examples from the fossil record are also based on particular fossil groups. Fortunately, as we discuss below, the factors controlling biotic distribution for one group will probably affect other groups in a similar manner, although certain groups, such as pelagic forms or those with a long pelagic larval stage, are likely to be less confined to provinces (i.e., they are *cosmopolitan*) than are benthic forms. In any event, recognizing biotic provinces on the basis of limited taxonomic groups is the standard practice. For clarity, however, such provinces should be prefixed to indicate their data base, as molluskan provinces, trilobite provinces, and so on.

FACTORS CONTROLLING GEOGRAPHIC DISTRIBUTION OF SPECIES

Each species has a potential geographic range that is determined by its habitat requirements, but few, if any species actually occur throughout their potential range. This is not to imply that there are vast areas of suitable habitat available for occupation that have unoccupied niches. The usual pattern is for similar species or similarly adapted species to occupy a given niche in areas separated by barriers. Using the centers of dispersal approach, one would argue that species have migrated to the barrier-separated areas from the center of dispersal by different routes and at different times, evolving slightly in the process so that different species ultimately occupy the areas. The vicariance approach would explain the patterns as resulting from development of barriers dividing a once-continuous area of suitable habitat. The separated populations would evolve *allopatrically* into different species, each with a more restricted geographic range than the parent species. Obviously, biogeography is closely tied to evolution, a fact recognized at an early date by both Wallace and Darwin.

Biogeography and biotic provinces are closely related to barriers and the geologic history of the barriers. Any feature of the environment that limits the distribution of a species is a barrier. The types of environmental parameters that will be of greatest importance in determining provinciality will be those that affect many species at the same or approximately the same locality. Also of importance is how systematically the environmental parameter varies geographically and how effective is the barrier produced by changes in that parameter. Sometimes, rather than being due to change in a single parameter, barriers may result from the interaction of two or more factors. Although practically any environmental parameter can be important in specific cases, the two most important are depth–elevation and temperature.

Depth–Elevation

As indicated previously, depth is not a pure environmental parameter because many parameters change with depth. The same could be said of the related concept of elevation. However, depth–elevation is a convenient concept in discussing barriers in biogeography. All marine organisms have a limited range of depths over which they can survive; hence water that is too deep or too shallow forms a barrier. A broad expanse of deep water is a very effective barrier to the distribution of shallow-water benthic organisms, and terrestrial organisms, too, have certain elevation limits. The most effective barrier to distribution of either terrestrial or marine organisms is sea level. Terrestrial organisms cannot exist below it (except obviously in terrestrial basins that go below sea level), and marine organisms cannot exist above it. Another way of expressing the same idea is the presence or absence of the life media, be it air or water. Even a relatively narrow expanse of elevation above sea level, such as the Isthmus of Panama, is an effective barrier to marine organisms. The boundaries of all modern marine provinces are in part determined by the depth–elevation parameter.

The depth–elevation parameter can be used to illustrate the importance of the geographic extent and effectiveness of barriers. Barriers obviously may differ in effectiveness, and effectiveness will not be the same for all species. One measure of the effectiveness of a barrier is the ease with which species circumvent it. Simpson (1940) classified dispersal routes around or through barriers into three categories on the basis of their effectiveness. The *corridor* is an open route for migration, with many species easily passing through the barrier. The broad shelf connecting the coastal waters of Siberia with those of Alaska would be a corridor for shallow marine organisms. A *filter bridge* is a more limited dispersal route available to some species but not all, at least in geologically short time periods. A relatively narrow strait of deep water between shoal water areas such as the Mozambique Channel between Africa and Madagascar would be an example. A *sweepstakes route* is a low-probability route that few species are able to take. The broad expanses of deep water around the Hawaiian Islands would be an example for the dispersal of both shallow marine and terrestrial forms.

Temperature

The boundaries of all modern biotic provinces are in part temperature controlled. All species, even eurythermal ones, are sensitive to temperature variation (see Chapter 2). [Jablonski et al. (1985) briefly review the reason for temperature control on biogeographical distributions.] But the reason temperature is so important in determining provincial boundaries is its systematic global variation. Year-round warm oceanic temperatures are restricted to the area relatively near the equator. Cooler, seasonally variable temperatures on either side of this zone form barriers. Not only is this low-temperature barrier effective, but there simply is no other side of the barrier! Hence warm-water taxa are restricted to the tropical zone.

Similarly, the tropical zone forms an effective barrier to cold-water taxa that might otherwise migrate from one temperate belt to that across the equator. However, the cold-water taxa have two alternatives not available to warm-water forms. First, there is cold water at depth even in the tropics; so a cold-water species that can tolerate depth can find suitable habitat. Many species do this, exhibiting the phenomenon of *submergence*, that is, occurring at greater depth in the low-latitude end of their range (Fig. 10.1). Surprisingly, a few species also submerge at the high-latitude end of their range because the surface waters may chill more during the winter than water slightly below the surface. The second alternative for cold-water forms is that if they can break

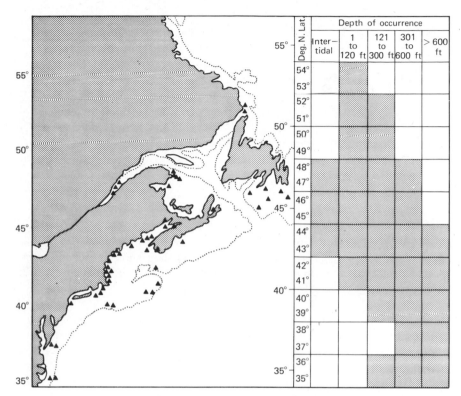

Figure 10.1 Depth and latitude distribution of the cool-water bivalve *Venericardia borealis* in the northwest Atlantic. Triangles on the map show the localities of living specimens as determined from published reports. The graph on the right summarizes the water depth of all known occurrences for each 2° of latitude. From McAlester and Rhoads (1967), by permission of Elsevier Scientific Publishing Co., Amsterdam, Netherlands.

through the tropical barrier, there is cold water at the surface in the other hemisphere.

A few species or closely related species do occur on either side of the equator, giving rise to *bipolar distributions* (Hesse et al., 1951); that is, the species occurs in middle or high latitudes in both hemispheres but not at low latitudes (Fig. 10.2). The method by which the species accomplish this distribution has been the subject of some controversy (Fig. 10.3; Briggs, 1987). One theory is that the species may submerge into deeper water in the tropics and occur at the surface in higher latitudes. No tropical deep-water forms of bipolar species have been described, however. Another theory is that in the past the temperature in the tropics was lower, so that the species ranged from one hemisphere to the other. Warming of the tropics caused elimination of the equatorial forms of the bipolar species. Valentine (1984) argues for this expla-

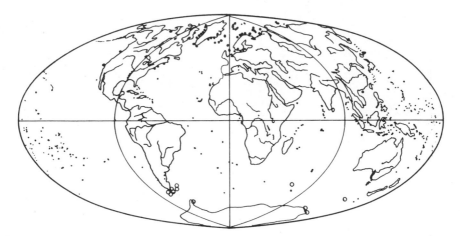

Figure 10.2 Geographic distribution of the species pair of priapulid worms *Priapulus caudatus* (dots) and *P. tuberculato-spinosus* (circles) showing a bipolar distribution. From Hedgpeth (1957b), Geological Society of America Memoir, *67*, published by the Geological Society of America.

nation and cites oxygen isotopic data in support of a Neogene warming of tropical surface waters. A third theory states that the species once lived throughout the tropics and into both temperate zones, but that a strictly tropical species evolved that was adapted for the same niche. The original species was eliminated from the tropics by competitive exclusion but remains in the temperate zones that the new species cannot invade. This theory is like the second in requiring evolution of new species in the tropics. Indeed, tropical species tend to be geologically younger than those at higher latitude. The tropics are a barrier that temperate species have a certain probability of penetrating, albeit perhaps a very low probability. But as Simpson's term *sweepstakes route* implies, the chance is there that given enough time the species may be able to cross the barrier and become bipolar.

Of the temperature-related barriers in the marine environment, none is more effective than the tropical–temperate barrier. A very high proportion of the species cannot cross the tropical–temperate boundary, which has a mean temperature of about 15 to 18°C for the coldest month of the year. Wherever that temperature boundary occurs in the shallow seas, there is a provincial boundary. If adequate data were available, the tropical–temperate boundary should be readily identified in the geologic record on the basis of distribution of ancient biotic provinces. As an example, the boundary between the Jurassic Tethyan and Boreal provinces in Europe probably corresponds to the tropical–temperate boundary.

Provincial boundaries often occur at other temperatures also, although the biotic change is usually not as great (Hall, 1964). Biotic changes usually occur

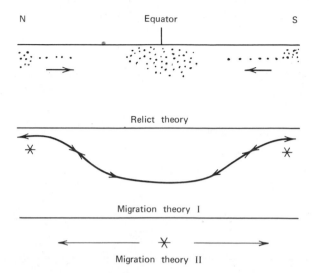

Figure 10.3   Schematic representation of three theories proposed to explain bipolarity. These three theories are discussed in detail in the text. Place of origin is indicated by an asterisk. From Hedgpeth (1957b), Geological Society of America Memoir, 67, published by the Geological Society of America.

in areas where there is a rapid temperature change over a relatively short distance. Such areas often correlate with changes of direction in the coastline, especially insofar as these are related to current patterns. For example, on the California coast one of the biggest changes in biota occurs at Point Conception (Fig. 10.4), south of which the coast runs approximately east-west for many miles, and north of which it runs north-south. North of Point Conception the water is uniformly cool as the result of the southward-flowing California Current immediately offshore and extensive upwelling of deep water caused by the surface waters moving offshore. South of Point Conception the temperature is warmer because of a northward-flowing countercurrent, the more offshore location of the California Current, and less extensive upwelling. Outliers of the more northerly fauna occur on the Channel Islands and at places far south into Baja California, where intense upwelling brings cold water to the surface.

The nature of the biota is controlled not only by the absolute temperature but also by the amount of temperature variation and the durations of low or high temperatures. The more stenothermal forms may not be able to survive in areas where the temperature is highly variable. This partly explains why some aspects of the shallow-water biotas of opposite sides of the same ocean are so different (Hedgpeth, 1957a). Generally, the major oceanic circulation patterns produce less temperature variation on the eastern sides of ocean

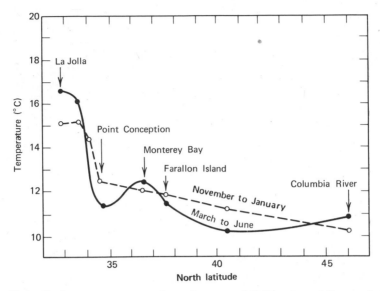

Figure 10.4   Surface temperatures along the coast of California and Oregon in March to June and November to January. Note the thermal break at Point Conception. From Sverdrup et al. (1942), from The Oceans, copyright 1942, renewed 1970, p. 724; reprinted by permission of Prentice-Hall, Inc., Englewood Cliffs, N.J.

basins than on the western sides. This also accounts for the greater diversity of the biota on the eastern sides of ocean basins.

Other Environmental Parameters

Other environmental parameters are capable of forming barriers that could mark provincial boundaries; however, they appear to be distinctly less important than depth–elevation or temperature. Currents are at least indirectly quite important in some areas because temperature patterns are strongly determined by the oceanic current pattern. Currents aid in dispersal of larvae (although they probably do not form an effective long-term barrier to dispersal of organisms), and they modify the odds in sweepstakes route dispersal.

The amount of rainfall is almost as important as temperature in controlling the distribution of terrestrial plants and animals (Good, 1974; Wolfe, 1978). Both plants and animals are highly adapted to the amount of water available in their environment, hence do not generally have a very broad range of rainfall tolerance. Maps of terrestrial biomes (geographically associated communities) clearly reflect precipitation patterns (Fig. 10.5) Rainfall, however, is not as important in determining the really large-scale global realms, which are likely to include a range of climatic conditions, including rainfall.

Salinity has been suggested as an important factor controlling provincial

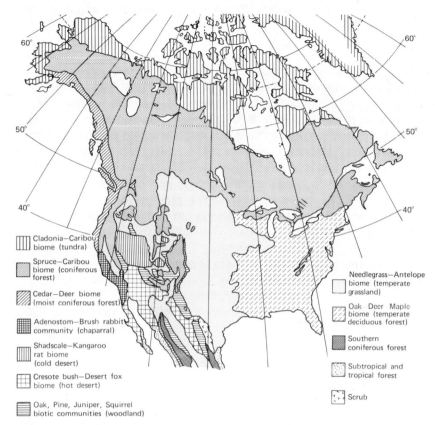

Figure 10.5 Distribution of terrestrial biomes in North America. From Shelford (1963), copyright 1963, the University of Illinois Press.

boundaries. It certainly is important in marking the boundary between freshwater and marine provinces, for few freshwater species are able to tolerate salinities above 1 to 2‰ (Kinne, 1971). Similarly, very few marine forms are able to live below that salinity barrier. In marine environments salinity is of minor importance in defining provincial boundaries because most of the oceans have a relatively uniform salinity of 35 ± 3‰ (Sverdrup et al., 1942). The rather irregular, smaller areas of lower or higher salinity are not distributed in such a way as to form barriers that cannot be penetrated by normal marine species. Salinity is important in controlling the local distribution of communities but not of provinces. Low- or high-salinity communities are simply part of the assemblage of communities that makes up the province.

The same types of statements can be made about the importance of other parameters, such as substrate or water turbulence. These parameters are highly variable at the provincial scale and, although they are important at the community level, they are not involved in forming barriers at provincial boundaries.

Geologic History

The distribution of species is greatly influenced by geologic history, particularly the geologic history of the environmental parameters that form barriers. The most important of these is the depth–elevation parameter or, more specifically, the relative position of land and sea. The major controlling factor on the distribution of land and sea is tectonic activity. The establishment and breakdown of barriers is often closely related to plate tectonic events. Provinciality is greatest during times when plate motion has produced a maximum number of separate continents (such as the present time). There are fewer barriers and thus less provinciality when plate motion has welded together continents, such as was the case at the end of the Paleozoic (Valentine, 1971a).

Several specific relationships between plate tectonics and biogeography can be stated (Fig 10.6 and Table 10.1):

1. When spreading ridges lie parallel to continents, they produce deep, ever-widening ocean basins and thus barriers to migration of terrestrial or shallow marine biota (Mid-Atlantic Ridge).
2. Transform faults parallel to continental margins also are usually associated with deep-water barriers (San Andreas fault, California).
3. Subduction zones parallel to and dipping toward the continent form deep-water barriers (Peru–Chili Trench).
4. Subduction zones dipping away from continents may have island arcs that aid in breaking down barriers (subduction zone from Burma to New Hebrides).
5. Midplate volcanoes, perhaps related to mantle plumes, help break down deep-water barriers (Hawaiian Islands).
6. Subduction zones, spreading ridges, and associated island arcs at high angles to continents may provide migration pathways breaking down barriers (Aleutian Islands) (Valentine, 1971a).

Some of the most obvious and best known examples of the effect of tectonic history are on terrestrial biotas. The distinctive biota of Australia is an excellent example of this effect (Keast, 1981). Before the advent of humans, only a few species had been able to penetrate the oceanic barrier around Australia since Cretaceous or early Cenozoic time. Most members of the biota of Australia seem to have evolved from Cretaceous ancestors, which probably entered Australia from Antarctica. In late Cretaceous to early Cenozoic time, Australia began moving northward, severing its land connection with Antarctica and the rest of the terrestrial world. Establishment of

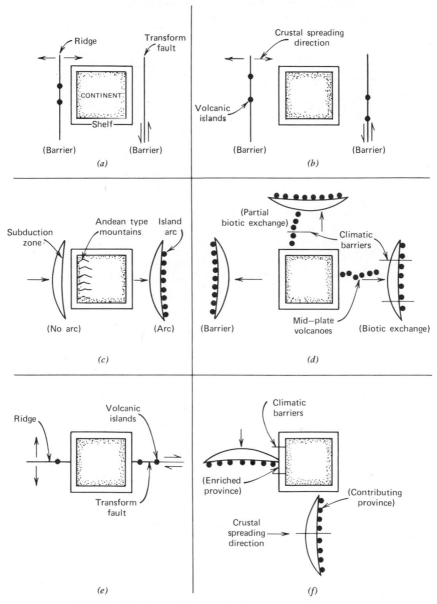

Figure 10.6 Relationships between crustal plates and continents as they affect biotic distributions. See Table 10.1 and text for a detailed explanation of each case. From Valentine (1971a); reprinted by permission from Systematic Zoology, *20*:261.

TABLE 10.1   Relationship of Crustal Plates to Continental Margins and Their Effect on Biogeography[a]

Geometry of Relation	Character of Margin	Distance	Biogeographic Implications for Continental Shelf	Fig. 10.6
Parallel	Ridge or transform	Near	Barrier but with depauperate provincial outliers on isolated islands	a
		Far	Barrier	b
	Subduction	Near	Barrier if truly marginal with no island arc; source of rich biota and dispersal route if arc present	c
	Zone	Far	No effect unless intervening region bridged by midplate volcanoes, then source of rich biota and dispersal route if no climatic barriers intervene	d
High angle	Ridge or transform	Near	Little effect, with depauperate provincial outliers on isolated islands	e
		Far	Not a case	
	Subduction	Near	N-S shelf, E-W arc: arc system a source of rich biota for local province. E-W shelf, N-S arc: proximal province of arc system a source of rich biota for entire shelf	f
	Zone	Far	Not a case	

[a]From Valentine (1971a).

deep-sea barriers surrounding Australia has been effective in preventing immigration of most outside species. Of course, humans have artificially broken down that barrier by the introduction of many outside species: for example, rabbits and dogs, in addition to humans.

South America had a similar history of isolation from the rest of the world beginning in Cretaceous or Early Cenozoic time, when spreading of the South Atlantic Ocean basin separated it from Africa. The isolation was not as complete as that of Australia. Primitive placental mammals had apparently already arrived and a few more advanced rodents and monkeys managed to penetrate the oceanic barrier, perhaps via volcanic islands (McKenna, 1980). Isolation of South America lasted until Pliocene time. With the development of the Isthmus of Panama there was a "great American interchange" of biota between North and South America (Marshall et al., 1982; Stehli and Webb, 1985). Some South American animals such as the opossum and the armadillo have migrated to North America and many North American animals have migrated to South America.

A related marine example is provided by the biotas of the Central American shelf and adjoining areas. Before the development of the Isthmus of Panama, no land barrier existed between the Caribbean and Atlantic on the one hand and the Pacific on the other. Since Pliocene time, distinctive biotas have developed on opposite coasts of Central America and adjoining areas of North and South America. Hallam (1973b) has pointed out that often the disappearance of a barrier to terrestrial migration is accompanied by the formation of a marine barrier, and vice versa (Fig. 10.7). An aspect of these faunas relating to their evolutionary history is that many species have a similar but slightly different counterpart on the opposite side of the land barrier. This is an excellent example of vicariance, evolution of the species from the same parent stock, which occurred in both areas before the formation of the land barrier. One might speculate about what would happen if the Panama land barrier were broken by the construction of a sea-level canal. The long-term effect might well be significant in terms of the extinction of many species and the spread of others.

Historical changes in temperature have also had pronounced effects on the distribution of organisms. The positions of temperature barriers are constantly moving through time, allowing organisms to change their distribution as the barriers move. This effect was especially pronounced during Pleistocene time. Many effects of fluctuating climatic patterns are apparent in the distribution of modern organisms. In terrestrial biotas there are numerous relict populations of species occurring far outside the main populations of that species. Hesse et al. (1951) cite several examples of glacial relicts in Europe and North America, such as the arctic ptarmigan and varying hare, which are now found in isolated populations in mountaintop localities such as the Alps, Pyrenees, and Caucasus. Many arctic plants and animals have been described far south of their normal range on Mt. Washington in New Hampshire. The populations have managed to continue to live in these relict locations by

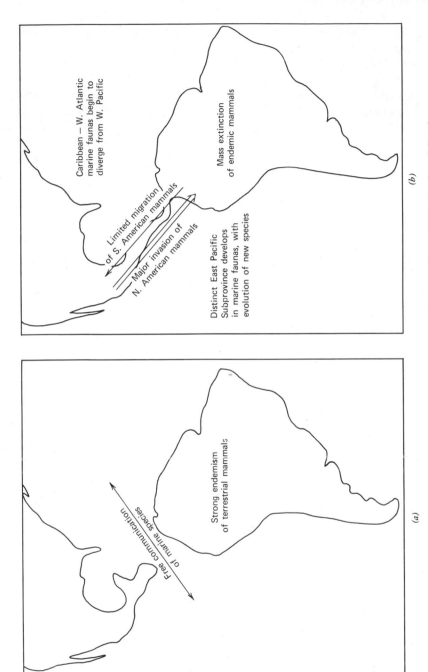

Figure 10.7 Changes in patterns of terrestrial and marine faunal distribution following creation of the Central America land bridge: (*a*) pre-Late Pliocene; (*b*) Late Pliocene–Pleistocene. From Hallam (1973b).

staying in areas such as on mountaintops, which are cooler than most of the surrounding areas.

Many similar marine examples have been described. Several species found in the Baltic Sea are relict populations of species that otherwise occur only far north in the Atlantic (Segerstråle, 1957). At one time, seawater temperatures were apparently low enough that the species extended through the Straits of Kattegat and throughout the surrounding area. Now the relict populations occur only in the northern portion of the sea far from the main area of distribution of the species.

Many shallow marine species on the southeastern U.S. coast show a disrupted distribution because of the high temperatures in south Florida (*disjunct endemism*). The species (e.g., *Mercenaria mercenaria, Mytilus edulis, Crassostrea virginica*) occur on the Atlantic coast and the Gulf of Mexico coast but not in south Florida, where the temperature is too warm (Valentine, 1963). The general similarity of the Gulf Coast and the Atlantic faunas suggests that the temperature in south Florida has not always been too warm to act as a barrier to distribution.

These are but a few of many examples of the historical effects. One of the most important contributions of paleoecologists to the study of biogeography is a better understanding of this process, which leads to a better reconstruction of earth's history.

ISLAND BIOGEOGRAPHY

The geographic range of species will change during its existence. Ideally, a species that develops by geographic speciation of a small, isolated population will start with a small geographic range. This range will increase with time as the species migrates into new areas and perhaps will eventually decline as the species becomes extinct. The processes of species migration and range changes are well suited for theoretical and statistical treatment. MacArthur and Wilson (1967) have published a much quoted study of this sort based largely on data on the populating of islands. Although this work is based on the biota of actual islands, it is more generally applicable. Areas of suitable habitat for a given group of species can be considered as *habitat islands* with intervening unsuitable areas as barriers. Freshwater lakes surrounded by a sea of land would be such an analogy.

MacArthur and Wilson reason that the number of species living on an island will be determined by a balance between immigration rate of new species and local extinction of species already on the island. This is shown graphically in Fig. 10.8, in which the descending curve represents immigration rate. The greater the number of species already on the island, the slower will be the rate of addition of new species. Ultimately, if all available species are on the island, the rate of immigration of new species must be zero. The ascending curve in Fig. 10.8 represents the local extinction rate. Of necessity,

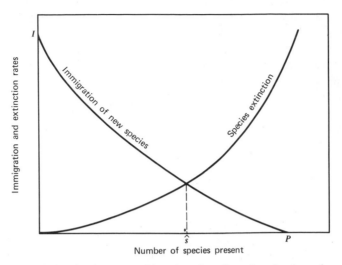

Figure 10.8 Variation in the rates of immigration, I, and extinction of species on an island as a function of the number of species on the island. *P* is the number of species available for immigration. The equilibrium species number is reached at the intersection point, $\hat{S}$, of the curves. See text for a detailed discussion of this model. From MacArthur and Wilson (1963), *Evolution*, 17:375.

that curve must start at zero because if there are zero species, there are none to become extinct. When there is a large number of species, the rate at which they are becoming extinct will be greater. At some point, *s*, immigration rate and extinction rate will be equal. The number of species corresponding to *s* is the equilibrium value for that particular island.

Each island will have its own characteristic immigration and extinction curves, but certain relationships should exist between the curves. For example, the immigration rate for distant islands should be lower than that for near islands because of the greater width of the barrier to be overcome (Fig. 10.9). Consequently, far islands should have fewer species than do near islands. Small islands should have less space for species and also a lesser variety of habitats and thus the extinction rate should be higher than for larger islands (Fig. 10.9). As a result, large islands should have more species than do small islands. In fact, MacArthur and Wilson have noted an excellent correlation between the size of an island and the number of species it supports. Limited data also indicate that far islands do have fewer species than near islands (Fig. 10.10).

Since the pioneering work of MacArthur and Wilson, many researchers have attempted to test their model with varying degrees of success [see Pielou (1979), Simberloff (1983), Cox and Moore (1985), and Case and Cody (1987) for discussions and reviews]. Various modifications of the original theory have been suggested. These modifications take into account factors such as (1)

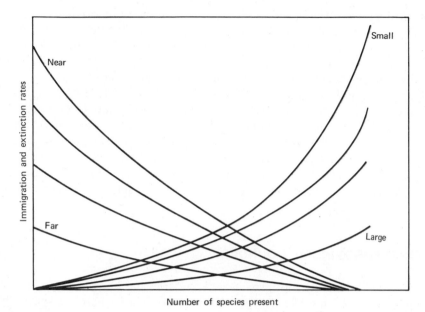

Figure 10.9 Equilibrium model as in Fig. 10.8 for biotas of islands of varying size and distance from the principal source area. From MacArthur and Wilson (1963), Evolution 17:376.

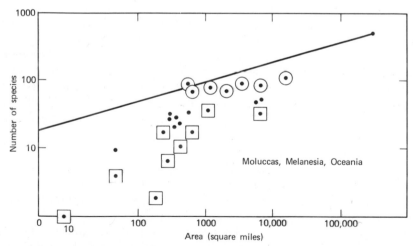

Figure 10.10 Number of land and freshwater bird species on Pacific islands as a function of the area of the islands. Islands that are near (less than 500 miles) a major landmass are circled. Islands that are far (more than 2000 miles) from major landmass are in squares. Islands at intermediate distances are dots. From MacArthur and Wilson (1963), Evolution, 17:377.

historical legacy of the island, (i.e., the nature of the biota originally occupying the island will affect its ultimate composition); (2) interaction between species, which will influence the migration and extinction patterns of those species; and (3) level and diversity of resources on the island, which also affect the diversity of species that can be maintained on the island. Based on a study of biotas on islands in the Gulf of California, Case and Cody (1987) emphasize the importance of colonization ability and persistence ability of the species in determining both species number and the degree of endemism of island biotas. The general outlines of the MacArthur and Wilson theory seem to apply, but the theory does not take into account many details that complicate the picture. Perhaps the main value of this approach is that it gives a conceptual framework for considering the process and results of colonization of islands, either real or habitat.

DEFINING PROVINCIAL BOUNDARIES

Traditionally, recognition of provincial boundaries has been largely a subjective process somewhat like the recognition of boundaries between species by taxonomists. Through the years, biogeographers have attempted to develop means for objectively determining natural boundaries between provinces. These efforts have met with varying degrees of success. Usually, the provinces recognized by these methods agree with those determined subjectively!

The paleobiogeographer often deals with data from a limited number of localities. In these cases precisely defining boundaries between provinces is not possible. Determining the degree of similarity between biotas of two or more different areas may be of more importance than determining the location of boundaries. Simpson (1960) suggested an approach to measure similarity (or provinciality). He did not attempt to define formal provinces by any quantitative measure but simply compared the degree of similarity of the biota between two areas by calculating a similarity coefficient, now called the *Simpson similarity coefficient* (or sometimes the *provinciality index*, PI)

$$S = C/E$$

in which C is the number of taxa in common between two areas and E is the number of taxa restricted to the area with fewer taxa. Many other similarity coefficients have been defined by various researchers and several have been used in paleobiogeographic studies; for example, Valentine (1966) used the Jaccard coefficient (see below). Johnson (1971) used the Simpson similarity coefficient to compare the Devonian brachiopod faunas of the Appalachian and Great Basin regions of the United States with the western and arctic Canada region. His analysis showed variation with time in the faunal similarity in both of these comparisons, suggesting that barriers existed part of the time but not all of the time between the regions.

Pielou (1979) discusses a number of objective methods of defining provincial boundaries. The method of *divisive information analysis* allows an area to be subdivided in a stepwise fashion into pairs of areas with a maximum of similarity within the subdivisions and minimal similarity between areas. This is done by calculating an *information content coefficient* which measures the similarity between samples within the area being investigated.

The method of *biogeographic co-ranges* is adapted for recognizing provinciality along a linear trend such as a coastline or continental shelf area. Species ranges along the trend (such as degrees of latitude) are plotted on *x-y* plots. Species with similar ranges will form clusters on these plots. The cluster may be regarded as a province, but as clusters overlap, objective definition of boundaries is not possible.

Perhaps the most effective means of defining provincial boundaries is by use of an *ordered similarity matrix* or OSM (Pielou, 1979, 1983). This method was first proposed by Valentine (1966) for molluskan provinces along the Pacific coast of North America (Fig. 10.11). He calculated a matrix of Jaccard similarity coefficients (an OSM) by considering the area bounded by each degree of latitude along the coast as a separate sample. Thus each 1-degree segment of coastline was compared to all other 1-degree segments on the basis of presence or absence data on the occurrence of molluskan species. Two segments that had many species in common, of course, had much higher Jaccard coefficients than segments with dissimilar biotas. Usually, the most similar coastal segments are those that are adjacent to each other. Widely separated segments have low similarities. Side-by-side segments in areas where marked biotic changes occur have lower similarity than side-by-side segments in areas with uniform biotas. This results in blocks of high similarity which are separated by breaks in similarity as shown by the Jaccard coefficient. Clusters can also be recognized by performing a *Q*-mode cluster analysis on the OSM (Fig. 10.12). Clusters recognized by this process correspond quite closely to previously subjectively defined molluskan provinces.

As originally used, this method has the advantage of objectively showing breaks in the biotic distribution. It is still subjective, however, because one has to decide at what level of similarity a cluster (or province) will be recognized. Pielou has attempted to overcome this problem by comparing the OSM or part of an OSM to one that would be generated by random mixing. Departure from randomness indicates an ordering of similarity within the matrix and presumably, provinciality. Pielou pointed out that the OSM method need not be restricted to biogeographical studies but can be used to recognize patterns of many sorts in both space and time. For example, Pielou (1983) used the method to study changes in benthic foram faunas in sediment cores from the Bay of Fundy.

None of these methods attempts to define provincial boundaries directly on the basis of the definition of biotic provinces as suggested by Valentine, that is, as a region in which communities maintain characteristic taxonomic compositions. Ideally, all communities should be studied and variations noted. Areas

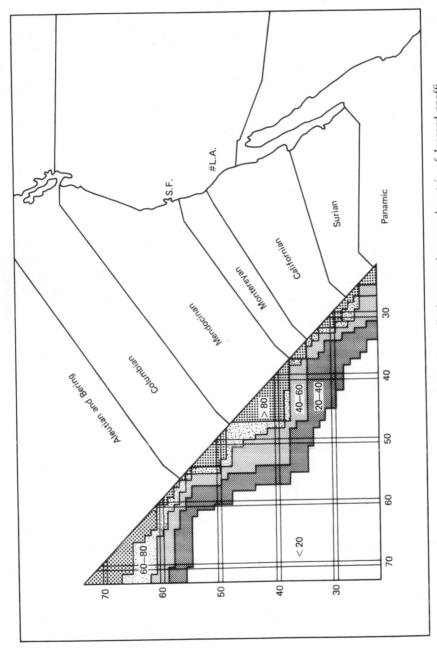

Figure 10.11 Modern northeast Pacific molluskan provinces and matrix of Jaccard coefficients of similarity for 1° latitude increments, based on shelled benthic gastropods and bivalves. From Valentine (1966).

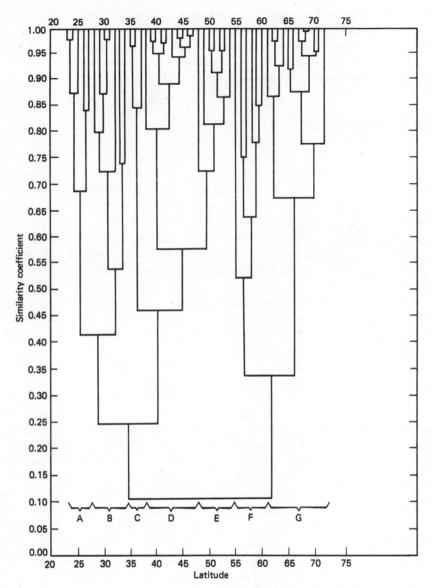

Figure 10.12 Dendrogram showing clustering of mollusks of each degree of latitude with those of other degrees of latitude with similarity determined by the Jaccard coefficient. The weighted pair-group method of clustering was used. A, Surian; B, Californian; C, Montereyan; D, Mendocinian; E, Columbian; F, Aleutian; G, Bering. From Valentine (1966).

where many communities changed in taxonomic composition would be provincial boundaries. The question remains as to how much change should be considered significant and how many communities have to show the change. Because provinces are by nature interpretive constructs and not really natural entities, the placing of boundaries probably will remain subjective.

MODERN PROVINCES

Wallace (1876) recognized five terrestrial zoogeographic regions (Fig. 10.13), and modern zoogeographers still use his basic classification. The northern regions, the Palearctic and Nearctic, are sometimes combined as the Holarctic region because they are less distinctive than the other regions. The Neotropical and the Ethiopian regions are peninsulas extending into the southern hemisphere and are semi-isolated by oceans and climates. The Oriental region is isolated by oceans on the south and mountains on the north. The Australian region is most isolated, being completely separated from other regions by water.

Floral regions which are essentially identical to the zoogeographic regions have also been recognized (Fig. 10.14; Good, 1974). Although different names are used for the floristic regions, most boundaries are nearly the same as those for the zoogeographic regions. A separate small region is recognized in South Africa. An Antarctic region is also recognized, and the Indo-Malaysian region is subdivided into three major subregions.

The geographic distribution of the terrestrial biota, especially the plants, can be shown as biomes (Fig. 10.5). These are not provinces in the usual sense but complexes of related biotas. Nevertheless, they are useful in studying the geographic distribution of terrestrial life. In fact, they are perhaps more useful in paleoecologic studies than are the biotic regions because their distribution is strongly controlled by environmental parameters, especially temperature and precipitation. Comparison between modern and ancient biomes has been used especially in palynological studies of Pleistocene sediments (e.g., Davis, 1969).

Modern marine shelf provinces (Fig. 10.15) clearly show the effect of the depth–elevation and temperature factors. Boundaries between provinces are between north-south coastal segments or are continental masses or deep oceans. Valentine (1973b) has pointed out that equatorial provinces are broad because the same temperature gradient is repeated north and south of the equator. The arctic and Antarctic provinces are relatively broad because water temperatures in these regions are uniformly near or at freezing. North-south coastlines in the temperate regions have many small provinces because of the steep temperature gradients (and concentration of biogeographers?). East-west coastlines have large provinces because they parallel isotherms (and thus have uniform temperatures).

Five shelf provinces have been recognized on the east coast of North America (Caribbean, Gulf, Carolinian, Virginian, and Nova Scotian). Five

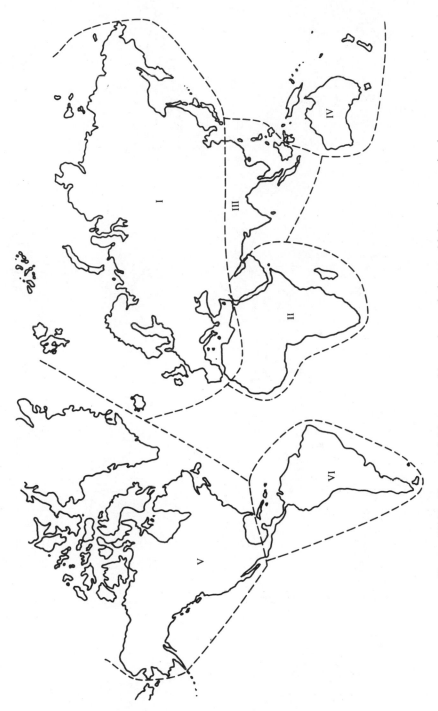

Figure 10.13 Terrestrial biogeographic regions of the world based on animals. Zoogeographic regions: I, Palearctic; II, Ethiopian; III, Oriental; IV, Australian; V, Nearctic; VI, Neotropical. From Good (1974).

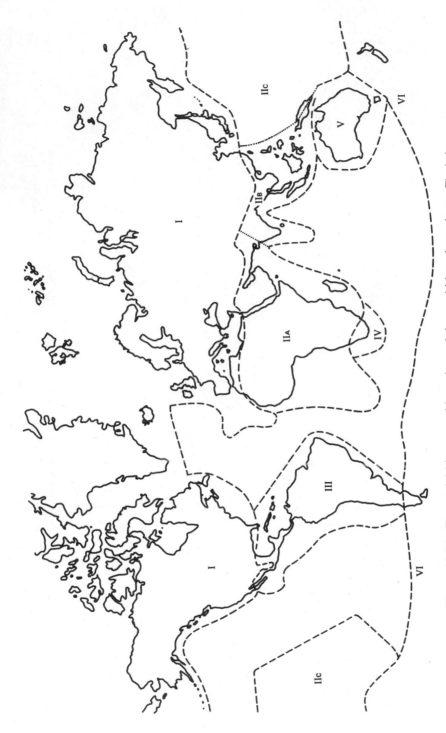

Figure 10.14 Terrestrial biogeographic regions of the world based on plants. Floristic regions: I, boreal; IIA, paleotropical (African); IIB, paleotropical (Indo-Malaysian); IIC, paleotropical (Polynesian); III, neotropical; IV, South African; V, Australian; VI, Antarctic. From Odum (1971).

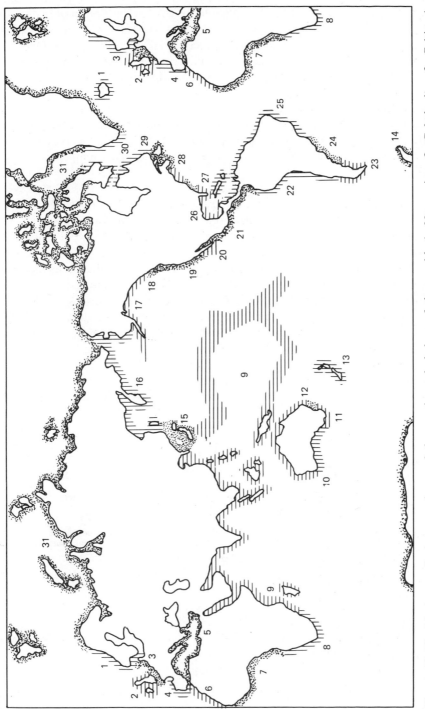

Figure 10.15 Marine molluskan provinces of the continental shelves of the world. 1, Norwegian; 2, Caledonian; 3, Celtic; 4, Lusitanian; 5, Mediterranean; 6, Mauritanian; 7, Guinean; 8, South African; 9, Indo-Pacific; 10, South Australian; 11, Maugean; 12, Peronian; 13, Zeolandian; 14, Antarctic; 15, Japonic; 16, Bering; 17, Aleutian; 18, Oregonian; 19, Californian; 20, Surian; 21, Panamic; 22, Peruvian; 23, Magellanic; 24, Patagonian; 25, Caribbean; 26, Gulf; 27, Carolinian; 28, Virginian; 29, Nova Scotian; 30, Labradorian; 31, Arctic. From James W. Valentine (1973a), Evolutionary paleoecology of the marine biosphere, copyright 1973, p. 356; reprinted by permission of Prentice-Hall, Inc., Englewood Cliffs, N.J.

411

provinces have also been named on the west coast (Panamic, Surian, Californian, Oregonian, and Aleutian). The Oregonian province has been further
subdivided into three subprovinces (the Columbian, Mendocinan, and Montereyan) by Valentine (1966). Six provinces are recognized in Europe and northern Africa (Mauritanian, Mediterranean, Lusitanian, Celtic, Caledonian, and
Norwegian). All of the North American and European provinces are clearly
determined by temperature gradients. The depth–elevation factor accounts
for the occurrence of three groups of provinces across the same temperature
gradient.

Provinciality can be recognized in pelagic biotas of the open oceans, but
formal provincial names have not been generally used (van der Spoel, 1983).
As indicated above, the pelagic biota is relatively uniform within a given
water mass, so that the biotic regions can simply be named for the water mass
with which each is associated. Many groups (e.g., the geologically important
planktonic foraminifera, diatoms, coccoliths) are much more diverse in the
warmer-water masses, with only a few cosmopolitan genera and species at
high latitudes. Many workers who have studied these groups discuss high-
latitude, low-latitude, and transitional biotas. Biogeographic studies of the
pelagic biota of Mesozoic and Cenozoic sediments have been especially common since the advent of the Deep Sea Drilling Project has made available
numerous deep-sea cores. These studies have especially emphasized the use
of paleobiogeography (1) to study past climates, as in the CLIMAP program,
which studied Pleistocene climates (Fig. 10.16; Cline and Hays, 1976); (2) to

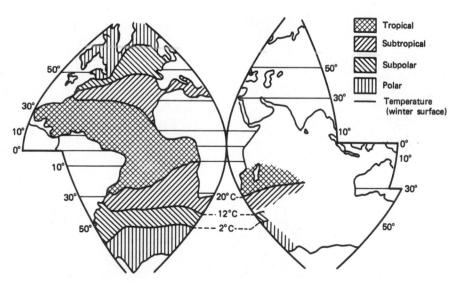

Figure 10.16 Biogeographic regions of the Atlantic Ocean based on the distribution
of planktic foraminifera on the sediment surface. From Imbrie and Kipp (1971), copyright 1971 by Yale University.

study temperature fluctuations (e.g., Haq et al., 1977); and (3) to study ancient circulation patterns. Based largely on the paleobiogeography of fossil plankton, Berggren and Hollister (1977) traced the development of oceanic circulation from a relatively tranquil state in the Mesozoic to the more vigorous system of today.

Knowledge of the modern benthic biota of the deep sea is still incomplete. A distinctive pattern of provinciality occurs below about 4000 m depth (Menzies et al., 1973). In shallower water the biota is very uniform, with many species occurring in all oceans. Below about 4000 m, however, the ocean floor is subdivided by oceanic rises and ridges that act as barriers to migration; temperature differences between different bottom-water masses may also help to establish provincial patterns. Bottom water which sinks from the surface in the Antarctic area extends northward into the Atlantic, Pacific, and Indian basins. These tongues of Antarctic water support a characteristic biota that can be traced far north of the Antarctic (Fig. 10.17). Menzies et al. (1973) have recognized five deep-water regions, corresponding to the five major oceans (Fig. 10.17). These have been further subdivided into provinces, and some of the provinces have been subdivided into areas, and in some cases, subareas.

ANCIENT BIOTIC PROVINCES

The composition of the Pliocene biota of the Kettleman Hills is very similar to that of the modern biota along the Pacific coast of North America. This allows a comparison of the fossil biota with that of the various modern provinces. The composition of the provinces has undoubtedly changed some since Pliocene time. In fact, Valentine (1961) used different terms for Pleistocene provinces that in composition are similar to but in locations slightly different from the modern ones. Temperature is the main factor forming the barriers that define provincial boundaries along the Pacific coast. Thus if the fossil biotas can be assigned to the modern province on the basis of the greatest similarity, an estimate can be made of the temperature conditions existing during the life of the fossils.

To determine which modern province most closely resembles that in which the fossil biota lived, we determined the modern provincial ranges of all the fossil genera for which data were available (Stanton and Dodd, 1970). Very wide ranging genera, those that live over the entire range from the Panamic to north of the Mendocinan province, were excluded from the analysis. The province where the maximum overlap of ranges occurs should correspond to the most similar modern province. A perplexing problem of this approach is that some of the ranges of the fossils that occur together do not presently overlap. That is, some genera that lived in the same province in Pliocene time no longer do so. This must indicate that some genera have evolved in terms of their temperature requirements (see Chapter 2). Perhaps in some cases they

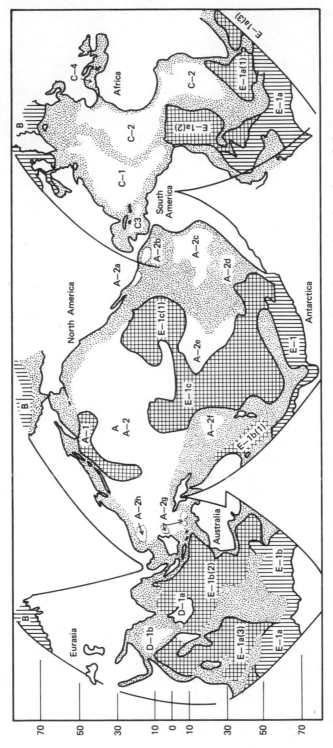

Figure 10.17 Proposed biogeographic subdivision of the lower abyssal regions of the oceans. Stippled areas are less than 4000 m deep and are thus above the lower abyssal depth. The letters and numbers refer to the regions and subdivisions as recognized by Menzies et al. (1973). From Menzies et al. (1973).

have become extinct in a given province because of other factors. Several examples of such changing environmental requirements have been identified.

We used this technique on each of the zones recognized by Woodring et al. (1940) in the Kettleman Hills (Fig. 10.18). The faunas of most of the units consist mainly of temperate species that extend southward. Warm-water genera ranging northward from the Panamic province are uncommon in most units except 9, 18, 21, and 22. These units all appear to have faunas indicating depositional climates that were somewhat warmer than the rest. A relatively small proportion of the faunas consists of genera that are endemic to the Montereyan subprovince or the Californian province. Most of the zones contain faunas that are most similar to those of the modern Montereyan sub-

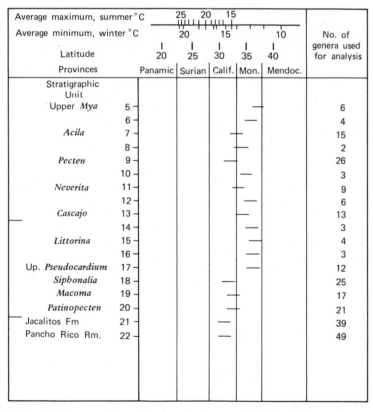

Average maximum, summer °C		25 20 15				
Average minimum, winter °C		20	15		10	No. of genera used for analysis
Latitude		20	25	30	35	40
Provinces		Panamic	Surian	Calif.	Mon.	Mendoc.
Stratigraphic Unit						
Upper *Mya*	5					6
	6					4
Acila	7					15
	8					2
Pecten	9					26
	10					3
Neverita	11					9
	12					6
Cascajo	13					13
	14					3
Littorina	15					4
	16					3
Up. *Pseudocardium*	17					12
Siphonalia	18					25
Macoma	19					17
Patinopecten	20					21
Jacalitos Fm	21					39
Pancho Rico Rm.	22					49

Figure 10.18 Modern provinces having fauna most similar to those of various units within the Neogene strata of the Kettleman Hills. The interpretations are based largely on data from Woodring et al. (1940) (Kettleman Hills); Nomland (1916) and Adegoke (1969) (Jacalitos Formation); and Durham and Addicott (1965) (Pancho Rico Formation). From Stanton and Dodd (1970). Journal of Paleontology, *44*:1100, by permission of Society of Economic Paleontologists and Mineralogists.

province of the Oregonian province. Several of the zones have faunas most similar to those of the Californian province, and some are intermediate. The paleotemperatures of individual stratigraphic units or zones were then determined by comparison with the temperature conditions found in the modern provinces (Fig. 10.18). This analysis indicates that the lower part of the stratigraphic section was deposited under slightly warmer conditions than that of most of the upper part of the section. A return to slightly warmer conditions occurred during *Pecten* zone time, followed by a cooling trend toward the end of deposition of the section.

The relationship between paleobiogeography and plate tectonics is vividly shown in studies of similarity between faunas on opposite sides of the Atlantic and Pacific Ocean basins (Fallaw, 1979, 1983). According to plate tectonic reconstruction, the North Atlantic Ocean began to open in early Jurassic time. At that time the shallow invertebrate marine faunas of western Europe and eastern North America were very similar. Similarity progressively decreased as the ocean basin widened, with the least similarity occurring today. The decreasing similarity is apparently due to the ever-increasing effectiveness of the deep Atlantic Ocean as a barrier to dispersal of shallow marine benthic organisms. Fallaw (1979) studied this progressive change by determining the Simpson similarity coefficient (or provinciality index; see above) for fossil faunas of seven ages, ranging from early Jurassic to Neogene. The similarity coefficient decreases linearly with increasing width of the Atlantic Ocean basin.

Although the Atlantic Ocean basin has been widening since mid-Jurassic time, the Pacific basin has been narrowing as the North American continent has moved westward. One might expect Pacific shallow marine faunas on opposite sides of the basin to become more similar with time. Indeed, Fallow (1983) has shown that they do. He plotted the Simpson similarity coefficient for invertebrate genera on opposite sides of the North Pacific Ocean for 11 times ranging from Early Jurassic to the present. The faunas become progressively more similar and show a strong negative correlation with the width of the Pacific Ocean basin as calculated from plate tectonics reconstruction (Fig. 10.19)

An interesting conclusion suggested by this pattern is that it appears to refute the expanding-earth hypothesis (Carey, 1976). This hypothesis states that tectonic and other features of the earth can be explained by a gradual expansion in the diameter of the earth since Triassic time. If the earth has been expanding, the Pacific basin should not be decreasing in size to compensate for opening of the Atlantic Ocean basin.

A well-known example of provinciality among fossils is in the molluskan faunas of Jurassic age. Arkell (1956), referring extensively to earlier work, recognized three major provinces (or faunal realms) for the Jurassic: Tethyan, Boreal, and Pacific. Much work has been done in more recent years, particularly by Hallam (1969, 1977, 1983) in delineating Jurassic biogeography. Hallam points out that although the Tethyan–Boreal boundary is well defined on the basis of ammonites and belemnites, it is not nearly as clear on the basis of

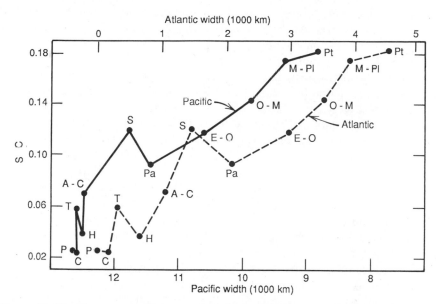

Figure 10.19 Simpson coefficient values (SC) for faunas on opposite sides of the Pacific Ocean basin plotted against average ocean basin width of the Atlantic (dashed line) and Pacific (solid line). P, Pliensbachian; C, Callovian; T, Tithonian; H, Hauterivian; A-C, Albian and Cenomanian averaged; S, Santonian; Pa, Paleocene; E-O, Eocene and Oligocene averaged; O-M, Oligocene and Miocene averaged; M-Pl, Miocene and Pliocene averaged; Pt, Pleistocene. From Fallaw (1983), reprinted by permission of American Journal of Science.

the more abundant bivalves. He further suggests that the boundary is emphasized by a widespread facies change between northern and southern Europe. This results in a different group of communities occupying the Boreal region than occupied the Tethyan region.

Hallam (1977) evaluated the relative importance of latitude (temperature) and sea level (the depth–elevation effect) in determining Jurassic biogeography. Temperature effects can be recognized in the distribution of many bivalve genera which are restricted to a belt within 30° of the paleoequator (the Jurassic tropics?). These include the rudists and other thick-shelled genera. The temperature effect does not seem to be as strong in controlling bivalve distribution as is sea level and tectonics. Hallam recognizes five Jurassic provinces (European, Ethiopian, East Asian, Southwest Pacific, and West American), which are largely controlled by distribution of landmasses.

Hallam also studied details of the distribution of bivalve genera in Europe and western North and South America. Early in Jurassic time differences between European and American faunas were high. At this time exchange between the areas was possible only via circuitous routes around the Pangean continent. Sea-level changes and incipient opening of the North Atlantic by middle Jurassic time allowed more direct interchange of faunas between the areas via the Central Atlantic Seaway. Simpson similarity coefficients between the areas become higher at this time, reflecting the breakdown of the land barrier and fragmentation of Pangea.

One of the most intriguing discoveries of paleobiogeography and plate tectonic reconstruction has been identification of suspect or accreted terranes. These are relatively small blocks of crust that are attached to larger crustal plates. The fossil biota of these exotic terranes may be markedly different from that of the larger plates to which they are accreted. The difference is due to the fact that the blocks were separated by broad expanses of deep water at the time the organisms were living. The amount of dissimilarity is in part a function of the distance between the crustal blocks. Sometimes the biota can be related to a distant paleobiogeographic province. Use of paleobiogeography in conjunction with paleomagnetic and stratigraphic techniques is especially valuable in studying the origin, movement, and timing of incorporation of these terranes.

A good example is provided by the several suspect terranes that have been identified on the western part of North America (Coney et al., 1980). They were attached to the North American Plate at various times during Mesozoic and early Cenozoic time. The Tethyan and mixed Tethyan–Boreal early Jurassic provinces can be recognized on three of these terranes (Fig. 10.20). Their northward displacement since early Jurassic time can be seen from the fact that the boundary between pure Tethyan and mixed faunas on the terranes is far north of its location on the craton. Smith and Tipper (1986) have measured up to 2400 km of displacement for the Wrangellian terrane by this method.

Although paleomagnetic data can be used to determine the paleolatitude of a suspect terrane, it cannot unequivocally distinguish between north and south latitude. Neither can paleomagnetic data be used to determine longitudinal position. Paleobiogeographical information can help solve these problems. In the Jurassic example from western North America discussed above, the Tethyan genera are found in both the northern and southern hemispheres. However, one Boreal genus, *Amaltheus,* is restricted to the northern part of the northern hemisphere, indicating that the suspect terranes were in the northern hemisphere during early Jurassic time. This restriction of high-latitude taxa to one hemisphere has been called the "polar bear principle." Another ammonite genus, *Fanninoceras,* is widespread in lower Jurassic strata of the eastern Pacific, but is not found in the western Pacific. Its occurrence in the suspect terranes of western North America indicates that these terranes were in the eastern Pacific during early Jurassic time.

An example of the use of paleobiogeography to study plate tectonic events in older rocks is provided by the study of the early Paleozoic Iapetus Ocean. The Iapetus Ocean occupied the approximate position of the present Atlantic Ocean relative to North America and Europe during early Paleozoic time. The North American landmass (including portions of present northern Ire-

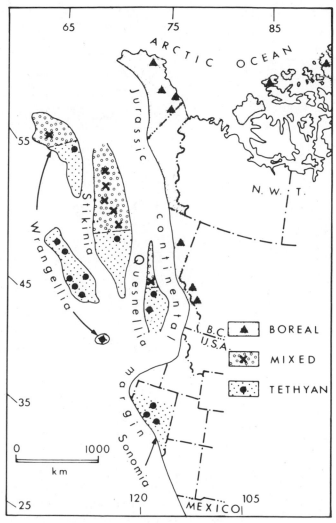

Figure 10.20 Biogeographic zonation in Pliensbachian strata in allochthonous terranes from western North America. From Smith and Tipper (1986). Used by permission of Society of Economic Paleontologists and Mineralogists.

land, Scotland, and Norway) lay to the north of the ocean and the European landmass (including portions of New England, Nova Scotia, and Newfoundland) lay to the south (Fig. 10.21). Indeed, the European continent may have been subdivided into a portion connected to Gondwana and a portion comprising a separate continent called Baltica, which was separated from Gondwana by Tornquist's Sea (Cocks and Fortey, 1982).

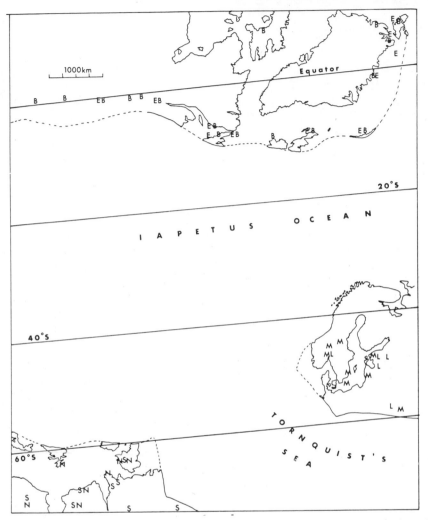

Figure 10.21 Distribution of selected shelf faunas in Arenig (Ordovician) time. B, bathyurid trilobites; E, the mollusk *Euchasma;* M, megistaspid trilobites; L, brachiopod *Lycophoria, Clitambonites, Antigonambonites;* N, the trilobite *Neseuretus;* S, the trilobite *Selenopeltis*. From Cocks and Fortey (1982), copyright Geological Society of London.

Paleomagnetic as well as sedimentologic and paleobiogeographic evidence indicates that North America was located near the equator. This is suggested by widespread apparently tropical carbonate deposits and diverse shallow benthic shelf communities. Gondwana and its included portion of Europe was located at high latitudes, as suggested by predominantly siliciclastic sediments and low-diversity shelf faunas. Baltica may have been located at intermediate latitudes as indicated by the presence of some carbonate sediments and benthic communities of moderate diversity. Pelagic faunas (graptolites, some trilobites, acritarchs, and chitinozoa) are less endemic but more diverse on the North American than the European continental margin, reflecting its lower-latitude position.

Due to subduction, the basin closed in late Ordovician and Silurian time. The "suture" zone of the closure can be seen in northeastern North America, Ireland, Great Britain, and Norway. McKerrow and Cocks (1976) and Cocks and Fortey (1982) note the gradual increase in similarity of the biota on opposite sides of the Iapetus Ocean during Ordovician and Silurian time. They point out that the oceanic barrier has a differential effectiveness in preventing dispersal of different groups of organisms. Provinciality first broke down among the pelagic forms, such as the graptolites. Early Ordovician brachiopod and trilobite faunas on opposite sides of this zone are quite distinct, as they were separated by a deep-water barrier. The faunas are much more similar in late Ordovician rocks, when only a remnant of the ocean remained. Differences in faunal composition persisted longest in the fresh-water fish and the ostracods, which lack a planktic larval stage.

Several puzzling faunal anomalies are not explained by this simple model, however. Neuman (1984) describes several brachiopod faunas in Ordovician rocks in the suture zone which do not clearly match either the North American, Gondwana, or Baltic faunas, suggesting that these are suspect terranes. They contain some species that are characteristic of each of the larger provinces and also a number of species that appear to be endemic to the local area. Neuman interprets these faunas as having lived on islands within the basin or on the margins of the basin. Various sedimentological and stratigraphic features of the associated rocks support this interpretation. The islands were swept into the suture zone when the major plates collided as the ocean basin closed. This distinctive fauna is called the Celtic fauna for its excellent development in Ireland and Scotland. The fauna has many characteristics of modern faunas living on isolated islands such as Hawaii.

11

Temporal Patterns

The paleoenvironmental interpretation of the fossil record is best carried out within the stratigraphic framework of time and space and within the facies mosaic of depositional units. This is because the absolute value of an environmental parameter can seldom be determined precisely, but the relative value of the parameter can be determined at one locality as being more or less than at an adjacent locality. Similarly, biotic gradients are usually sought and interpreted within the stratigraphic framework because the reconstruction of the local environment on the basis of a single fossil assemblage is difficult if the assemblage is isolated by limitations of sampling or exposure from other samples. *Walther's law,* which states that the vertical sequence of facies at a single locality formed by the horizontal shift through time of a laterally arranged pattern of depositional environments and resulting facies, is an indication of the strong role that stratigraphic patterns may play in paleoenvironmental reconstruction. The most evident expression of Walther's law is provided by the lateral and vertical facies distributions during a transgressive–regressive cycle (Fig. 11.1). The potential interpretation of any specific locality in this facies mosaic is both determined and constrained by its stratigraphic position relative to that of other localities.

Lateral and vertical gradients may exist for any paleontologic parameter, as they do for present-day biotic characteristics. Commonly described ones, for example, are taxonomic composition, diversity, and morphology. Lateral paleontologic gradients must reflect differences in the original environment and/or subsequent taphonomic, preservational conditions. Primary lateral paleontologic differences between samples must represent contemporaneous differences in the original environment of deposition. The taphonomic processes that modify the original community of living organisms are environmentally determined, and thus have discrete distributions that can be identi-

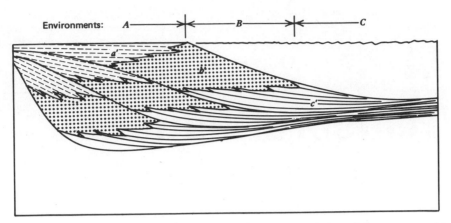

Figure 11.1 Cross section through deltaic sediments consisting of three asymmetric cycles of rapid transgression with little deposition and prolonged regression with major deposition to form a single, primarily regressive, genetic unit. *a', b',* and *c'* are facies deposited in environments *A, B,* and *C.* After Selley (1976), with permission from *An introduction to sedimentology,* Fig. 128, copyright Academic Press, Inc. (London) Ltd.

fied and analyzed as taphofacies. Vertical paleontologic differences within the stratigraphic framework are not as easily explained, however. They may represent changes in the depositional and/or taphonomic environment, as in the case of lateral patterns, but in addition, they may be the result of evolutionary causes more-or-less independent of the external environment. The temporal pattern represented by a vertical sequence in the stratigraphic record is particularly important in paleoenvironmental studies for several reasons: (1) Certainly in core data, and commonly in outcrop, it is better preserved and more likely to be continuously complete than is the lateral record; (2) its environmental interpretation is fairly well established because the vertical sequence has been well studied by sedimentologists, which provides a strong independently based environmental interpretation with which to compare paleontologic conclusions; and (3) the temporal dimension encompasses a wide range of important ecologic and paleontologic phenomena, ranging from short-term succession, driven by internal biological and external physical changes, to longer-term replacement, driven by external environmental changes, and to even longer-term evolutionary phenomena.

The nature of a vertical sequence of fossils, and thus its interpretation, are strongly controlled by the rate of sedimentation. In describing stratigraphic intervals, depositional rate is commonly expressed in terms of centimeters or meters per million years. In paleoecology, however, rate of deposition is important in the much more immediate sense of whether it is great enough to inhibit, overwhelm, or kill, bury, and preserve the community or the individual organism. If sedimentation was very rapid, the fossil assemblage may be essentially a mass-mortality census of the living community, containing only the individuals living at that particular time (see Chapter 8). Two important attributes of such a census population would be that it includes all the organisms that were present, with few having escaped burial, and that it does not include admixtures from previous or subsequent communities. Alternatively, if sedimentation was slow, an individual fossil assemblage may be time-averaged—that is, it may have accumulated over a long period of time and consist of an incomplete and probably biased sampling of many temporally distinct populations and communities (Fürsich, 1978).

Sedimentation is generally slow compared to the life span of most organisms that are potentially preservable as fossils (Olsen, 1978; Glass and Rosen, 1978). Typical rates of deposition on the seafloor distant from land and terrigenous influx are on the order of fractions of a millimeter per year, and on the continental shelf away from shore, deposition since the Pleistocene has been negligible and the seafloor in many places consists of relict Pleistocene sediment. Even in terrestrial and shallow marine settings the rate of deposition is apparently low if calculated by dividing sediment thickness by the time interval represented by the sediments. One might conclude, then, that the resulting fossil record is the time-averaged representation of successive communities. At one extreme, with slow sedimentation the seafloor should be paved with shells and fossils should be the dominant component of sedimentary rocks. From the other extreme point of view, with slow sedimentation, little fossil material should accumulate because destructive taphonomic processes at the substrate should be completely effective before the skeletal material could be buried.

A more reasonable model of the formation of the fossil record is that apparent depositional rate are net and average rates; sedimentation actually is in relatively rapid pulses separated by longer intervals of nondeposition or even erosion. Consequently, the fossil record may more closely approximate sequences of census assemblages buried by distinct but infrequent sedimentation events than it would approximate assemblages of gradually accumulating and time-averaged fragmentary representations of successive communities. If so, the fossil record in vertical sequence largely consists of discrete and disjunct bits of the record of life at that locality. The rare instances of excellent preservation, Lagerstätten, show how good the fossil record can be, and thus emphasize, correspondingly, how poor it is as a result of the usual conditions of poor to nonpreservation during slow deposition, as represented by non- to poorly fossiliferous strata.

In this chapter we concentrate on the shorter-term phenomena of succession and replacement, in which evolution, although it cannot be eliminated, is a minor factor. Examples of these are listed in Table 11.1 and discussed in the latter part of the chapter.

TABLE 11.1   Examples of Temporal Paleoecologic Change, Listed in Order of Increasing Duration

I. Primary Autogenic Succession

A. Benthic marine invertebrates on soft substrate; pioneer community provides firm substrate for succeeding community: Walker and Parker (1976), Johnson (1977), Wilson (1982)

B. Growth of reefs from quiet water below wave base into shallow rough water, and corresponding changes in community of reef builders and reef dwellers: Lowenstam (1957), Nicol (1962), Shaver (1974), Walker and Alberstadt (1975), Hoffman and Narkiewicz (1977)

C. Swamp fills and progrades, with change from marine-brackish to fresh water to land: Raymond (1988)

II. Allogenic Succession or Replacement

A. Biota changes during progressive lithification of the substrate as represented by differences in the life habits of the benthos consisting of burrowing and boring infauna, and epifauna: Goldring and Kazmierczak (1974), Walker and Diehl (1986)

B. Climatic cycles control Triassic lacustrine environment, sedimentation, and fossil assemblage: Olsen et al. (1978)

C. Community changes within marginal marine bay–lagoon system as bar development reduced communication between bay and open ocean: Miller (1986)

D. Community changes within large embayment as communication with open ocean fluctuated due to tectonic and/or eustatic causes: Stanton and Dodd (1972)

E. Biotic changes in Carboniferous cyclothems attributed to eustatic sea-level fluctuations: Rollins et al. (1979), Boardman et al. (1987)

SUCCESSION

Succession has been an important topic in ecology for many years. Consequently, if it can be recognized in the fossil record, ecologic concepts associated with it can be tested and applied to paleoecology. *Succession* refers to the predictable temporal changes that take place within a community that has become established in an area newly opened for colonization. The starting point for succession is most commonly a natural disturbance that removes the preexisting biota—for example, a fire that destroys the vegetation, a flood and rapid sedimentation that buries the marine or lacustrine benthos, or a storm that sweeps away the plants and animals living on a rocky coast. The starting point for succession may also be on new ground rather than following a disturbance—a new surface formed by a volcanic eruption, new land exposed at the edge of a retreating ice sheet, or a new seafloor or land surface resulting from transgression or regression. Much of the study of succession by ecologists, however, has been concerned with changes in the biota following human intervention—how does the original native vegetation reestablish itself on cleared and once-farmed lands, or how can timber be harvested and then the natural succession be modified or controlled to replenish the timber resources most rapidly?

The concept of succession became an important topic in ecology with the work of Clements (1916). He noted that distinctive communities replace one another in a predictable sequence, he characterized these communities as pioneer, mature, and climax stages (or seres) in a succession, and he inferred that these seres were analogous to the young, adult, and old age ontogenetic stages of an individual. Thus the community was viewed as a superorganism in which the constituent species are interacting and tightly interdependent. A corollary of this view of succession as a regular and predictable phenomenon is that community attributes such as diversity, biomass, and productivity, should also change in a predictable way (Fig. 11.2; Odum, 1969). Clements's concept of succession was based on the floral changes that occur on disturbed land, and the reality of this succession is not in doubt. Succession is, however, no longer generally perceived in terms of the superorganism analogy or of marked and discrete changes from sere to sere, but as a continuum of changes determined by adaptations of the individual component species.

In *autogenic succession,* the sequence of biotic changes is the result of internally driven biologic processes within the community, in the absence of external environmental changes. In *allogenic succession,* on the other hand, external changes in the physical environment cause the changes within the community.

Autogenic succession is commonly subdivided into *primary autogenic* and *secondary autogenic* succession. The distinction between the two is based on whether or not the organisms permanently modify their own physical environment. In primary autogenic succession, the biota does modify its habitat and creates new habitat conditions that are more or less suitable both for that community itself and for other communities that may be available to invade the site. The substrate is perhaps the environmental characteristic that is most typically modified in ways that persist beyond the duration of a particular sere. In the classical example of primary autogenic succession in a terrestrial ecosystem, the pioneer and subsequent communities alter the chemical and physical characteristics of the soil through the processes of plant growth and animal activity. The time span for such a succession is much longer than the life span of any of the individual plants because it involves significant and slowly developing changes in soil parameters that concomitantly lead to changes in the biota. Because in primary autogenic succession the biota induces changes in the microenvironment, if the succession is interrupted the starting point for the subsequent succession would theoretically be an intermediate sere rather than the initial starting point.

In secondary autogenic succession, the biotic changes occur independently

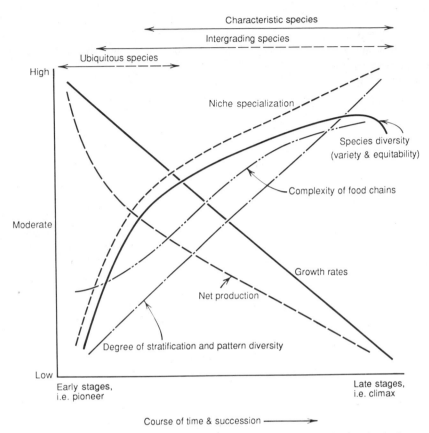

Figure 11.2 Some expected changes in community structure during ecological succession. From Walker and Alberstadt (1975).

of or in the absence of any permanent modification of the environment. Connell and Slatyer (1977) have proposed three ways in which the biota control the succession (Fig. 11.3). In facilitation, the classical explanation, the initial abiotic setting is colonized by a pioneer community that because of the rigorous initial conditions, is composed of species that mature and reproduce rapidly, have a great many offspring, and have relatively short life spans. That is, they are opportunistic or have an *r*-selected life strategy. The diversity and equitability of the pioneer community are low because few species are adapted to the initial conditions, but those that are present may be abundant. The pioneer community alters its local or microenvironment by damping fluctuations in the microclimate, by changing the moisture or organic content of the substrate, and by creating new niches for other organisms as it provides

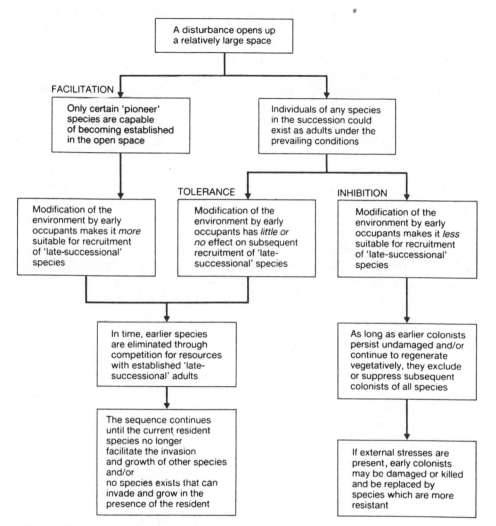

Figure 11.3 Mechanisms of secondary autogenic succession according to Connell and Slater (1977). From Begon et al. (1986).

nutrients and new physical and biological dimensions to the environment. As succession continues, however, early dominants are replaced by others; diversity, equitability, and the proportion of opportunistic (r-selected) species to equilibrium or K-selected species should all change. (In contrast to r-selected species, K-selected species are more specialized or have narrower niches, reproduce at lower rates, and have longer life spans.) The final or climax stage of succession is theoretically stable and will persist until disturbed. In contrast to the facilitation mechanism, in the inhibition mechanism, the early organisms inhibit others from becoming established or, if present, from achieving any prominent role in the community. The later seres of the succession are only able to develop as the earlier organisms disappear. In the tolerance model, all of the organisms that are going to be present in the succession are able to tolerate one another, but the dominants at each stage are those best adapted to those particular conditions, and thus will prosper and outcompete the others.

When a secondary autogenic succession is disturbed, it is commonly by a short-term catastrophic event such as fire, storm, or epidemic. Because in theory the physical environment has not been permanently changed by the biota, as during a primary succession, the community that reestablishes itself after the disturbance will be an early sere in the succession. If the disturbance has been drastic, as after a severe fire, the new community might be the pioneer one; if the disturbance is mild, the existing community might be replaced by a sere only a stage or two earlier in the succession. If the environment is disturbed frequently relative to the time span required for the complete succession to develop, the communities characteristic of the later stages of the succession may never or only rarely occur. Thus, in seasonal environments, succession may be of annual duration, beginning in the spring, running through the summer and fall, and terminating with the cold weather of winter. Examples of settings in which annual succession occurs would be high-latitude arctic tundra and strongly seasonal temperate bays, lakes, and terrestrial habitats with communities not dominated by perennial plants. Succession may be an even shorter-term phenomenon, as in ephemeral ponds. In this case, the usually attained last community is as close as the succession comes to a climax condition, and it is not possible to define the climax community to which the succession might develop if there were time. If disturbance is infrequent, however, secondary succession may continue to the climax sere, which will remain as the equilibrium condition for an indefinite period. The time span to reach the climax stage should not be longer than several times the life span of the dominant species in the climax community. This is because the succession consists of several seres, and because the life spans of early-stage species tend to be shorter than those of later-stage species. Thus the duration of most secondary autogenic successions will be measured in years or tens of years. In an extreme case, such as a *Sequoia* forest, it will be measured in thousands of years.

Allogenic succession consists of subevolutionary biotic changes caused by changes in the external environment. An example would be the community changes associated with the gradual filling of an estuary or lake with terrigenous sediments. Conceptually, allogenic succession is indistinguishable from replacement, but generally, replacement is more commonly used for longer time spans, allogenic succession for shorter ones.

These definitions are straightforward, but actual examples may not fall neatly into one another of these pigeonholes. If a shallow lake or swamp fills up, becoming progressively a marsh, bog, and eventually dry land, the accompanying succession of plant communities would be an example of allogenic succession or replacement if the filling were largely through the influx of detrital terrigenous sediment. If the filling was from autochthonous plant material, on the other hand, the biotic sequence would be an example of primary autogenic succession because the microenvironmental change was caused by the biota itself, and would be comparable to the process of soil formation in the classic example of primary autogenic succession in terrestrial habitats. To the extent that the filling is a combination of terrigenous allochthonous sediment and authochthonous plant material, the effect would be intermediate between allogenic and primary autogenic succession. At any point in the overall sequence, if the biota were removed by a disturbance such as fire or storm, the short-term sequence as the biota became fully reestablished would represent secondary autogenic succession.

As another example, if localized and prolific productivity of marine organisms resulted in a buildup or reef through the accumulation of skeletal material, community changes as the reef grew into shallower and more agitated water would represent primary autogenic succession. Again, in contrast to the overall biotic sequence of the primary succession, a shorter-term secondary autogenic succession could occur at any point of reef growth.

TEMPORAL PROCESSES IN PALEONTOLOGY

The recognition of temporal processes in paleontology depends on the likelihood that they will be preserved in the fossil record. This likelihood depends on whether the particular process is of long enough duration to be preserved within the normally slowly accumulating stratigraphic record. That is, if very little sediment has accumulated during the short time span of succession, the fossil record will be so condensed that the seres within the succession cannot be identified.

Secondary autogenic succession is not likely to be preserved in the geologic record because the duration of a secondary succession is short relative to geologic time and to the average rate of sedimentation, but long relative to the short term. Secondary autogenic succession in plant communities is not likely to be preserved by plant macrofossils because of their limited preservability and because of the high rate of sedimentation necessary to bury and preserve the successive communities. It is more likely to be preserved in the pollen record if sedimentation is rapid enough so that sequential events are not combined into a condensed horizon, and if bioturbation is absent so that they are not intermixed. Fürsich (1978) has described in detail the conditions of high sedimentation rate and low bioturbation rate that are necessary to ensure that temporally distinct communities will be vertically separated in the sedimentary rock. The absence of unequivocal examples of secondary succession in the fossil record indicates how uncommon these conditions are.

Primary autogenic succession, allogenic succession, and replacement represent increasingly long time spans and thus have increasingly better chances of being preserved. They also are likely to be recognized because allogenic succession and replacement are in response to external environmental changes, so that the biotic changes are commonly correlated with corresponding lithologic changes, and because primary autogenic succession permanently modifies the environment, and thus it too should be correlated with recognizable lithology changes.

Although primary autogenic succession may be a short-duration phenomenon in terms of geologic time, it may be recognizable through the changes made in the environment at each stage of the succession. One of the more common categories is that of a pioneer community becoming established on a soft substrate and creating a hard substrate for subsequent organisms as its shells accumulate. For example, in the Silurian Hopkinton Dolomite of Iowa, the pioneer community was dominated by orthotetacean brachiopods that were able to populate the soft lime mud and crinodial sand substrates, and thus it formed a suitable substrate for the subsequent community dominated by pentameracean brachiopods. Once the apparently necessary firm shelly substrate had been established by the orthotetacean brachiopods, the pentameracean community would persist as the climax stage until the succession was interrupted (Fig. 11.4; Johnson, 1977).

A more specific role of substrate in controlling succession is exemplified by the organisms commonly found encrusting present-day shells on the beaches of the Texas Gulf coast (Fig. 11.5). The encrusting bryozoan uses the empty pectinid valve; the scleractinian coral is apparently unable to attach itself directly to the bivalve but does very well on the bryozoan. Host–substrate specificity of this type has been described as well from the fossil record, and emphasizes that succession may be highly predictable because of constraints imposed by specific organism interactions.

The succession of benthic communities in the Ordovician Chickamauga Group of Tennessee, beginning with strophomenid brachiopods as the pioneer population on a soft substrate, was apparently more complex (Walker and Parker, 1976). Seres that followed the pioneer community were characterized by encrusting bryozoans, ramose bryozoans, and a diverse late-stage, or perhaps climax, community dominated by a rhynchonellid brachiopod and including gastropods, bivalves, bryozoans, and pelmatozoan echinoderms. The succession was apparently terminated by an influx of argillaceous sediment, but was then repeated many times, forming in each instance a limestone bed 2 to 6 cm thick interbedded with the mudstone. In both the Ordovician and Silurian examples, only the very local, substrate aspect of the microenvironment was modified by the pioneer community, and after each disturbance the environment returned to its pre-succession condition.

The shelly benthos of present-day marine soft substrate is largely composed of organisms living in or on the sediment. However, as the shells accumulate on the seafloor, the relative abundance of epifaunal organisms utilizing the shells as an attachment substrate increases. This is both because the shells provide the requisite substrate for the epifauna and because the accumulating shells inhibit infaunal organisms that require a soft-sediment substrate. This relation between epifaunal/infaunal proportion and the substrate has long been recognized—the examples above are special cases of it. Kidwell and Jablonski (1983) have elaborated on it in their thorough discussion and through their appellation of it as *taphonomic feedback*.

Present-day swamps or lakes may gradually be filled by the accumulation of the plant material produced in situ. Consequently, a vertical sequence within the resulting peat will consist of plants adapted to an aqueous habitat, water-logged swamp, drier swamp or marsh, and finally dry land when the lake is filled. The sequence is preserved in the sediment or peat infill by differences in the composition of the plant communities at different levels and by cross-cutting age relationship of the roots from the plants in the successive communities (Raymond, 1988). Building on this model and associated criteria derived from modern examples, Raymond has similarly demonstrated a succession in a Carboniferous coal. The sequence from bottom to top of (1) cordaitaleans, (2) *Psaronius* tree ferns, (3) medulosan seed ferns, and (4) lycopods preserved only as stigmarian rootlets indicates that a brackish marine swamp changed to a freshwater swamp and then built up to form a dry surface. Because there was no contemporaneous terrigenous deposition in the swamp, this is a clear example of primary autogenic succession.

Another important example of primary autogenic succession is the construction of carbonate buildups and reefs, which permanently modify both the substrate and local topography. As organisms, through the production of carbonate sediment and, sometimes, a skeletal framework, create local topographic relief on the seafloor within an otherwise apparently uniform physical environment, they modify the local environment so that lateral differentiation of the benthos occurs. If the reef or buildup is able to grow into progressively shallower water, a succession should occur of communities suited to the progressively shallower and more turbulent environment.

This model of successional change associated with reef growth from deeper and quieter to shallower and rougher water has been widely described. An early example is the description by Lowenstam (1950, 1957) of the growth of Silurian reefs in Illinois and Indiana. Nicol (1962) phrased Lowenstam's description in explicitly successional terminology, and Shaver (1974) and Shaver and Sunderman (1989) have added further detail. In the initial growth stage, these reefs were mounds of lime mud containing a low-diversity assemblage of benthic invertebrates. Through in situ accumulation of biogenic carbonate sediment, the reefs grew upward from deeper, quiet-water conditions into shallower rougher water within an external environment that otherwise was apparently not changing. The corresponding biotic changes (Fig. 11.6) represent primary autogenic succession because (1) at each stage of the succession, activity of the existing community permanently modified the environment for the following community, (2) the duration of the succession was orders of magnitude longer than the life span of any of the species, and (3) the environment was being continuously and permanently modified. Consequently, if the

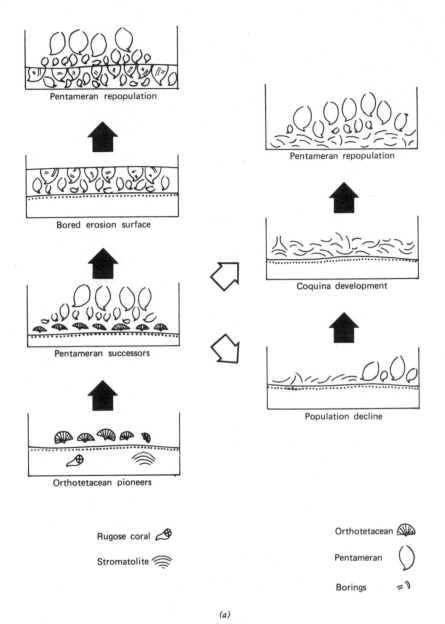

(a)

Figure 11.4 Succession of brachiopod communities in Silurian strata, Iowa: (1) Orthotetacean pioneers form hard shelly substrate for pentameran successors, which remain dominant even after interruptions as long as substrate is suitable; (b) variations in successor communities during a long period of time as a result presumably of environmental differences. After Johnson (1977), Lethaia *10*.

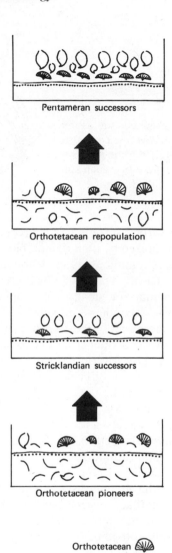

Pentameran successors

Orthotetacean repopulation

Stricklandian successors

Orthotetacean pioneers

Orthotetacean

Stricklandian

Pentameran

(b)

Figure 11.4 (Continued)

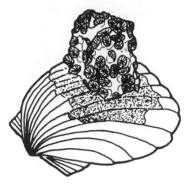

Figure 11.5 Substrate-determined succession. The bryozoan encrusts the *Pecten,* the coral then encrusts the bryozoan. Corals do not attach to *Pecten* shells directly, only where bryozoan substrate is present. Length of shell: 3.8 cm.

environment had been disturbed at some stage of reef growth, the reef surface would not have been recolonized by the original, pioneer, deeper-water community but by some pioneer community of the secondary autogenic succession for that stage of the reef growth. Thus as the reef grew, the normally present biota was probably the climax or steady-state community for that habitat condition and a short-term, secondary autogenic succession that would lead back to the normal community after short-term disturbances. The recognizable succession consists of the sequence of climax communities as the reef environment changed; the shorter-term secondary autogenic successions resulting from disturbance would be difficult to recognize because of the slow rate of sedimentation relative to the duration of the secondary successions at each stage in the primary succession.

This shoaling sequence in reefs and reef communities has been widely recognized and has been subdivided by Walker and Alberstadt (1975) into four successional stages: stabilization, colonization, diversification, and domination. Hoffman and Narkiewicz (1977) have further expanded on this model of succession. However, if the characteristic shoaling pattern of reef development is coupled with the fact that reefs and buildups are narrowly restricted to a limited number of discrete intervals of reef growth, it suggests instead that the model may not represent autogenic succession, but be controlled by external conditions and thus represent allogenic succession or replacement. In particular, the documentation of Devonian reefs within eustatic transgressive/regressive cycles by Hladil (1986), suggests that the apparently self-generated growth into rough water within stable conditions may instead represent reef growth into falling sea level during the regressive limbs of eustatic cycles.

ENVIRONMENTAL CHANGE: ALLOGENIC SUCCESSION OR REPLACEMENT

Primary autogenic succession (change in the environment caused by the organisms) and allogenic succession or replacement (biotic change in response to

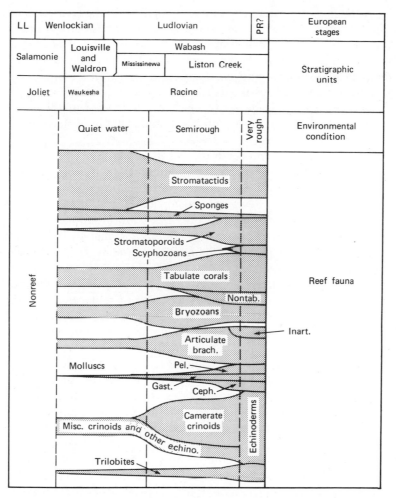

Figure 11.6 Successional changes in community of Silurian reefs of Indiana and Illinois. Inferred cause of change was growth of the reefs from quiet, deeper water into shallow, rough water. Relative abundances are based on volumetric contributions to the reef sediment. Stromatactids are enigmatic structures that are largely probably inorganic. PR, Pridolian; LL, Liandoverian. After Shaver (1974), published with permission of The American Association of Petroleum Geologists.

environmental change caused by extrinsic factors) may be difficult to distinguish. However, only by recognizing and taking into account primary autogenic succession can the biotic changes resulting from allogenic succession and replacement be used to recognize changes in the external environment. In general, environmental interpretations of temporal paleontologic changes have assumed that the paleontologic change resulted from environmental change without taking succession into account. Use of the term *succession* for both autogenic and allogenic phenomena leads to confusion in both terminology and thought [but for the contrary point of view see Walker and Diehl (1986) and Copper (1988)]. As the examples in Table 11.1 show, succession in the restricted, secondary autogenic sense is only rarely a definite factor in temporal change in the fossil record, and the biotic changes normally are examples of replacement, analyzed and interpreted by the methods and criteria described in other chapters of this book, and result in explanations involving a changing environment.

Progressive changes in the benthos as a soft substrate becomes increasingly rigid and finally lithified to form a hardground may be preserved in the fossil record. Such instances most commonly represent a decrease or a cessation of sedimentation followed by dewatering, compaction, and cementation (Walker and Diehl, 1986). The biological processes of bioturbation, which destroy existing bedding, generate biogenic bedding, and change the sedimentary chemistry, will also progressively change the sediment, but the direction of change is not as easily predicted (Aller, 1982[a]). If the change in biota is caused by changes in sedimentation or physical chemical properties of the sediment, this would be an example of allogenic succession or replacement. If the sedimentary changes are the result or organism activity, this would be an example of primary succession. Trace fossils, whether they are burrows in soft sediments or borings in lithified sediment, provide the primary criteria for determining sediment rigidity, and have been discussed in Chapter 6. Additional criteria are listed in Table 11.2, modified from Goldring and Kazmierczak (1974). The full succession that may be correlated with these lithologic changes would start with an initial infauna able to move through soft fluid sediments and an epifauna with specialized morphologic features or life habit adapted for the soft substrate. In both cases, the ability to maintain a stable position and to avoid being smothered by the sediments are prime requirements. With increasing sediment rigidity, however, it becomes less of a problem to maintain a stable position on or in the sediment, burrows become more distinct, and at the hardground end of the spectrum are replaced by borings that cut across grains, fossils, and previous burrows and borings within the sediment. Throughout this progression, epifauna should increase in relative abundance and comprise the bulk of the community at the hardground of the succession. The complete succession is theoretically possible, but is seldom recorded. The sequence more commonly consists of (1) soft sediment with a burrowing infauna, (2) lithification and development of a hardground, and (3) a boring and epifaunal community.

Fluctuations within a marginal marine bay/lagoon system resulting from variation in the degree of isolation of the water body from the open ocean have been described by Miller (1986) in Pleistocene terrigenous strata in North Carolina. A correlated vertical sequence of communities can be described in terms of taxonomic composition, relative and rank abundance of species, dominance and diversity trends, trophic structure, shell borings and

TABLE 11.2   Criteria for Determining the Degree of Consolidation[a]

1. Burrow-in-burrow structure, where successive generations of burrows show progressively sharper margins, less distortion, and increasingly circular cross sections, indicating that the sediment was undergoing an increase in degree of consolidation.

2. Deformed crypts indicate that the organism penetrated firm but not lithified sediment and deformation occurred with subsequent compaction of the sediment.

3. Borings truncating evenly across shells, ooids, oncoids, and older crypt fills indicate that the matrix was as hard as the shells and other clasts.

4. Discontinuity surfaces evenly truncating clasts, shells, and matrix, likewise indicate full lithification of the surface.

5. The hardness of the substrate at the time of penetration may be estimated from the form of the shell and borings of pholads.

6. The absence from a sedimentary unit of burrows penetrating down from the overlying unit may indicate that an increase in consolidation had occurred before the overlying unit was deposited, if it can be shown that nonpenetration was unlikely to have been because of other factors (e.g., depth of penetration required).

7. Shells and other objects of known hardness introduced above the discontinuity surface in the smothering layer and pressed into the surface show that the discontinuity surface was sufficiently plastic to take an impression.

8. Toolmarks scratched on the discontinuity surface provide information, especially if the tool can be identified. A. Pszczolkowski has observed, on the top of a Lower Kimmeridgian discontinuity surface penetrated by bivalves, prod or impact marks probably made by the small calcareous algae *Marinella*.

[a]After Goldring and Kazmierczak (1974).

incrustations, and evidences of predation. By comparison of the Pleistocene communities with present-day communities from the western Atlantic, Miller has concluded that the environment changed from an initial open bay to restricted bay to closed lagoon, and then fluctuated between open and closed lagoons. This detailed interpretation was possible because of the similarity of the Pleistocene fauna to the present-day fauna and because of the well-preserved stratigraphic and sedimentologic control. In general, paleontologic data are not so readily interpretable. Replacement sequences are particularly useful if they are cyclic because then turning points of the cycles, both lithologic and paleontologic, provide reference points from which to analyze intermediate replacement stages.

The oscillation chart of Israelsky (1949) was a pioneer use of cyclic paleontologic patterns. Correlation within the thick basinal Tertiary mudstones of the Gulf coast was difficult then because of the lithologic uniformity and the subtle vertical changes in the foraminiferal fauna. In addition, first or last appearances of species did not seem to be useful for correlation because of strong lateral environmental differences. However, Israelsky determined paleontologic characteristics such as relative abundances of depth-diagnostic assemblages and ratios of pelagic to benthic and arenaceous to calcareous foraminifers for a series of wells along a dip section. He integrated these into what he termed an oscillation chart, which showed from the faunal data the cyclic changes in environment and bathymetry. To construct an oscillation chart, the vertical distribution of environmentally significant foraminiferal assemblages were determined for each well. By arranging the wells in a dip section and tying environmental intervals together from well to well, the cyclic pattern of transgression and regression becomes evident (Fig. 11.7). The oscillation chart provides a paleoenvironmental framework that can be used for further analysis of the sections. In addition, the point of maximum water depth in each well provides a reliable horizon for correlation because the point of maximum water depth is the turning point from the transgression to regression in the cycle, and as such is correlative between wells even though the water depth at each site at that time was different (Fig. 11.6). Thus Israelsky's oscillation chart anticipates the concepts embodied in the coenocorrelation method of Cisne and Rabe (1978) and in present-day ecostratigraphy. The fundamental theme of ecostratigraphy is that biofacies recognizable in the stratigraphic record establish a temporal pattern of community distributions within ecologic space. If communities replace one another because of discrete environmental changes

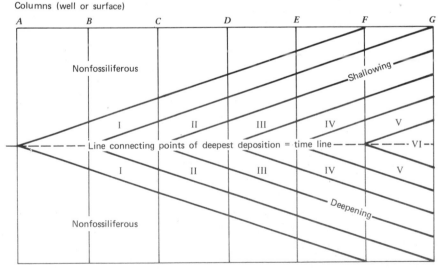

Figure 11.7 Depositional dip section of a marine cycle. Each column is simplified from the data by indicating at any elevation only the dominant assemblage. After Israelsky (1949), published with permission of The American Association of Petroleum Geologists.

or because of some process of community evolution, biostratigraphic datum planes may be generated. Community replacement can be observed readily in the fossil record, and the common assumption is that it reflects environmental change. The reality of community evolution is not as clear and is widely debated. The paleoenvironmental significance in the context of this chapter is that the analysis of the faunal sequence within the stratigraphic framework reveals a cyclic pattern and that this pattern provides the means both for paleoenvironmental reconstruction and for correlation.

Another example of the value of the vertical sequence in paleoecology is provided by the Cretaceous sedimentary rocks and fossils of the western interior of the United States (Kauffman, 1977). The key step in reconstructing the Cretaceous paleoenvironments has been to group both sediments and fossils into cycles (Fig. 11.8) such as those portrayed by Israelsky's oscillation

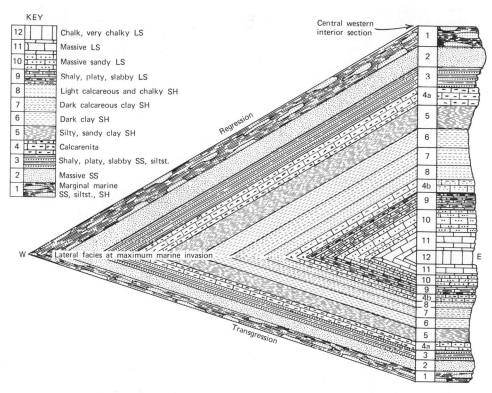

Figure 11.8 Model of sedimentation patterns within a Cretaceous marine cycle. Western Interior United States, showing: distribution and repetition of 12 major lithologic types; complete lithologic development of cycles in the central part of the basin and thinning toward strand (actual sections thicken from basin center to strand of maximum transgression); thinner transgressive than regressive sequences of rocks. After Kauffman (1969), reprinted with the approval of the author and Allen Press, Inc., copyright 1970.

charts. The interpretation of the cyclic pattern of rock types and fossils is in terms of a transgressive–regressive model that serves as a hypothesis to be tested and as a framework for more detailed analysis. For example, in general, cycles within a basin should be correlative as in the oscillation chart unless the geography and bathymetry is being modified by tectonism as well as eustatic sea-level changes. Thus a cyclic model developed in the western interior puts constraints on the interpretation of more seaward Cretaceous strata and fossils in New Mexico or Texas (Fig. 11.9), just as the cycles Israelsky recognized at any one locality had to be compatible with the cycle pattern in the adjacent well.

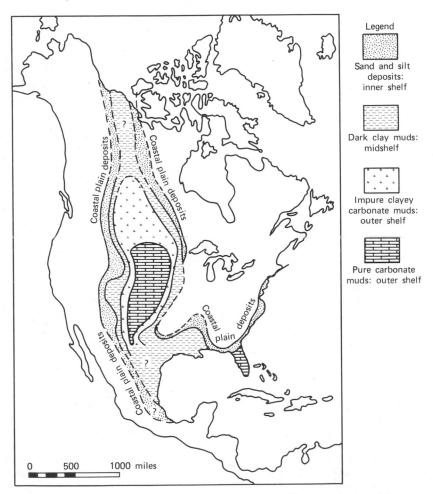

Figure 11.9 Early Late Cretaceous paleogeography of North America. After Kauffman (1969), reprinted with the approval of the author and Allen Press, Inc., copyright 1970.

The model presented by Fig. 11.8 must be tested as well as used. The symmetry of the cyclic model as portrayed in Figs. 11.7 and 11.8 may be misleading, for commonly, deposition during transgression may be much less than during regression, or even absent. Asymmetric cycles are the general rule in carbonate strata (Wilson, 1975b) and are also characteristic of clastic deposits in tectonic areas in basins that attain their maximum depth rapidly but then fill gradually during the regressive phase.

Caution must be exercised, however, in extrapolating too broadly with a general model of this sort, because the classical Triassic Lofer cycles of the Northern Calcareous Alps in Austria and Bavaria record primarily the transgressive limbs of the cycles, with each regressive limb recorded only by a solution surface and residual argillaceous red clay (Fig. 11.10; Fischer, 1964b).

Repeated transgressive–regressive cycles in Pliocene terrigenous strata of the Kettleman Hills of the California coast ranges are represented by distinctive lithologic and corresponding biotic replacement sequences (Fig. 11.11). Each of these cycles, described as a zone in the literature of the area, grades laterally from relatively marine at the northern end of the Kettleman Hills to brackish or nonmarine at the southern end. Within each zone, fossils are distributed vertically in a systematic and cyclical way. Diversity (number of genera) and abundance (number of specimens) of typically marine macrofossils are greatest at the base of each cycle and decrease rapidly upward. These fossils are rare to absent in the upper part of the cycle, where they are replaced by crab remains, fish bones, teeth, and dermal plates, terrestrial plant remains (leaves and stem fragments), and freshwater mollusks. Fragmentary remains of terrestrial vertebrates are most common either more or less in place in the upper part of the cycle or as reworked residual material in the basal gravel of the cycle.

The vertical variations in lithology parallel those of the biota but are not as evident because the pattern is less uniform from the cycle to cycle. The base of each cycle is generally marked by a bored and irregular erosion surface overlain by a thin basal pebble conglomerate or pebbly sandstone. The pebbles are mudstone clasts derived locally from erosion of the underlying strata, and igneous and metamorphic clasts derived from basement sources to the west. Dominant rock types consist of laminated or cross-bedded sandstone low in the cycle, overlain by bioturbated sandy mudstone and muddy sandstone. Strata in the upper parts of the cycles are predominantly dark green to gray claystone and thin-bedded to laminated tan siltstone with locally abundant plant remains; lenticular channel deposits of muddy, medium- to fine-grained sandstone with locally abundant plant remains are less common. Most of the channels were probably fluvial, but some contain shark teeth and marine fossils in the basal lag gravel and may have been tidal or estuarine. Probable aeolian dune deposits are found only in upper parts of the cycles, as are thin discontinuous beds of massive to laminated argillaceous dolomite. The dolomite beds contain rare ostracods, and adjacent mudstones may contain freshwater mollusks; the dolomite probably formed in shallow lakes on, or isolated lagoons bordering, a deltaic plain.

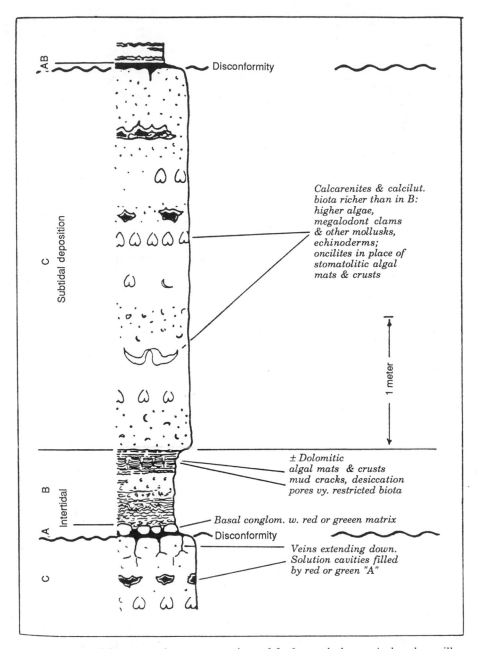

Calcarenites & calcilut.
biota richer than in B:
higher algae,
megalodont clams
& other mollusks,
echinoderms;
oncilites in place of
stomatolitic algal
mats & crusts

± Dolomitic
algal mats & crusts
mud cracks, desiccation
pores vy. restricted biota

Basal conglom. w. red or greeen matrix

Veins extending down.
Solution cavities filled
by red or green "A"

Figure 11.10 Diagrammatic representation of Lofer cyclothem. *A*, basal, argillaceous member, representing reworked residue of weathered material (red or green),
commonly confined to cavities in underlying limestone; *B*, intertidal member of
"loferites" with algal mats and abundant desiccation features; *C*, subtidal "megalodont
limestone" member, with cavities produced by desiccation and solution during succeeding drop in sea level. From Fischer (1964b), Kamnan Geological Survey, Bulletin 169.

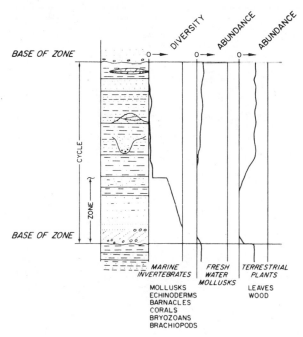

Figure 11.11 General lithologic and paleontologic characteristics of Pliocene cycles. Kettleman Hills, California. After Stanton and Dodd (1972).

The cyclic pattern as described above and illustrated in Fig. 11.11 is a composite derived from the study of outcrops throughout the exposed section of all three domes of the Kettleman Hills. A typical section (the *Siphonalia* zone and small intervals of the adjacent cycles) is described here to add detail to the generalized description and to indicate the range of variability present (Fig. 11.12). Underlying the *Siphonalia* zone is siltstone and very fine sandstone of the *Macoma* zone, with abundant plant fossils that probably were deposited in a very restricted marine-to-swamp environment. A lenticular sandstone at approximately the same stratigraphic level 1½ miles to the northwest of the section plotted is probably an aeolian dune, in harmony with the restricted marine to nonmarine interpretation. In the *Siphonalia* zone, gradients upward include decreasing grain size, sorting, amount of inclined and crossed bedding, faunal diversity and abundance, and increasing abundance of plant fossils. Consequently, the sediments and fossils at the top of the cycle are very similar to those at the top of the underlying cycle. These overall gradients, along with trends and variations in the composition of the fauna, are interpreted as reflecting fluctuations in a general long-term change in the environment from relatively open-marine or outer-bay with normal salinity to highly restricted marine of probably reduced salinity.

The cycle overlying the *Siphonalia* zone consists of siltstone very much like that in the upper part of the *Siphonalia* zone plus lenticular bodies of blue

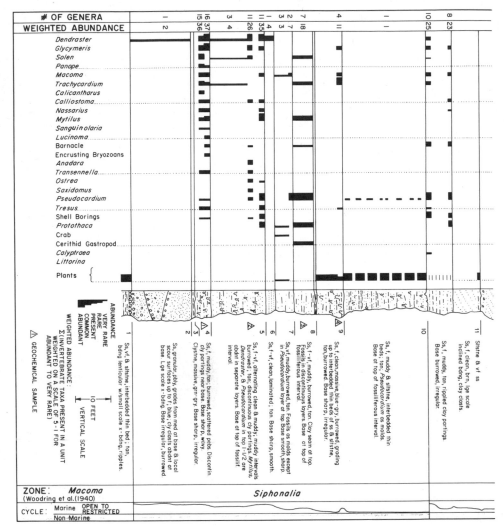

Figure 11.12 Detailed lithologic and paleontologic characteristics of the *Siphonalia* zone, a single cycle in the Pliocene Etchegoin Formation, Kettleman Hills, California. After Stanton and Dodd (1972).

sandstone of diverse genesis. It begins with an erosional bored basal surface of a blue sandstone, then grades upward through cross-bedded blue sandstone, muddier and finer brown sandstone, and into plant-bearing siltstone. Marine fossils are confined to the lower part of the cycle, associated with the relatively clean high-energy, cross-bedded deposits.

The individual cyclic pattern as described in general and in more detail for the *Siphonalia* zone is repeated throughout the Pliocene section, resulting in a cyclic stratigraphic package (Fig. 11.13) grading laterally in each cycle from

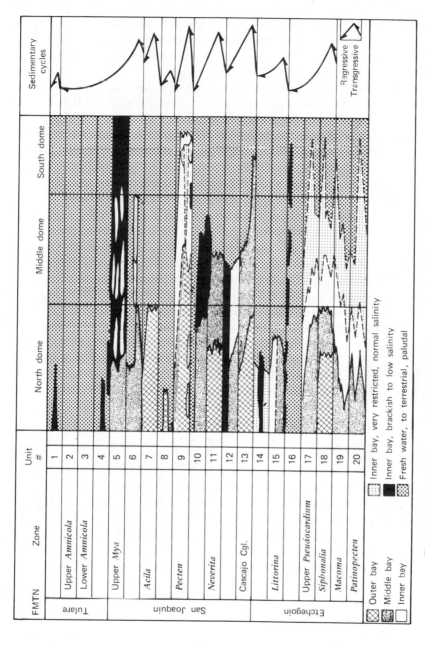

Figure 11.13 Diagrammatic north-south cross section illustrating cyclic sedimentation and environmental facies pattern. After Stanton and Dodd (1970), Journal of Paleontology, Society of Economic Paleontologists and Mineralogists.

447

more-or-less open marine to nonmarine, and establishing datum planes that have a greater refinement in this fluctuating marginal marine setting than any datum plane based on the first appearance of a species. These replacement units, which represent replacement in the replacement and succession terminology, are punctuated aggregational cycles (PACs) within a terrigenous depositional regime. PACs have been more commonly described in carbonate strata because the typically horizontal depositional surface results in more widespread cyclic units and, as noted above, a generally asymmetric cycle because carbonate sedimentation is strongly determined by sea-level and eustatic fluctuations.

In the Pliocene example, the cause of the cyclic changes within the section is not known, but in addition to eustatic fluctuations, climatic fluctuations, pulses in tectonic movement controlling subsidence of the basin and the degree of communication with the open ocean, are probably most important. The climatic mechanisms that have been proposed are considered unlikely because they do not provide a comprehensive explanation of the sedimentologic as well as paleontologic patterns, and because biogeochemical and faunal evidence indicate that climatic fluctuations were minor. The great number of widespread thin shallow marine-to-nonmarine cycles points to eustatic changes or to episodic basin subsidence under tectonic control as prime causes of the cycles. Episodic lateral movement of the San Andreas Fault, located along the western margin of the San Joaquin Basin, may also have been important because the basin was largely landlocked during the middle and Late Pliocene with only a narrow connection across the San Andreas Fault to the Pacific Ocean (Fig.

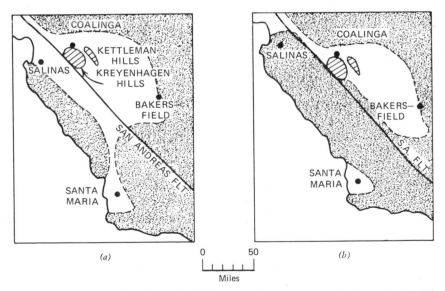

Figure 11.14   Early (*a*) and Late (*b*) Pliocene paleogeography of west-central California. After Galehouse (1967). Geological Society of American Bulletin.

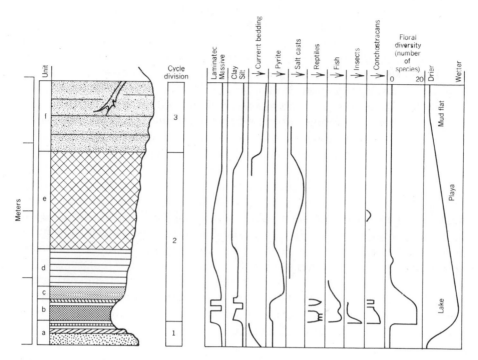

Figure 11.15 (right). Stratigraphic section of cycle CB1-2 showing the change in certain key properties (arrow indicating the direction of increase). "Cycle division" refers to the divisions of the modal cycle and "unit" refers to the following: *a*, black and dark gray micaceous siltstone and calcareous fine siltstone with some graded bedding; *b*, black, microlaminated calcareous siltstone: coarse graded siltstone near the top; *c*, black pyritic, well-bedded siltstone, intensely slickensided and disturbed; *d*, black, pyritic, well-bedded siltstone with common slickensided bedding planes; *e*, microlaminated siltstone with numerous crumpled casts of a salt: and *f*, gray, well-bedded siltstone with scour marks on bedding planes and large plant stems in growth position. From Olsen, et al. (1978), Science 201:732. Copyright 1978 by the AAAS.

11.14). Thus variations in the width of the inlet as lateral movement occurred along the fault would result in changes in salinity and degree of circulation but not necessarily in water depth. The upper part of a cycle would have been deposited in brackish to fresh water when connection with the open ocean was closed, and subsequent apparent transgression would have occurred when communication with the Pacific Ocean was increased or reestablished and more normal marine conditions were reintroduced to the basin.

A cyclic replacement sequence that was apparently determined by paleoclimate has been described by Olsen et al. (1978). Repeated sedimentary cycles in upper Triassic strata in North Carolina formed in large meromictic lakes. The lithologic and biotic replacement sequence for a typical cycle is illustrated in Figure 11.15. As the lake deepened, the shallow and relatively oxygenated conditions of unit A were replaced by the stratified water mass and anoxic bottom conditions of unit B. Both the abundance of fossils and the laminated sediments of unit B reflect the excellent conditions for preservation in the anoxic lake bottom. During deposition of units C and D, the water mass was less stratified and the benthic fauna that appeared destroyed sediment laminations by bioturbation. Progressive shallowing led to playa conditions during deposition of unit E and mudflat conditions during unit F. The maximum floral diversity, during units B and C, is the result of greater preservation potential within the stratified lake and anoxic bottom waters, and of amelioration of the adjacent terrestrial climate coincident with the maximum extent of the lake.

In each of these examples of replacement, the paleontologic changes are closely linked to changes in the external physical environment. Because they are cyclic, they are commonly tied at reversal points to some environmental condition that is perhaps fairly easily established, such as sea level in many of these cases, or restricted and relatively anoxic conditions as in the black core shales of Carboniferous cyclothems. In any case, although replacement sequences provide valuable temporal patterns, it is important to distinguish carefully between them and allogenic succession on the one hand, and autogenic succession on the other. The differences in origin and process lead to great differences in interpretation of the resulting fossil record.

REFERENCES

Abel, O., 1935, Vorzeitliche Lebensspuren: Gustav Fischer, Jena, East Germany, 644 p.

Adams, F. D., 1938, The birth and development of the geological sciences: Dover, New York, 506 p.

Adams, D. P. and G. J. West, 1983, Temperature and precipitation estimates through the last glacial cycle from Clear Lake, California, pollen data: Science, *219:* 168–170.

Adegoke, O. S., 1969, Stratigraphy and paleontology of the Neogene formations of the Coalinga region, California: Univ. Calif. Publ. Geol. Sci., *80,* 269 p.

Adey, W. H. and I. G. Macintyre, 1973, Crustose coralline algae: A reevaluation in the geological sciences: Geol. Soc. Am. Bull., *84:* 883–904.

Ager, D. V., 1963, Principles of paleoecology: McGraw-Hill, New York, 371 p.

Ager D. V., 1965, The adaptation of Mesozoic brachiopods to different environments: Palaeogeogr. Palaeoclimatol. Palaeoecol., *1:* 143–172.

Ager, D. V., 1967, Brachiopod paleoecology: Earth Sci. Rev., *3:* 157–179.

Aharon, P., 1985, Carbon isotope record of late Quaternary coral reefs: Possible index of sea surface paleoproductivity: *In* E. T. Sundquist and W. S. Broecker (eds.), The carbon cycle and atmospheric CO_2: Natural variations Archean to present: Am. Geophys. Un., Geophys. Monogr. Scr., *32:* 343–355.

Ahr, W. M., 1971, Paleoenvironment, algal structures, and fossil algae in the Upper Cambrian of central Texas: J. Sediment. Petrol., *41:* 205–216.

Ahr, W. M. and R. J. Stanton, Jr., 1973, The sedimentologic and paleoecologic significance of *Lithotrya,* a rock boring barnacle: J. Sediment. Petrol., *43:* 20–23.

Aigner, T., 1977, Schalenpflaster im unteren Hauptmuschelkalk bei Crailsheim (Württ., Trias, mol)—Stratinomie, Ökologie, Sedimentologie: Neues Jahrb. Geol. Palaeontol. Abh., *153:* 193–217.

Alexander, R. R., 1974, Morphologic adaptations of the bivalve *Anadara* from the Pliocene of the Kettleman Hills, California: J. Paleontol., *48:* 633–651.

Alexander, R. R., 1975, Phenotypic lability of the brachiopod *Rafinesquina alternata* (Ordovician) and its correlation with the sedimentologic regime: J. Paleontol., *49:* 607–618.

Alexander, R. R., 1977, Growth morphology and ecology of Paleozoic and Mesozoic opportunistic species of brachiopods from Idaho–Utah: J. Paleontol., *51:* 1133–1149.

Alexander, R. R., 1984, Comparative hydrodynamic stability of brachiopod shells on current-scoured arenaceous substrates: Lethaia, *17:* 17–32.

Alexander, R. R., 1986, Life orientation and post-mortem reorientation of Chesterian brachiopod shells by paleocurrents: Palaios, *1:* 303–311.

Ali, O. E., 1984, Sclerochronology and carbonate production in some Upper Jurassic reef corals: Palaeontology, *27:* 537–548.

Aller, R. C., 1982a, The effects of macrobenthos on chemical properties of marine sediment and overlying water: *In* P. L. McCall and M. J. S. Tevesz (eds.), Animal—sediment relations: Plenum Press, New York, p. 53–102.

Aller, R. C., 1982b, Carbonate dissolution in nearshore terrigenous muds: The role of physical and biological reworking: J. Geol., *90:* 79–95.

Aller, R. C. and R. E. Dodge, 1974, Animal–sediment relations in a tropical lagoon Discovery Bay, Jamaica: J. Mar. Res., *32:* 209–232.

Allison, P. A., 1988, Konservat-Lagerstätten: Cause and classification: Paleobiology, *14:* 331–344.

Anderson, O., 1983, Radiolaria: Springer-Verlag, New York, 328 p.

Arkell, W. J., 1956, Jurassic geology of the world: Oliver & Boyd, Edinburgh, 806 p.

Ausich, W. I., 1980, A model for niche partitioning in Lower Mississippian crinoid communities: J. Paleontol., *54:* 273–288.

Ausich, W. I., 1983, Functional morphology and feeding dynamics of the Early Mississippian crinoid *Barycrinus asteriscus:* J. Paleontol., *57:* 31–41.

Ausich, W. I. and D. J. Bottjer, 1985, Phanerozoic tiering in suspension-feeding communities on soft substrate: Implications for diversity: *In* J. W. Valentine (ed.), Phanerozoic diversity patterns: Princeton Univ. Press, Princeton, N.J., p. 255–274.

Axelrod, D. I., 1938, A Pliocene flora from the Mount Eden beds, southern California: Carnegie Inst. Wash. Publ., *476:* 127–183.

Ayala, F. J., D. Hedgcock, G. S. Zumwalt, and J. W. Valentine, 1973, Genetic variation in *Tridacna maxima,* an ecological analog of some unsuccessful evolutionary lineages: Evolution, *27:* 177–191.

Ayala, F. J., J. W. Valentine, T. E. Delaca, and G. S. Zumwalt, 1975, Genetic variability of the Antarctic brachiopod *Liothyrella notorcadensis* and its bearing on mass extinction hypotheses: J. Paleontol., *49:* 1–9.

Bader, R. G., 1954, The role of organic matter in determining the distribution of pelecypods in marine sediments: J. Mar. Res., *13:* 32–47.

Badgley, C., 1986, Counting individuals in mammalian fossil assemblages from fluvial environments: Palaios, *1:* 328–338.

Bailey, I. W. and E. W. Sinnott, 1915, A botanical index of Cretaceous and Tertiary climates: Science, *41:* 831–834.

Baker, R. G., 1986, Sangamonian (?) and Wisconsinan paleoenvironments in Yellowstone National Park: Geol. Soc. Am. Bull., *97:* 717–736.

Bakker, R. T., 1975, Dinosaur renaissance: Sci. Am., *232:* 59–79.

Bambach, R. K., 1983, Ecospace utilization and guilds in marine communities through the Phanerozoic: *In* M. J. S. Tevesz and P. L. McCall (eds.), Biotic interactions in recent and fossil benthic communities: Plenum Press, New York, p. 719–746.

Bandel, K. and H. Knitter, 1983, Litho-und biofazielle Untersuchungen eines Posidonienschieferprofils in Oberfranken: Geol. Bl. Nordost-Bayern, *32:* 95–129.

Bandy, O. L., 1960a, The geologic significance of coiling ratios in the foraminifer *Globigerina pachyderma* (Ehrenberg): J. Paleontol., *34:* 671–681.

Bandy, O. L., 1960b, General correlation of foraminiferal structure with environment: Int. Geol. Congr., 21st, *22:* 7–19.

Bandy, O. L., 1964, General correlation of foraminiferal structure with environment: *In* J. Imbrie and N. D. Newell (eds.), Approaches to paleoecology: Wiley, New York, p. 75–90.

Barker, R. M., 1964, Microtextural variation in pelecypod shells: Malacologia, *2:* 69–86.

Barnes, R. D., 1974, Invertebrate zoology, 3rd ed.: Saunders, Philadelphia, 870 p.

Barron, J. A., 1973, Late Miocene–Early Pliocene paleotemperatures for California from marine diatom evidence: Palaeogeogr. Palaeoclimatol. Palaeoecol., *14:* 277–291.

Barth, R. H. and R. E. Broshears, 1981, The invertebrate world: Saunders, Philadelphia, 646 p.

Bate, R. H., E. Robinson, and L. M. Sheppard (eds.), 1982, Fossil and Recent Ostracods: Ellis Horwood, Chichester, West Sussex, England, 493 p.

Bathurst, R. G. C., 1975, Carbonate sediments and their diagenesis: Elsevier, Amsterdam, 658 p.

Bayer, U., 1977a, Cephalopoden-Septen. 1. Konstruktionsmorphologie des Ammoniten-Septums: Neues Jahrb. Geol. Palaeontol. Abh., *154:* 290–366.

Bayer, U., 1977b, Cephalopoden-Septen. 2. Regelmechanismen im Gehuse- und Septenbau der Ammoniten: Neues Jahrb. Geol. Palaeontol. Abh., *155:* 162–215.

Bayer, U., 1978, Morphogenetic programs, instabilities, and evolution: A theoretical study: Neues Jahrb. Geol. Palaeontol. Abh., *156:* 226–261.

Bé, A. W. H., 1968, Shell porosity of Recent planktonic foraminifera as a climatic index: Science, *161:* 881–884.

Beadle, S. C., 1988, Dasyclads, cyclocrinitids, and receptaculitids: Lethaia, *21:* 1–12.

Beatty, J., 1980, Optimal-design models and the strategy of model building in evolutionary biology: Philos. Sci. *47:* 532–561.

Begon, M., J. L. Harper, and C. R. Townsend, 1986, Ecology, individuals, populations and communities: Blackwell Scientific, Oxford, 876 p.

Behrens, E. W. and R. L. Watson, 1969, Differential sorting of pelecypod valves in the swash zone: J. Sediment. Petrol., *39:* 159–165.

Behrensmeyer, A. K., 1975, The taphonomy and paleoecology of Plio-Pleistocene vertebrate assemblages east of Lake Rudolf, Kenya: Bull. Mus. Comp. Zool. Harv. Univ., *146:* 473–578.

Behrensmeyer, A. K. and A. P. Hill, (eds.), 1980, Fossils in the making: Univ. Chicago Press, Chicago, 338 p.

Behrensmeyer, A. K. and D. E. Schindel, 1983, Resolving time in paleobiology: Paleobiology, *9:* 1–8.

Benson, R. H., 1961, Ecology of ostracode assemblages: *In* R. C. Moore (ed.), Treatise on invertebrate paleontology, pt. Q, Arthropoda 3: Geol. Soc. Am., Boulder, Colo., p. 56–63.

Benson, R. H., 1975, Morphologic stability in Ostracod: *In* F. M. Swain (ed.), Biology and paleobiology of Ostracoda: Bull. Am. Paleontol., *65:* 11–46.

Benson, R. H., 1984, Estimating greater paleodepths with ostracodes, especially in past thermospheric oceans: Palaeogeogr. Palaeoclimatol. Palaeoecol., *48:* 107–141.

Berger, W. H., 1970, Planktonic foraminifera: Selective solution and the lysocline: Mar. Geol., *8:* 111–138.

Berger, W. H. and J. S. Killingley, 1977, Glacial–Holocene transition in deep-sea carbonates: Selective dissolution and the stable isotope signal: Science, *197:* 563–556.

Berger, A., J. Imbrie, J. Hays, G. Kukla, and B. Saltzman, 1984, Milankovitch and climate, pt. 1: D. Reidel, Dordrecht, The Netherlands, 895 p.

Berggren, W. A. and C. D. Hollister, 1977, Plate tectonics and paleocirculation-commotion in the ocean: Tectonophysics, *38:* 11–48.

Berglund, B. E. (ed.), 1986, Handbook of Holocene palaeoecology and palaeo-hydrology: Wiley-Interscience, Chichester, West Sussex, England, 869 p.

Bergquist, P. R., 1978, Sponges: Univ. California Press, Berkeley, Calif., 268 p.

Berman, D. B., 1976, Occurrence of *Gnathorhiza* (Osteichthyes Dipnoi) in aestivation burrows in the Lower Permian of New Mexico with discription of a new species: J. Paleontol., *50:* 1034–1039.

Berner, R. A., 1975, The role of magnesium in the crystal growth of calcite and aragonite from sea water: Geochim. Cosmochim. Acta, *39:* 489–504.

Bidder, G. P., 1923, The relation of the form of a sponge to its currents: Q. J. Microsc. Soc., *67:* 293–325.

Binyon, J., 1966, Salinity tolerance and ion regulation: *In* R. A. Boolootian (ed.), Physiology of Echinodermata: Interscience, New York, p. 359–377.

Birks, H. J. B. and H. H. Birks, 1980, Quaternary palaeoecology: Univ. Park Press, Baltimore, 289 p.

Blome, C. D. and N. R. Albert, 1985, Carbonate concretions: An ideal sedimentary host for microfossils: Geology, *13:* 212–215.

Boardman, R. S. and A. H. Cheetham, 1987, Phylum bryozoa: *In* R. S. Boardman, A. H. Cheetham, and A. J. Rowell (eds.), Fossil invertebrates: Blackwell Scientific, Boston, p. 297–549.

Boardman, R. S., A. H. Cheetham, and A. J. Rowell, 1987, Fossil invertebrates: Blackwell Scientific, Boston, 720 p.

Boersma, A., 1978, Foraminifera: *In* B. V. Haq and A. Boersma (eds.), Introduction to marine micropaleontology: Elsevier, New York, p. 19–77.

Bøggild, O. B., 1930, The shell structure of the mollusks: K. Dan. Vidensk. Selsk. Skr. naturvidensk. Math. Afd. (Acad. R. Sci. Lett. Dan.) Mem. Ser., *9:* 233–326.

Boltovskoy, E. and R. Wright, 1976, Recent Foraminifera: Dr. W. Junk, Publishers, The Hague, The Netherlands, 515 p.

Bonner, J. T., 1965, Size and cycle: An essay on the structure of biology: Princeton Univ. Press, Princeton, N.J., 219 p.

Bosence, D. W. J., 1976, Ecological studies on two unattached coralline algae from western Ireland: Palaeontology, *19:* 365–395.

Bosence, D. W. J., 1979a, Live and dead faunas from coralline algal gravels, Co. Galway: Palaeontology, *22:* 449–478.

Bosence, D. W. J., 1979b, Trophic analysis of communities and death assemblages: Lethaia, *12:* 120.

Bosence, D. W. J., 1985, The morphology and ecology of a mound-building coralline alga (*Neogoniolithon strictum*) from the Florida Keys: Palaeontology, *28:* 189–206.

Bottjer, D. J. and J. G. Carter, 1980, Functional and phylogenetic significance of project-ing periostracal structures in the Bivalvia (Mollusca): J. Paleontol., *54:* 200–216.

Bottjer, D. J., C. S. Hickman, and P. D. Ward (eds.), 1985, Mollusks: Notes for a short course: Univ. Tenn., Dep. Geol. Sci., Stud. Geol., *13,* 306 p.

Boucot, A. J., 1953, Life and death assemblages among fossils: Am. J. Sci., *251:* 25–40.

Boudreau, B. P., 1987, A steady-state diagenetic model for dissolved carbonate species and pH in the porewaters of oxic and suboxic sediments: Geochim. Cosmochim. Acta, *51:* 1985–1996.

Bourget, E., 1977, Shell structure in sessile barnacles: Nat. Can., *104:* 281–323.

Bowen, R., 1966, Paleotemperature analysis: Elsevier, New York, 265 p.

Boyd, D. W. and N. D. Newell, 1972, Taphonomy and diagenesis of a Permian fossil assemblage from Wyoming: J. Paleontol., *46:* 1–14.

Boyle, E. A., 1988, Cadmium: Chemical traces of deepwater paleoceanography: Palaeoceanography, *3:* 471–489.

Brand, U., 1981a, Mineralogy and chemistry of the lower Pennsylvanian Kendrick fauna, eastern Kentucky. 1. Trace elements: Chem. Geol., *32:* 1–16.

Brand, U., 1981b, Mineralogy and chemistry of the lower Pennsylvanian Kendrick fauna, eastern Kentucky. 2. Stable isotopes: Chem. Geol., *32:* 17–28.

Brand, U., 1986, Paleoenvironmental analysis of Middle Jurassic (Callovian) ammonoids from Poland: Trace elements and stable isotopes: J. Paleontol., *60:* 293–301.

Brand, U., 1987a, Biogeochemistry of nautiloids and paleoenvironmental aspects of Buckhorn seawater (Pennsylvanian), southern Oklahoma: Palaeogeogr. Palaeoclimatol. Palaeoecol., *61:* 255–264.

Brand, U., 1987b, Depositional analysis of the Breathitt Formation's marine horizons, Kentucky: Trace elements and stable isotopes: Chem. Geol., *65:* 117–136.

Brand, U. and J. Veizer, 1981, Chemical diagenesis of multicomponent carbonate systems. 1. Trace elements: J. Sediment. Petrol., *50:* 1219–1236.

Braiser, M.D., 1975, An outline history of seagrass communities: Palaeontology, *18:* 681–702.

Brasier, M. D., 1980, Microfossils: Allen & Unwin, London, 193 p.

Brass, G. W., J. R. Southam, and W. H. Peterson, 1982, Warm saline bottom water in the ancient ocean: Nature, *296:* 620–623.

Breimer, A., 1969, A contribution to the paleoecology of Palaeozoic stalked crinoids: K. Ned. Akad. Wet. Proc., *B72:* 139–150.

Breimer, A. and N. G. Lane, 1978, Ecology and paleoecology: *In* R. C. Moore and C. Teichert (eds.), Treatise on invertebrate paleontology, Pt. T, Echinodermata 2, v. 1: Geol. Soc. Am., Boulder, Colo., p. 316–347.

Brenner, K. and G. Einsele, 1976, Schalenbruch im Experiment: Zentralbl. Geol. Palaeontol., *2:* 349–354.

Bretsky, P. W., 1968, Evolution of Paleozoic marine invertebrate communities: Science, *159:* 1231–1233.

Bretsky, P. W., 1969, Central Appalachian Late Ordovician communities: Geol. Soc. Am. Bull., *80:* 193–212.

Brett, C. E. and G. C. Baird, 1986, Comparative taphonomy: A key to paleoenvironmental interpretation based on fossil preservation: Palaios, *1:* 207–227.

Brewer, P. G., 1975, Minor elements in sea water: *In* J. P. Riley and G. Skirrow (eds.), Chemical oceanography, 2nd ed. v. 1: Academic Press, London, p. 415–496.

Briggs, J. C., 1974, Marine zoogeography: McGraw-Hill, New York, 475 p.

Briggs, J. C., 1987, Biogeography and plate tectonics: Elsevier, Amsterdam, 204p.

Britton, E. R. and R. J. Stanton, Jr., 1973, Origin of "dwarfed" fauna in the Del Rio Formation, Lower Cretaceous, East Central Texas (abs.): Geol. Soc. Am. Abstr. Programs, 5: 248–249.

Broecker, W. S., 1974, Chemical oceanography: Harcourt Brace & World, New York, 214 p.

Broecker, W. S. and J. van Donk, 1970, Insolation changes, ice volumes, and the O-18 record in deep-sea cores: Rev. Geophys. Space Phys., 8: 169–198.

Bromley, R. G., 1972, On some ichnotaxa in hard substrates with a redefinition of Trypanites magdefrau: Paleontol. Zh., 46: 93–98.

Bromley, R. G. and A. A. Ekdale, 1984a, Trace fossil preservation in flint in the European chalk: J. Paloentol., 58: 298–311.

Bromley, R. G. and A. A. Ekdale, 1984b, Chondrites: A trace fossil indicator of anoxia in sediments: Science, 224: 872–874.

Bromely, R. G. and A. A. Ekdale, 1986, Composite ichnofacies and tiering of burrows: Geol. Mag., 123: 59–65.

Bromley, R. G., S. G. Pemberton, and R. A. Rahmani, 1984, A Cretaceous woodground: The Teredolites ichnofacies: J. Paleontol., 58: 488–498.

Brongersma-Sanders, M., 1957, Mass mortality in the sea: In J. W. Hedgpeth (ed.), Treatise on marine ecology and paleoecology, v. 1, Ecology: Geol. Soc. Am. Mem., 67(1): 941–1010.

Brood, K., 1972, Cyclostomatous bryozoa from the Upper Cretaceous and Danian in Scandinavia: Stockholm Contrib. Geol., 26: 1–464.

Buchardt, B., 1977, Oxygen isotope ratios from shell material from the Danish Middle Paleocene (Selandian) deposits and their interpretation as paleotemperature indicators: Palaeogeogr. Palaeoclimatol. Palaeoecol., 22: 209–230.

Buchardt, B., 1978, Oxygen isotope paleotempertures from the Tertiary period in the North Sea area: Nature, 275: 121–123.

Buchardt, B. and P. Fritz, 1978, Strontium uptake in shell aragonite from the freshwater gastropod Limnaea stagnalis: Science, 199: 291–292.

Buchardt, B., and H. J. Hansen, 1977, Oxygen isotope fractionation and algal symbiosis in benthic foraminifera from the Gulf of Elat, Israel: Bull, Geol. Soc. Den., 26: 185–194.

Buchbinder, B. T., 1977, The coralline algae from the Miocene Ziglar Formation in Israel and the environmental significance: In E. Flügel (ed.), Fossil algae: Recent results and developments: Springer-Verlag, West Berlin, p. 279–285.

Buddemeier, R. W., J. E. Maragrs, and D. W. Knutson, 1974, Radiographic studies of reef coral exoskeletons: Rates and patterns of coral growth: J. Exp. Mar. Biol. Ecol., 14: 179–200.

Buffrenil, V. de, J. D. Farlow, and A. De Ricqles, 1986, Growth and function of Stegosaurus plates: Evidence from bone histology: Paleobiology, 12: 459–473.

Bullock, T. H., 1955, Compensation for temperature in the metabolism and activity of poikilotherms: Biol. Rev. Cambridge Philos. Soc., 30: 311–342.

Burckle, L. H., 1978, Marine diatoms: In B. U. Haq and A. Boersma (eds.), Introduction to marine micropaleontology: Elsevier, New York, p. 245–266.

Burckle, L. H., 1984, Ecology and paleoecology of the marine diatom *Eucanipia antarctica* (Castr.) Mangin: Mar. Micropaleontol., *9:* 77–86.

Buzas, M. A. and B. K. Sen Gupta (eds.), 1982, Foraminifera: Notes for a short course: Univ. Tenn., Dep. Geol. Sci., Stud. Geol., *6,* 219 p.

Buzas, M. A., R. C. Douglas, and C. C. Smith, 1987, Kingdom Protista: *In* R. S. Boardman, A. H. Cheetham, and A. J. Rowell (eds.), Fossil invertebrates: Blackwell Scientific, Boston, p. 67–106.

Byers, C. W. and L. E. Stasko, 1978, Trace fossils and sedimentologic interpretation: McGregor Member of Plattville Formation (Ordovician) of Wisconsin: J. Sediment. Petrol., *48:* 1303–1310.

Cadée, G. C., 1968, Molluscan biocoenoses and thanatocoenoses in the Ria de Arosa, Galicia, Spain: Zool. Verh. Rijksmus. Nat. Hist. Leiden, *95:* 1–121.

Cadée, G. C., 1982, Low juvenile mortality in fossil brachiopods, some comments: Interne Verslagen, Nederlands Instituut voor, Onderzoek der Zee, Texel, The Netherlands, p. 1–29.

Campana, S. E., 1984, Lunar cycles of otolith growth in the juvenile starry flounder *Platichthys stellatus:* Mar. Biol., *80:* 239–246.

Carey, S. W., 1976, The expanding earth: Developments in geotectonics, v. 10, Elsevier, New York, 488 p.

Carpenter, W. B., 1844, On the microstructure of shells: Rep. Bri. Assoc. Adv. Sci., 1–24.

Carriquiry, J. D., M. J. Risk, and H. P. Schwarcz, 1988, Timing and temperature record from stable isotopes of the 1982–83 El Niño warming event in eastern Pacific corals: Palaios, *3:* 359–364.

Carter, R. W. G., 1974, Feeding sea birds as a factor in lamellibranch valve sorting patterns: J. Sediment. Petrol., *44:* 689–692.

Carter, J. G., 1980, Environmental and biological controls of bivalve shell mineralogy and microstructure: *In* D. C. Rhoads and R. A. Lutz (eds.), Skeletal growth of aquatic organisms: Plenum Press, New York, p. 69–113, 645–673.

Carter, J. G. and G. R. Clark II, 1985, Classification and phylogenetic significance of molluscan shell microstructure: *In* T. W. Broadhead (ed.), Mollusks: Notes for a short course, organized by D. J. Bottjer, C. S. Hickman, and P. D. Ward: Univ. Tenn., Dep. Geol. Sci., Stud. Geol., *13:* 50–71.

Case, T. J. and M. L. Cody, 1987, Testing theories of Island biogeography: Am. Sci., *75:* 402–411.

Casey, R. E., 1972, Neogene radiolarian biostratigraphy and paleotemperatures: Southern California, the experimental Mohole Antarctic core E 14-8: Palaeogegr. Palaeoclimatol. Palaeoecol., *12:* 115–130.

Casey, R. E., 1977, The ecology and distribution of Recent radiolaria: *In* A. T. S. Ramsey (ed.), Oceanic micropaleontology: Academic Press, London, p. 809–845.

Chaloner, W. G. and W. S. Lacey, 1973, the distribution of Late Palaeozoic floras: *In* N. F. Hughes (ed.), Organisms and continents through time: Palaeontol. Assoc. Spec. Pap., *12:* 271–289.

Chamberlain, J. A., 1976, Flow patterns and drag coefficients of cephalopod shells: Palaeontology, *19:* 539–563.

Chamberlain, C. K., 1978, Recognition of trace fossils in cores: *In* P. W. Basan (ed.), Trace fossil concepts, Soc. Econ. Paleontol. Mineral. Short Course Notes, *5:* 123–183.

Chamberlain, J. A. and R. B. Chamberlain, 1986, Is cephalopod septal strength index an index of cephalopod septal strength?: Alcheringa, *10:* 85–97.

Chang, Y.-M., 1967, Accuracy of fossil percentage estimations: J. Paleontol., *41:* 500–502.

Chappell, J., 1981, Coral morphology, diversity and reef growth: Nature, *286:* 249–252.

Chave, K. E., 1954, Aspects of the biogeochemistry of magnesium. I. Calcareous marine organisms: J. Geol., *62:* 266–283.

Chave, K. E., 1964, Skeletal durability and preservation: *In* J. Imbrie and N. Newell (eds.), Approaches to paleoecology: Wiley, New York, p. 377–387.

Chesher, R. H., 1969, Destruction of Pacific corals by the sea star *Acanthaster planci:* Science, *165:* 280–283.

Chivas, A. R., P. DeDeckker, and J. M. G. Shelley, 1985, Strontium content of ostracods indicates lacustrine palaeosalinity: Nature, *316:* 251–253.

Cisne, J. L. and B. D. Rabe, 1978, Coenocorrelation: Gradient analysis of fossil communities and its application in stratigraphy: Lethaia, *11:* 341–364.

Clark, G. R. II, 1968, Mollusk shell: Daily growth lines: Science, *161:* 800–802.

Clarke, F. W. and W. C. Wheeler, 1922, The inorganic constituents of marine invertebrates: U.S. Geol. Surv. Prof. Pap., *124,* 62 p.

Clements, F. E., 1916, Plant succession: An analysis of the development of vegetation: Carnegie Inst. Wash. Publ., *242.*

Clements, F. E. and V. E. Shelford, 1939, Bio-ecology: Wiley, New York, 425 p.

CLIMAP Project Members, 1984, The last interglacial ocean: Quaternary Research, 21:123–224.

Cline, R. M. and J. D. Hays (eds.), 1976, Late Quanternary paleoceanography and paleoclimatology: Geol. Soc. Am. Mem., *145:* 464 p.

Cocks, L. R. M. and R. A. Fortey, 1982, Faunal evidence for oceanic separations in the Palaeozoic of Britain: J. Geol. Soc. London, *139:* 465–478.

Cody, M. L., 1974, Optimization in ecology: Science, *183:* 1156–1164.

Cody, M. L. and J. M. Diamond (eds.), 1975, Ecology and evolution of communities: Harvard Univ. Press–Belknap Press, Cambridge, Mass., 545 p.

Coe, W. R., 1957, Fluctuations in littoral populations: *In* J. W. Hedgpeth (ed.), Treatise on marine ecology and paleoecology, v. 1, Ecology: Geol. Soc. Am. Mem., *67*(1): 935–940.

Cohen, A. S. and C. Nielsen, 1986, Ostracodes as indicators of paleohydrochemistry in lakes: A late Quaternary example from Lake Elmenteita, Kenya: Palaios, *1:* 601–609.

Colbert, E. H., 1964, Climatic zonation and terrestrial faunas: *In* A. E. M. Nairn (ed.), Problems in paleoclimatology: Wiley-Interscience, New York, p. 617–639.

Colgan, M. W., 1987, Coral reef recovery on Guam (Micronesia) after catastrophic predation by *Acanthaster planci:* Ecology, *68:* 1592–1605.

Coney, P. J., D. L. Jones, and J. W. H. Monger, 1980, Cordilleran suspect terranes: Nature, *288:* 329–333.

Connell, J. H., 1961, Effects of competition, predation by *Thais lapillus,* and other factors on natural populations of the barnacle *Balanus balanoides:* Ecol. Monogr., *31:* 61–104.

Connell, J. H. and R. O. Slatyer, 1977, Mechanisms of succession in natural communities and their role in community stability and organization: Am. Nat., *111:* 1119–1144.

Conrad, M. A., 1977, The Lower Cretaceous calcareous algae in the area surrounding Geneva (Switzerland): Biostratigraphy and depositional environments: *In* E. Flügel (ed.), Fossil algae: Recent results and developments: Springer-Verlag, West Berlin, p. 295–300.

Copper, P., 1988, Ecological succession in Phanerozoic reef ecosystems: Is it real?: Palaios, *3:* 136–151.

Corliss, B. H., 1985, Microhabitats of benthic foraminifera within deep-sea sediments: Nature, *314:* 435–438.

Corliss, B. H. and C. Chen, 1988, Morphotype patterns of Norwegian Sea deep-sea benthic foraminifera and ecological implications: Geology, *16:* 716–719.

Coutts, P. J. F., 1970, Bivalve-growth patterning as a method for seasonal dating in archaeology: Nature, *226:* 874 p.

Coutts, P. J. F., 1975, The seasonal perspective of marine-oriented prehistoric hunter-gatherers: *In* G. D. Rosenberg and S. K. Runcorn (eds.), Growth rhythms and the history of the earth's rotation: Wiley-Interscience, New York, p. 243–252.

Cowen, R., R. Gertman, and G. Wiggett, 1973, Camouflage patterns in *Nautilus,* and their implications for cephalopod paleobiology: Lethaia, *6:* 201–213.

Cox, C. B. and P. D. Moore, 1985, Biogeography an ecological and evolutionary approach: Blackwell Scientific, Boston, 244 p.

Cracraft, J., 1983, Cladistic analysis and vicariance biogeography: Am. Sci., *71:* 273–281.

Craig, H., 1957, Isotopic standards for carbon and oxygen corrections factors for mass spectrometric analysis of CO_2: Geochim. Cosmochim. Acta, *12:* 133–149.

Craig, H., 1961, Standard for reporting concentrations of deuterium and oxygen-18 in natural waters: Science, *133:* 1833–1834.

Craig, H., 1965, The measurement of oxygen isotope paleotemperatures: *In* Stable isotopes in oceanographic studies and paleotemperatures: Speleto, Italy, Consiglio Nationale delle Ricerdy Lab. di Geol. Mus., Pisa, p. 3–24.

Craig, G. Y. and A. Hallam, 1963, Size-frequency and growth-ring analyses of *Mytilus edulis* and *Cardium edule,* and their paleoecological significance: Palaeontology, *6:* 731–750.

Craig, G. Y. and G. Oertel, 1966, Deterministic models of living and fossil populations of animals: Q. J. Geol. Soc. London, *122:* 315–355.

Creer, K. M., 1975, On a tentative correlation between changes in the geomagnetic polarity bias and reversal frequency and the earth's rotation through Phanerozoic time: *In* G. D. Rosenberg and S. K. Runcorn (eds.), Growth rhythms and the history of the earth's rotation: Wiley-Interscience, New York, p. 293–317.

Crick, R. E. and K. O. Mann, 1987, Biomineralization and systematic implications: *In* W. B. Saunders and N. H. Landman (eds.), *Nautilus:* The biology and paleobiology of a living fossil: Plenum Press, New York, p. 115–134.

Crick, R. E., and V. M. Ottensman, 1983, Sr, Mg, Ca and Mn chemistry of skeletal components of a Pennsylvanian and Recent nautiloid: Chem. Geol., *39:* 147–163.

Crocker, K. C., M. J. DeNiro, and P. D. Ward, 1985, Stable isotope investigations of early development in extant and fossil chambered cephalopods. I. Oxygen isotopic composition of eggwater and carbon isotopic composition of siphuncle organic matter in *Nautilus:* Geochim. Cosmochim. Acta, *49:* 2527–2532.

Cronin, T. M., 1983, Bathyl ostracodes from the Florida–Hatteras Slope, the Straits of Florida, and the Blake Plateau: Mar. Micropaleontol., *8:* 89–119.

Cuffey, R. J. and F. G. McKinney, 1982, Reteporid cheilostome bryozoans from the modern reefs of Eniwetok Atoll, and their implications for Paleozoic fenestrate bryozoan paleoecology: Pac. Geol., *16:* 7–13.

Cummins, H., E. N. Powell, H. J. Newton, R. J. Stanton, Jr., and G. Staff, 1986a, Assessing transportation by the covariance of species with comments on contagious and random distributions: Lethaia, *19:* 1–22.

Cummins, H., E. N. Powell, R. J. Stanton, Jr., and G. Staff, 1986b, The rate of taphonomic loss in modern benthic habitats: How much of the potential preservable community is preserved?: Palaeogeogr. Palaeoclimatol. Palaeoecol., *52:* 291–320.

Cummins, H., E. N. Powell, R. J. Stanton, Jr., and G. Staff, 1986c, The size-frequency distribution in palaeoecology: Effects of taphonomic processes during formation of molluscan death assemblages in Texas bays: Palaeontology, *29:* 495–518.

Curran, H. A. (ed.), 1985, Biogenic structures: Their use in interpreting depositional environments: Soc. Econ. Paleontol. Mineral. Spec. Publ., *35:* 347 p.

Currey, J. D., 1980, Mechanical properties of mollusc shell: Symp. Soc. Exp. Biol., *34:* 74–97.

Dafni, J., 1986, A biomechanical model for the morphogenesis of regular echinoid tests: Paleobiology, *12:* 143–160.

Dansgaard, W., 1964, Stable isotopes in precipitation: Tellus, *16:* 436–468.

Dansgaard, W. and H. Tauber, 1969, Glacier oxygen-18 content and Pleistocene ocean temperatures: Science, *166:* 499–502.

Dansgaard, W., S. J. Johnson, J. Moller, and C. C. Langway, Jr., 1969, One Thousand centuries of climatic record from Camp Century on the Greenland Ice Sheet: Science, *166:* 377–381.

Darlington, P. J., 1957, Zoogeography: The geographical distribution of animals: Wiley, New York, 675 p.

Davenport, C. B., 1938, Growth lines in fossil pectens as indicators of past climates: J. Paleontol., *12:* 514–515.

Davies, T. T., 1972, Effect of environmental gradients in the Rappahannock River Estuary on the molluscan fauna: *In* B. W. Nelson (ed.), Environmental framework of coastal plain estuaries: Geol. Soc. Am. Mem., *133:* 263–290.

Davis, W. M., 1926, The value of outrageous geological hypotheses: Science, *63:* 463–468.

Davis, M. B., 1969, Palynology and environmental history during the Quaternary Period: Am. Sci., *57:* 317–332.

De Deckker, P., A. R. Chivas, J. M. G. Shelley, and T. Torgersen, 1988, Ostracod shell chemistry: A new palaeoenvironmental indicator applied to a regressive–

transgressive record from the Gulf of Carpentaria, Australia: Palaeogeogr. Palaeoclimatol. Palaeoecol., *66:* 231–241.

Deevey, E. S., 1947, Life tables for natural populations of animals: Q. Rev. Biol., *22:* 283–314.

Degens, E. T., 1969, Biogeochemistry of stable carbon isotopes: *In* G. Eglinton and M. T. J. Murphy (eds.), Organic geochemistry: Methods and results: Springer-Verlag, New York, p. 304–329.

Degens, E. T. and D. A. Ross (eds.), 1974, The Black Sea: Geology, chemistry, and biology: Am. Assoc. Pet. Geol. Mem., *20,* 633 p.

Delaney, M. L. and E. A. Boyle, 1987, Cd/Ca in late Miocene benthic foraminifera and changes in the global organic carbon budget: Nature, *330:* 156–159.

deLaubenfels, M. W., 1957, Marine sponges: *In* J. W. Hedgpeth (ed.), Treatise on marine ecology and paleoecology, v. 1, Ecology: Geol. Soc. Am. Mem. *67*(1): 1083–1086.

Delcourt, P. A. and H. R. Delcourt, 1980, Pollen preservation and Quarternary environmental history in the southeastern United States: Palynology, *4:* 215–231.

Delorme, L. D., 1971, Paleoecological determinations using Pleistocene freshwater ostracodes: *In* H. J. Oertl (ed.), Paleoecologie ostracodes: Bull. Cent. Rech. Pau-SNPA, *5* (suppl.): 341–347.

DeNiro, M. J., 1987, Stable isotopy and archaeology: Am. Scientist, *75:* 182–191.

Dennison, J. M. and W. W. Hay, 1967, Estimating the needed sampling area for subaquatic ecologic studies: J. Paleontol., *41:* 706–708.

Denton, E. J. and J. B. Gilpin-Brown, 1973, Flotation mechanisms in modern and fossil cephalopods: Adv. Mar. Biol., *11:* 197–268.

Dill, R. F., E. A. Shinn, A. T. Jones, K. Kelly, and R. P. Steinen, 1986, Giant subtidal stromatolites forming in normal salinity water: Nature, *324:* 55–58.

Dodd, J. R., 1963, Paleoecological implications of shell mineralogy in two pelecypod species: J. Geol., *71:* 1–11.

Dodd, J. R., 1964, Environmentally controlled variation in the shell structure of a pelecypod species: J. Paleontol., *38:* 1065–1071.

Dodd, J. R., 1965, Environmental control of strontium and magnesium in *Mytilus:* Geochim. Cosmochim. Acta, *29:* 385–398.

Dodd, J. R., 1966, Diagenetic stability of temperature-sensitive skeletal properties in *Mytilus* from the Pleistocene of California: Geol. Soc. Am. Bull., *77:* 1213–1224.

Dodd, J. R., 1967, Magnesium and strontium in calcareous skeletons: A review: J. Paleontol., *41:* 1313–1329.

Dodd, J. R. and E. L. Crisp, 1982, Non-linear variation with salinity of Sr/Ca and Mg/Ca ratios in water and aragonitic bivalve shells and implications for paleosalinity studies: Palaeogeogr. Palaeoclimatol. Palaeoecol., *38:* 45–56.

Dodd, J. R. and T. J. M. Schopf, 1972, Approaches to biogeochemistry: *In* T. J. M. Schopf (ed.), Models in paleobiology: Freeman, Cooper, San Francisco, p. 46–60.

Dodd, J. R. and R. J. Stanton, Jr., 1975, Paleosalinities within a Pliocene Bay, Kettleman Hills, California: A study of the resolving power of isotopic and faunal techniques: Geol. Soc. Am. Bull., *86:* 51–64.

Dodd, J. R. and R. J. Stanton, Jr., 1976, Paleosalinities within a Pliocene Bay, Kettleman Hills, California: A study of the resolving power of isotopic and faunal techniques (reply to discussion): Geol. Soc. Am. Bull., *87:* 160.

Dodd, J. R., R. J. Stanton, Jr., and M. Johnson, 1984, Oxygen isotopic composition of Neogene molluscan fossils from the Eel River Basin of California: Geol. Soc. Am. Bull., *95:* 1253–1258.

Dodd, J. R., R. R. Alexander, and R. J. Stanton, Jr., 1985, Population dynamics in *Dendraster, Merriamaster,* and *Anadara* from the Neogene of the Kettleman Hills, California: Palaeogeogr. Palaeoclimatol. Palaeoecol., *52:* 61–76.

Dodge, R. E. and J. R. Vaismya, 1980, Skeletal growth chronologies of Recent and fossil corals: *In* D. C. Rhoads and R. A. Lutz (eds.), Skeletal growth of aquatic organisms: Plenum Press, New York, p. 493–517.

Doguzhaeva, L., 1982, Rhythms of ammonoid shell secretion: Lethaia, *15:* 385–394.

Donnay, G. and D. L. Pawson, 1969, X-ray diffraction studies of echinoderm plates: Science, *166:* 1147–1150.

Dorman, F. H. and E. D. Gill, 1959, Oxygen isotope paleotemperature determinations of Australian Cainozoic fossils: Science, *130: 1576.*

Drake, C. L., J. Imbrie, J. A. Knauss, and K. Turekian, 1978, Oceanography: Holt, Rinehart and Winston, New York, 447 p.

Driscoll, E. G., 1967, Experimental field study of shell abrasion: J. Sediment. Petrol., *37:* 1117–1123.

Driscoll, E. G., 1969, Animal–sediment relationships of the Coldwater and Marshall Formations of Michigan: *In* K. S. W. Campbell (ed.), Stratigraphy and paleontology: Essays in honor of Dorothy Hill: Australian National Univ. Press, Canberra, p. 337–352.

Driscoll, E. G., 1970, Selective bivalve shell destruction in marine environments, a field study: J. Sediment. Petrol., *40:* 898–905.

Driscoll, E. G. and D. E. Brandon, 1973, Mollusc–sediment relationships, northwestern Buzzards Bay, Massachusetts: Malacologia, *12:* 13–46.

Driscoll, E. G. and T. P. Weltin, 1973, Sedimentary parameters as factors in abrasive shell reduction: Palaeogeogr. Palaeoclimatol. Palaeoecol., *13:* 275–288.

Droser, M. L. and D. J. Bottjer, 1986, A semiquantitative field classification of ichnofabric: J. Sediment. Petrol., *56:* 558–559.

Druffel, E. M., 1982, Banded corals: Changes in oceanic carbon-14 during the Little Ice Age: Science, *218:* 13–19.

Druffel, E. M. and H. E. Suess, 1983, on the radio-carbon record in banded corals: Exchange parameters and net transport of $^{14}CO_2$ between atmosphere and surface ocean: J. Geophys. Res., *88:* 1271–1280.

Dullo, W. C., 1983, Fossildiagenese im miozenen Leitha-Kalk der Paratethys von Österreich: Ein Beispiel für Faunenverschiebungen durch Diageneseunterschiede: Facies, *8:* 1–112.

Duplessy, J-C., C. Lalou, and A. C. Vinot, 1970, Differential isotopic fractionation in benthic foraminifera and paleotemperatures reassessed: Science, *168:* 250–251.

Durham, J. W., 1967, The incompleteness of our knowledge of the fossil record: J. Paleontol., *14:* 559–565.

Durham, D. L. and W. O. Addicott, 1965, Pancho Rico Formation, Salinas Valley, California: U.S. Geol. Survey, Prof. Pap., *524-A*, 22 p.

Eckman, J. E., A. R. M. Nowell, and P. A. Jumars, 1981, Sediment destabilization by animal tubes: J. Mar. Res., *39:* 361–374.

Eichler, R. and H. Ristedt, 1966, Isotopic evidence on the early life history of *Nautilus pompilius* (Linné): Science, *153:* 734–736.

Eisma, D., 1965, Shell-characteristics of *Cardium edule* L. as indicators of salinity: Neth. J. Sea Res., *2:* 493–540.

Eisma, D., W. G. Mook, and H. A. Das, 1976, Shell characteristics, isotopic composition and trace element contents of some euryhaline molluscs as indicators of salinity: Palaeogeogr. Palaeoclimatol. Palaeoecol., *19:* 39–62.

Ekdale, A. A., 1977, Abyssal trace fossils in worldwide deep sea drilling project cores: Geol. J. Spec. Issue, *9:* 163–182.

Ekdale, A. A., 1980, Graphoglyptid burrows in modern deep-sea sediment: Science, *207:* 304–306.

Ekdale, A. A., 1985, Paleoecology of the marine endobenthos: Palaeogeogr. Palaeoclimatol. Palaeoecol., *50:* 63–81.

Ekdale, A. A. and T. R. Mason, 1988, Characteristic trace fossil associations in oxygen-poor sedimentary environments: Geology, *16:* 720–723.

Ekdale, A. A., R. G. Bromley, and S. G. Pemberton, 1984, Ichnology: Trace fossils in sedimentology and stratigraphy: Soc. Econ. Paleontol. Mineral. Short Course Notes, *15,* 317 p.

Ekman, S., 1953, Zoogeography of the sea: Sidgwick and Jackson, London, 417 p.

Elias, M. K., 1937, Depth of deposition of the Big Blue (late Paleozoic) sediments in Kansas: Geol. Soc. Am. Bull., *48:* 403–432.

Elliott, G. F., 1984, Climatic tolerance in some aragonitic green algae of the post-Paleozoic: Palaeogeogr. Palaeoclimatol. Palaeoecol., *48:* 163–169.

Emery, K. O. and R. E. Stevenson, 1957, Estuaries and lagoons: *In* J. W. Hedgpeth (ed.), Treatise on marine ecology and paleoecology, v. 1, Ecology: Geol. Soc. Am. Mem., *67*(1): 673–749.

Emig, C. C., 1981, Implications de données veceutes sur les lingules actuelles dans les interprétations paléoécologiques: Lethaia, *14:* 151–156.

Emig, C. C., J. C. Gall, D. Pajand, and J.-C. Plaziat, 1978, Réflexions critiques sur l'écologie et la systématique des lingules actuelles et fossiles: Geobios, *11:* 573–609.

Emiliani, C., 1954, Temperatures of Pacific bottom waters and polar superficial waters during the Tertiary: Science, *119:* 853–855.

Emiliani, C., 1955, Pleistocene temperatures: J. Geol., *63:* 538–578.

Emiliani, C., 1966, Paleotemperature analysis of Caribbean cores P6304-8 and P6304-9 and a generalized temperature curve for the past 425,000 years: J. Geol., *74:* 109–126.

Emrich, K., D. H. Ehhalt, and J. C. Vogel, 1970, Carbon isotope fractionation during the precipitation of calcium carbonate: Earth Planet. Sci. Lett., *8:* 363–371.

Epstein, S., 1959, The variation of the O^{18}/O^{16} ratio in nature and some geologic implications: *In* P. H. Abelson (ed.), Researches in geochemistry: Wiley, New York, p. 217–240.

Epstein, S., R. Buchsbaum, H. Lowenstam, and H. C. Urey, 1951, Carbonate–water isotopic temperature scale: Geol. Soc. Am. Bull., *62:* 417–426.

Epstein, S. and T. Mayeda, 1953, Variation of O^{18} content of waters from natural sources: Geochim. Cosmochim. Acta, *4:* 213–224.

Epstein, S., R. Buchsbaum, H. Lowenstam, and H. C. Urey, 1953, Revised carbonate–water isotopic temperature scale: Geol. Soc. Am. Bull., *64:* 1315–1326.

Erez, J., 1978, Vital effect on stable-isotope composition seen in foraminifera and coral skeletons: Nature, *273:* 199–202.

Fager, E. W., 1963, Communities of organisms: *In* M. N. Hill (ed.), The sea, v. 2: Wiley-Interscience, New York, p. 415–437.

Fager, E. W., 1964, Marine sediments: Effects of a tube-building polychaete: Science, *143:* 356–359.

Fagerstrom, J. A., 1964, Fossil communities in paleoecology: Their recognition and significance: Geol. Soc. Am. Bull., *75:* 1197–1216.

Fagerstrom, J. A., 1987, The evolution of reef communities: Wiley-Interscience, New York, 600 p.

Fallow, W. C., 1979, Trans-North Atlantic similarity among Mesozoic and Cenozoic invertebrates correlated with widening of the ocean basin: J. Geol., *88:* 723–727.

Fallow, W. C., 1983, Trans-Pacific faunal similarities among Mesozoic and Cenozoic invertebrates related to plate tectonic processes: Am. J. Sci., *283:* 166–172.

Farlow, J. O., 1987, Lower Cretaceous dinosaur tracks, Paluxy River Valley, Texas: Baylor Univ. Press, Waco, Tex., 50 p.

Farlow, J. O., C. V. Thompson, and D. E. Rosner, 1976, Plates of the dinosaur *Stegosaurus:* Forced convection heat loss fins?: Science, *192:* 1123–1125.

Faure, G., 1977, Principles of isotope geology: Wiley, New York, 464 p.

Fell, H. B., 1954, Tertiary and Recent Echinoidea of New Zealand: N. Z. Geol. Surv. Paleontol. Bull., *23*, 62 p.

Fell, H. B., 1966, Ecology of the crinoids: *In* R. A. Boolootian (ed.), Physiology of Echinodermata: Wiley-Interscience, New York, p. 49–62.

Fenchel, T., 1970, Studies on the decomposition of organic detritus derived from the turtle grass *Thalassia testudinum:* Limnol. Oceanogr., *15:* 14–20.

Finks, R. M., 1970, The evolution and ecological history of sponges during Paleozoic times: *In* W. G. Fry (ed.), The biology of Porifera: Academic Press, New York, p. 3–22.

Fischer, A. G, 1964a, Growth patterns of Silurian Tabulata as palaeoclimatologic and palaeogeographic tools: *In* A. E. M. Nairn (ed.), Problems in paleoclimatology: Wiley-Interscience, New York, p. 608–615.

Fischer, A. G., 1964b, The Lofer cyclothems of the Alpine Triassic: Kans. Geol. Surv. Bull., *169:* 107–149.

Fisher, D. C., 1984, Mastodon butchery by North American paleo-indians: Nature, *308:* 271–272.

Fisher, D. L., 1985, Evolutionary morphology: Beyond the analogous, the anecdotal, and the ad hoc: Paleobiology, *11:* 120–138.

Fisher, R. A., A. S. Corbet, and C. B. Williams, 1943, The relation between the number of species and the number of individuals in a random sample of an animal population: J. Anim. Ecol., *12:* 42–58.

Fleming, R. H., 1957, General features of the oceans: *In* J. W. Hedgpeth (ed.), Treatise on marine ecology and paleoecology, v. 1: Geol. Soc. Am. Mcm., *67*(1): 87–108.

Flessa, K. W. and T. J. Brown, 1983, Selective solution of macroinvertebrate calcareous hard parts: A laboratory study: Lethaia, *16:* 193–205.

Flügel, E., 1982, Microfacies analysis of limestones: Springer-Verlag, West Berlin, 633 p.

Forbes, E., 1843, Report on the Mollusca and Radiata of the Aegean Sea and their distribution, considered as bearing on geology: Br. Assoc. Adv. Sci. Rep., *13,* p. 130–193.

Forbes, E. and R. Godwin-Austen, 1859, The natural history of the European seas: J. Van Voorst, London, 306 p.

Foster, M. W., 1974, Recent antarctic and subantarctic brachiopods: Antarct. Res. Ser., *21,* Am. Geophys. Union, Washington, D.C., 189 p.

Frey, R. W., 1987, Hermit crabs: Neglected factors in taphonomy and paleoecology: Palaios, *2:* 313–322.

Frey, R. W. and S. W. Henderson, 1987, Left–right phenomena among bivalve shells: Examples from the Georgia coast: Senckenbergiana Marit., *19:* 223–247.

Frey, R. W. and S. G. Pemberton, 1985, Biogenic structures in outcrops and cores. 1. Approaches to ichnology: Bull. Can. Pet. Geol., *33:* 72–115.

Frey, R. W., J. D. Howard, and W. A. Pryor, 1978, *Ophiomorpha:* Its morphologic, taxonomic, and environmental significance: Palaeogeogr. Palaeoclimatol. Palaeoecol., *23:* 199–229.

Frey R. W., S. G. Pemberton, and J. A. Fagerstrom, 1984, Morphological, ethological, and environmental significance of the ichnogenera *Scoyenia* and *Ancorrichnus:* J. Paleontol. *58:* 511–528.

Frey, R. W., P. B. Basan, and J. M. Smith, 1987, Rheotaxis and distribution of oysters and mussels, Georgia tidal creeks and salt marshes, U.S.A.: Palaeogeogr. Palaeoclimatol. Palaeoecol., *61:* 1–16.

Friedman, G. M., C. D. Geblein, and J. E. Sanders, 1971, Micrite envelopes of carbonate grains are not exclusively of photosynthetic algal origin: Sedimentology, *10:* 89–96.

Fritz, L. W. and R. A. Lutz, 1986, Environmental perturbations reflected in internal shell growth patterns of *Corbicula fluminea* (Mollusca: Bivalvia): Veliger, *28:* 401–417.

Fritz, P., A. V. Morgan, U. Eicher, and J. H. McAndrews, 1987, Stable isotope, fossil coleoptera and pollen stratigraphy in late Quanternary sediments from Ontario and New York state: Palaeogeogr. Palaeoclimatol. Palaeoecol., *58:* 183–202.

Frost, S. H., 1977, Cenozoic reef systems of the Caribbean: Prospects for Paleoecological synthesis: *In* S. H. Frost (ed.), Reefs and related carbonates: Ecology and sedimentology: Am. Assoc. Pet. Geol. Stud. Geol., *4:* 93–110.

Frost, S. H. and R. L. Langenheim, Jr., 1974, Cenozoic reef biofacies: Northern Illinois Univ. Press, De Kalb, Ill., 388 p.

Fry, W. G. (ed.), 1970, The biology of porifera: Academic Press, New York, 512 p.

Füchtbauer, H. and L. A. Hardie, 1976, Experimentally determined homogeneous distribution coefficients for precipitated magnesian calcites: Geol. Soc. Am. Abstr. Programs, *8*(6): 877.

Fürsich, F. T., 1978, The influence of faunal condensation and mixing on the preservation of fossil benthic communities: Lethaia, *11:* 243–250.

Fürsich, F. T. and K. W. Flessa, 1987, Taphonomy of tidal flat molluscs in the northern Gulf of California: Paleoenvironmental analysis despite the perils of preservation: Palaios, *2:* 543–559.

Fürsich, F. T. and J. M. Hurst, 1974, Environmental factors determining the distribution of brachiopods: Palaeontology, *17:* 879–900.

Fürst, M. J., 1981, Boron in siliceous material as a paleosalinity indicator: Geochim. Cosmochim. Acta, *45:* 1–13.

Futterer, E., 1978a, Studien über die Einregelung, Anlagerung und Einbettung biogener Hartteile im Strömungskanal: Neues Jahrb. Palaeontol. Abh., *156:* 87–131.

Futterer, E., 1978b, Untersuchungen über die Sink- und Transportgeschwindigkeit biogener Hartteile: Neues Jahrb. Palaeontol. Abh., *155:* 318–359.

Fyfe, W. S., and J. L. Bischoff, 1965, The calcite–aragonite problem: *In* L. C. Pray and R. C. Murray (eds.), Dolomitization and limestone diagenesis: Soc. Econ. Paleontol. Mineral. Spec. Publ., *13:* 3–13.

Galehouse, J. S., 1967, Provenance and paleocurrents of the Paso Robles Formation, California: Geol. Soc. Am. Bull., *78:* 951–978.

Gastaldo, R. A., D. P. Douglass, and S. M. McCarroll, 1987. Origin, characteristics and provenence of plant macrodetritus in a Holocene crevasse splay, Mobile delta, Alabama: Palaios, *2:* 229–240.

Gat, J. R., 1980, The isotopes of hydrogen and oxygen in precipitation: *In* P. Fritz and J. C. Fontes (eds.), Handbook of environmental isotope geochemistry, v. 1, The terrestrial environment: Elsevier, Amsterdam, p. 21–47.

Geitzenauer, K. R., 1969, Coccoliths as Late Quaternary paleoclimatic indicators in the subantarctic Pacific: Nature, *223:* 170–172.

Genin, A., P. K. Dayton, P. F. Lonsdale, and F. N. Spiess, 1986, Corals on sea mount peaks provide evidence of current acceleration over deep-sea topography: Nature, *322:* 59–61.

Gennett, J. A. and E. L. Grossman, 1986, Oxygen and carbon isotope trends in a late glacial-holocene pollen site in Wyoming, U.S.A.: Geogr. Phys. Quat. *40:* 161–169.

Gibson, T. G. and M.A. Buzas, 1973, Species diversity: Patterns in modern and Miocene Foraminifera of the eastern margin of North America: Geol. Soc. Am. Bull., *84:* 217–238.

Ginsburg, R. N., 1956, Environmental relationships of grain size and constituent particles in some south Florida carbonate sediments: Bull. Am. Assoc. Pet. Geol., *40:* 2384–2427.

Gladfelter, E. H., 1983, Skeletal development in *Acropora cervicornis:* Coral Reefs, *2:* 91–100.

Glass, B. P. and L. J. Rosen, 1978, Sedimentation rates, deep sea: *In* R. W. Fairbridge and J. Bourgeois (eds.), The encyclopedia of sedimentology: Dowden, Hutchinson & Ross, Stroudsburg, Pa., p. 692–694.

Goldberg, E. D., 1957, Biogeochemistry of trace metals: In J. W. Hedgpeth (ed.), Treatise on marine ecology and paleoecology, v. 1, Ecology: Geol. Soc. Am. Mem., 67(1): 345–358.

Goldring, R. and J. Kazmierczak, 1974, Ecological succession in intraformational hardground formation: Palaeontology, 17: 949–962.

Golubic, S., S. E. Campbell, K. Drobne, B. Cameron, W. L. Balsam, F. Cimerman, and L. Dubois, 1984, Microbial endoliths: A benthic overprint in the sedimentary record, and a paleobathymetric cross-reference with Foraminifera: J. Paleontol., 58: 351–361.

Good, R., 1974, The geography of the flowering plants: Longman Group, Harlow, Essex, England, 557 p.

Good, S. C., 1987, Mollusc-based interpretations of lacustrine paleoenvironments of the Sheep Pass Formation (latest Cretaceous to Eocene) of central Nevada: Palaios, 2: 467–478.

Goodfriend, G. A. and M. Magaritz, 1987, Carbon and oxygen isotope composition of shell carbonate of desert land snails: Earth Planet. Sci. Lett., 86: 377–388.

Goreau, T. F., 1959, The physiology of skeleton formation in corals. I. A method of measuring the rate of calcium deposition by corals under different conditions: Biol. Bull. Mar. Biol. Lab. Woods Hole, 116: 59–75.

Goreau, T. F., 1963, Calcium carbonate deposition by coralline algae and corals in relation to their roles as reef-builders: Ann. N.Y. Acad. Sci., 109: 127–167.

Goreau, T. F., N. I. Goreau, and T. J. Goreau, 1979, Corals and coral reefs: Sci. Am., 241: 124–136.

Gould, S. J., 1965, Is uniformitarianism necessary? Am. J. Sci., 263: 223–228.

Gould, S. J. and R. C. Lewontin, 1979, The spandrels of San Marco and the Panglossian paradigm: A critique of the adaptationist programme: Proc. R. Soc. London, B205: 147–164.

Gould, S. J. and E. S. Vrba, 1982, Exaption: A missing term in the science of form: Paleobiology, 8: 4–15.

Graham, D. W., M. L. Bender, D. F. Williams, and L. D. Keigwin, Jr., 1982, Strontium–calcium ratios in Cenozoic planktonic foraminifera: Geochim. Cosmochim. Acta, 46: 1281–1292.

Grant, R. E., 1972, The lophophore and feeding mechanism of the productidina (Brachiopoda): J. Paleontol., 46: 213–249.

Graus, R. R. and I. G. Macintyre, 1976, Light control of growth form in colonial reef corals: Computer simulation: Science, 193: 895–897.

Graus, R. R., J. A. Chamberlain, Jr., and A. M. Baker, 1977, Structural modification of corals in relation to waves and currents: Am. Assoc. Pet. Geol. Stud. Geol., 4: 135–153.

Gray, J. and A. J. Boucot (eds.), 1979, Historical biogeography, plate tectonics, and the changing environment: Oregon State Univ. Press, Corvallis, Oreg., 500 p.

Greiner, G. O. G., 1969, Recent benthonic foraminifera: Environmental factors controlling their distribution: Nature, 223: 168–170.

Greiner, G. O. G., 1974, Environmental factors controlling the distribution of Recent benthonic foraminifera: Breviora Mus. Comp. Zool. Harv. Univ., 420: 1–35.

Grossman, E. L., 1984, Stable isotope fractionation in live benthic foraminifera from the southern California borderland: Palaeogeogr. Palaeoclimatol. Palaeoecol., *47:* 301–327.

Grossman, E. L., 1987, Stable isotopes in modern benthic foraminifera: A study of vital effect: J. Foraminiferal Res., *17:* 48–61.

Grossman, E. T. and T. L. Ku, 1981, Aragonite–water isotopic paleotemperature scale based on the benthic foraminifera *Hoeglundia elegans:* Geol. Soc. Am. Abstr. Programs, *13:* 464.

Grossman, E. L., and T. L. Ku, 1986, Oxygen and carbon isotope fractionation in biogenic aragonite: Temperature effects: Chem. Geol., *59:* 59–74.

Haack, S. C., 1986, A thermal model of the sailback pelycosaur: Paleobiology, *12:* 450–458.

Haggerty, J. A., J. N. Weber, R. J. Cuffey, and P. Deines, 1980, Environment-related morphologic and geochemical variability in the modern reef corals *Favia pallida* and *Favia stelligera* on Enewetak Atoll: Pac. Geol. *14:* 226–234.

Hallam, A., 1961, Brachiopod life assemblages from the Marlstone rock-bed of Leicestershire: Palaeontology, *4:* 653–657.

Hallam, A., 1967, The interpretation of size-frequency distributions in molluscan death assemblages: Palaeontology, *10:* 25–42.

Hallam, A., 1969, Faunal realms and facies in the Jurassic: Palaeontology, *12:* 1–18.

Hallam, A., 1972, Models involving population dynamics: *In* T. J. M. Schopf (ed.), Models in paleobiology: Freeman, Cooper, San Francisco, p. 62–80.

Hallam, A. (ed.), 1973a, Atlas of paleobiogeography: Elsevier, Amsterdam, 531 p.

Hallam, A., 1973b, Distributional patterns in contemporary terrestrial and marine animals: Palaeontol. Assoc. Spec. Pap., *12:* 93–105.

Hallam, A., 1977, Jurassic bivalve biogeography: Paleobiology, *3:* 58–73.

Hallam, A., 1983, Early and mid-Jurassic molluscan biogeography and the establishment of the central Atlantic seaway: Palaeogeogr. Palaeoclimatol. Palaeoecol., *43:* 181–193.

Hallam, A. and M. J. O'Hara, 1962, Aragonitic fossils in the Lower Carboniferous of Scotland: Nature, *195:* 273–274.

Hallock, P. and E. C. Glenn, 1986, Large foraminifera: A tool for paleoenvironmental analysis of Cenozoic carbonate depositional facies: Palaios, *1:* 55–64.

Hallock, P. and W. Schlager, 1986, Nutrient excess and the demise of coral reefs and coral platforms: Palaios, *1:* 389–398.

Hansen, H. J., 1979, Test structure and evolution in the Foraminifera: Lethaia, *12:* 173–182.

Häntzschel, W., 1975, Trace fossils and problematica: *In* C. Teichert (ed.), Treatise on invertebrate paleontology, pt. W. Miscellanea, suppl. 1: Geol. Soc. Am., Boulder, Colo., 269 p.

Haq, B. U., 1978, Calcareous nannoplankton: *In* B. U. Haq and A. Boersma (eds.), Introduction to marine micropaleontology: Elsevier, New York, p. 79–107.

Haq, B. U., I. Primoli-Silva, and G. P. Lohmann, 1977, Calcareous plankton paleobiogeograaphic evidence for major climatic fluctuations in the Early Cenozoic Atlantic Ocean: J. Geophys. Res., *82:* 3861–3876.

Harman, R. A., 1964, Distribution of foraminifera in the Santa Barbara Basin, California: Micropaleontology, *10:* 81–96.

Harmelin, J. G., 1975, Relations entre la forme zoariale et l'habitat chez les bryozoaires cyclostomes: Conséquences taxonomiques: *In* S. Pouyet (cd.), Bryozoa, 1974, Proc. 3rd Conference International Bryozoology Assoc., Doc. Lab. Geol. Fac. Sci. Lyon, *3*(1): 369–384.

Harmon, R. S., H. P. Schwarcz, and D. C. Ford, 1978, Stable isotope geochemistry of speleothems and cave water from the Flint Ridge–Mammoth Cave System, Kentucky: Implication for terrestrial climate change during the period 230,000 to 100,000 years B.P.: J. Geol., *86:* 373–384.

Harper, C. W., Jr., 1977, Groupings by locality in community ecology and paleoecology: Tests of significance: Lethaia, *11:* 251–257.

Hartman, W. D., 1983, Modern and Ancient Sclerospongiae: *In* J. K. Rigby and C. W. Stearn (eds.), Sponges and spongiomorphs: Notes for a short course: Univ. Tenn., Dep. Geol. Sci., Stud. Geol., *7:* 116–129.

Hartman, W. T. and T. F. Goreau, 1970, Jamaican coralline sponges: Their morphology, ecology, and fossil relatives: *In* W. G. Fry (ed.), The biology of Porifera: Academic Press, New York, p. 205–240.

Hay, W. H., K. M. Towe, and R. C. Wright, 1963, Ultra-microstructure of some selected foraminiferal tests: Micropaleontology, *9:* 171–195.

Hayami, I. and T. Okamoto, 1986, Geometric regularity of some oblique sculptures in pectinid and other bivalves: Recognition by computer simulation: Paleobiology, *12:* 433–449.

Haynes, J., 1965, Symbiosis, wall structure, and habitat in foraminifera: Cushman Found. Foraminiferal Res. Contrib., *16:* 40–43.

Haynes, J. R., 1981, Foraminifera: Halsted Press, New York, 433 p.

Hayward, B. W. and M. A. Buzas, 1979, Taxonomy and paleoecology of early Miocene benthic foraminifera of northern New Zealand and the north Tasman Sea: Smithsonian Contrib. Paleobiol., *36,* 154 p.

Hazel, J. E., 1971, Ostracode biostratigraphy of the Yorktown Formation (Upper Miocene and Lower Pliocene) of Virginia and North Carolina: U.S. Geol. Surv. Prof. Pap., *704,* 13 p.

Hazel, J. E., 1975, Patterns of marine ostracode diversity in the Cape Hatteras, North Carolina, area: J. Paleontol., *49:* 731–744.

Hecht, A. D., 1976, The oxygen isotopic record of foraminifera in deep-sea sediment: *In* R. H. Hedley and C. G. Adams (eds.). Foraminifera, v. 2: Academic Press, London, p. 1–43.

Heckel, P. H., 1972, Recognition of ancient shallow marine environments: Soc. Econ. Paleontol. Mineral. Spec. Publ., *16:* 26–286.

Heckel, P. H., 1977, Origin of phosphatic black shale facies in Pennsylvanian cyclothems of mid continent North America: Am. Assoc. Pet. Geol. Bull., *61:* 1045–1068.

Hedgpeth, J. W. (ed.), 1957a, Treatise on marine ecology and paleoecology, v. 1, Ecology: Geol. Soc. Am. Mem., *67*(1): 1296.

Hedgpeth, J. W., 1957b, Marine biogeography: *In* J. W. Hedgpeth (ed.), Treatise on marine ecology and paleoecology, v. 1, Ecology: Geol., Soc. Am. Mem., *67*(1): 359–382.

Hedgpeth, J. W., 1967, Ecological aspects of the Laguna Madre, a hypersaline estuary: *In* G. H. Lauff (ed.), Estuaries: Am. Assoc. Adv. Sci., Publ., *83,* 408–419.

Hedrick, P. W., 1984, Population biology: The evolution and ecology of populations: Jones and Bartlett Publishers, Boston, 445 p.

Hennebert, M. and A. Lees, 1985, Optimized similarity matrices applied to the study of carbonate rocks: Geol. J., *20:* 123–131.

Henrich, R. and G. Wefer, 1986, Dissolution of biogenic carbonates: Effects of skeletal structure: Mar. Geol., *71:* 341–362.

Herman, Y., 1978, Pteropods: *In* B. U. Haq and A. Boersma (eds.), Introduction to marine micropaleontology: Elsevier, New York, p. 151–159.

Herz, N. and N. E. Dean, 1986, Stable isotopes and archaeological geology: The Carrara marble, northern Italy: Appl. Geochim., *1:* 139–151.

Hesse, R., W. C. Allee, and K. P. Schmidt, 1951, Ecological animal geography: Wiley, New York, 715 p.

Hessler, R. R. and H. L. Sanders, 1967, Faunal diversity in the deep sea: Deep-Sea Res., *14:* 65–79.

Hester, K., and E. A. Boyle, 1982, Water chemistry control of cadmium content in Recent benthic foraminifera: Nature, *298:* 260–262.

Heusser, L. E. and N. J. Shackleton, 1979, Direct marine–continental correlation: 150,000-year oxygen isotope–pollen record for the north Pacific: Science, *204:* 837–839.

Hewitt, R. A. and G. E. G. Westermann, 1986, Functions of complexly fluted septa in ammonoid shells. I. Mechanical principles and functional models: Neues Jahrb. Geol. Palaeontol. Abh., *172:* 47–69.

Hickman, C. S., 1984, Composition, structure, ecology, and evolution of six Cenozoic deep-water communities: J. Paleontol., *58:* 1215–1234.

Hickman, C. S., 1985, Gastropod morphology and function: Univ. Tenn., Dep. Geol. Sci., Stud. Geol., *13:* 138–156.

Hill, M. N. (ed.), 1962 (etc.), The sea: Ideas and observations of progress in the study of the sea: Wiley-Interscience, New York (numerous vols.).

Hill, M. O., 1973, Diversity and evenness: A unifying notation and its consequences: Ecology, *54:* 427–432.

Hiller, N., 1988, The development of growth lines on articulate brachiopods: Lethaia, *21:* 177–188.

Hladil, J., 1986, Trends in the development and cyclic patterns of Middle and Upper Devonian buildups: Facies, *15:* 1–34.

Hoffman, A., 1977, Synecology of macrobenthic assemblages of the Korytnica Clays (Middle Miocene; Holy Cross Mountains, Poland): Acta Geol. Pol., *27:* 227–280.

Hoffman, A., 1979, Community paleoecology as an epiphenomenal science: Paleobiology, *5:* 357–379.

Hoffman, A., 1981, Stochastic versus deterministic approach to paleontology: The question of scaling or metaphysics: Neues Jahrb. Geol. Palaeontal. Abh., *182:* 80–96.

Hoffman, A. and M. Narkiewicz, 1977, Developmental pattern of Lower to Middle Paleozoic banks and reefs: Neues Jahrb. Geol. Palaeontol. Monatsh., *5:* 272–283.

Hoffman, A., A. Pisera, and W. Studencki, 1978, Reconstruction of a Miocene kelp-associated macrobenthic ecosystem: Acta Geol. *28:* 377–387.

Holland, H. D., J. J. Holland, and J. L. Munoz, 1964, The co-precipitation of cations with CaCO$_3$. II. The co-precipitation of Sr^{2+} with calcite between 90° and 100°C: Geochim. Cosmochim. Acta, *28:* 1287–1301.

Hollmann, R., 1968, Zur Morphologie rezenter Molluskenbruchschille: Palaeontol. Z., *42:* 217–235.

Hopson, J. A., 1975, The evolution of cranial display structures in hadrosaurans: Paleobiology, *1:* 21–44.

Horowitz, A. S. and P. E. Potter, 1971, Introductory petrography of fossils: Springer-Verlag, New York, 302 p.

Hsu, K. J., 1986, Environmental changes in times of biotic crisis: *In* D. M. Raup and D. Jablonski (eds.), Patterns and processes in the history of life: Springer-Verlag, West Berlin, p. 297–312.

Hubbard, J. A. E. B., 1970, Sedimentological factors affecting the distribution and growth of Visian caninioid corals in northwest Ireland: Palaeontology, *13:* 191–209.

Hudson, J. D., 1977, Oxygen isotope studies on Cenozoic temperatures, oceans, and ice accumulation: Scott. J. Geol., *13:* 313–325.

Hudson, J. H. and E. A. Shinn, 1977, Stress banding in corals: Normal and abnormal periodicity (abs.): J. Paleontol., *51*(2, pt. III): 15.

Hudson, J. H., E. A. Shinn, R. B. Halley, and B. Lidz, 1976, Sclerochronology: A tool for interpreting past environments: Geology, *4:* 361–364.

Humphries, C. J. and L. R. Parenti, 1986, Cladistic biogeography: Clarendon Press, Oxford, 98 p.

Hunt, R. M., Jr., 1978, Depositional setting of a Miocene mammal assemblage, Sioux County, Nebraska (U.S.A.): Palaeogeogr. Palaeoclimatol. Palaeoecol., *24:* 1–52.

Hurlbert, S. H., 1971, The nonconcept of species diversity: A critique and alternative parameters: Ecology, *52:* 577–586.

Huston, M., 1985, Variation in coral growth rates with depth at Discovery Bay, Jamaica: Coral Reefs, *4:* 19–25.

Hutchinson, G. E., 1959, Homage to Santa Rosalia or Why are there so many kinds of animals?: Am. Nat., *93:* 145–159.

Hutchinson, G. E., 1978, An introduction to population ecology: Yale Univ. Press, New Haven, Conn., 260 p.

Imbrie, J. and N. G. Kipp, 1971, A new micropaleontological method for quantitative paleoclimatology: Application to a Late Pleistocene Caribbean core: *In* K. K. Turekian (ed.), The Late Cenozoic glacial ages: Yale Univ. Press, New Haven, Conn., p. 71–181.

Inman, D. L., 1963, Ocean waves and associated currents: *In* F. P. Shepard, Submarine geology, 2nd ed.: Harper & Row, New York, p. 49–81.

Israelsky, M. C., 1949, Oscillation chart: Bull. Am. Assoc. Pet. Geol., *33:* 92–98.

Jaanusson, V., 1984, Functional morphology of the shell in platycope ostracodes: A study of arrested evolution: Lethaia, *18:* 73–84.

Jablonski, D., K. W. Flessa, and J. W. Valentine, 1985, Biogeography and paleo-biology: Paleobiology, *11:* 75–90.

Jackson, J. B. C., 1968, Bivalves: Spatial and size-frequency distribution of two inter-tidal species: Science, *161:* 479–480.

Jackson, J. B. C., T. F. Goreau, and W. D. Hartman, 1971, Recent brachiopod–coralline sponge communities and their paleoecological significance: Science, *173:* 623–625.

Jannasch, H. W. and M. J. Mottl, 1985, Geomicrobiology of deep-sea hydrothermal vents: Science, *229:* 717–725.

Johansen, H. W., 1981, Coralline algae: A first synthesis: CRC Press, Boca Raton, Fla., 239 p.

Johnson, J. H., 1961, Limestone-building algae and algal limestones: Colorado School of Mines Press, Golden, Colo., 297 p.

Johnson, R. G., 1964, The community approach to paleoecology: *In* J. Imbrie and N. Newell (eds.), Approaches to paleoecology: Wiley, New York, p. 107–134.

Johnson, R. G., 1965a, Pelecypod death assemblages in Tomales Bay, California: J. Paleontol., *39:* 80–85.

Johnson, R. G., 1965b, Temperature variation in the infaunal environment of a sand flat: Limnol. Oceanogr., *10:* 114–120.

Johnson, J. G., 1971, A quantitative approach to faunal province analysis: Am. J. Sci., *270:* 257–280.

Johnson, T. C., 1976, Biogenic opal preservation in pelagic sediments of a small area in the eastern tropical Pacific: Geol. Soc. Am. Bull., *87:* 1273–1282.

Johnson, M. E., 1977, Succession and replacement in the development of Silurian brachiopod populations: Lethaia, *10:* 83–93.

Johnson, A. L. A., 1981, Detection of ecophenotypic variation in fossils and its applica-tion to a Jurassic scallop: Lethaia, *14:* 277–285.

Johnson, G. A. L. and J. R. Nudds, 1975, Carboniferous coral geochronometers: *In* G. D. Rosenberg and S. K. Runcorn (eds.), Growth rhythms and the history of the earth's rotation: Wiley-Interscience, New York, p. 27–42.

Jones, D. S., 1980, Annual cycle of shell growth increment formation in two continen-tal shelf bivalves and its paleoecological significance: Paleobiology, *6:* 331–340.

Jones, D. S., 1981, Annual growth increments in shells of *Spisula solidisima* record marine temperature variability: Science, *211:* 165–167.

Jones, D. S., 1983, Sclerochronology: Reading the record of the molluscan shell: Am. Sci., *71:* 384–391.

Jones, D. S., D. F. Williams, M. A. Arthur, and D. E. Krantz, 1984, Interpreting the paleoenvironmental, paleoclimatic and life history records in mollusc shells: Geo-bios Mem., *8,* 333–339.

Jones, D. S., D. F. Williams, and C. S. Romanek, 1986, Life-history of symbiont-bearing giant clams from stable isotope profiles: Science, *231:* 46–48.

Jouzel, J., C. Lorius, J. R. Petit, C. Genthon, N. I. Barkov, V. M. Kotlyakov, and V. M. Petrov, 1987, Vostok ice core: A continuous isotope temperature record over the last climatic cycle (160,000 years): Nature, *329:* 403–408.

Juillet-Leclerc, A. and H. Schrader, 1987, Variations of upwelling intensity recorded in varved sediment from the Gulf of California during the past 3,000 years: Nature, *329:* 146–149.

Kahn, P. G. K. and S. M. Pompea, 1978, Nautiloid growth rhythms and dynamical evolution of the earth–moon system: Nature, *275:* 606–611.

Kammer, T. W., 1979, Paleosalinity, paleotemperature, and oxygen isotopic fractionation records of Neogene foraminifera from DSDP site 173 and the Centerville Beach section, California: Mar. Micropaleontol., *4:* 45–60.

Karhu, J. and S. Epstein, 1986, The implication of the oxygen isotope records in coexisting cherts and phosphates: Geochim. Cosmochim. Acta, *50:* 1745–1755.

Kaston, B. J., 1978, How to know the spiders: Wm. C. Brown, Dubuque, Iowa, 272 p.

Kato, M., 1963, Fine skeletal structure in Rugosa: J. Fac. Sci. Hokkaido Univ. Ser. 4, *11:* 571–630.

Kauffman, E. G., 1969, Cretaceous marine cycles of the Western Interior: Mt. Geol., *6:* 227–245.

Kauffman, E. G., 1977, Geological and biological overview: Western Interior Cretaceous Basin: Mt. Geol., *14:* 75–99.

Kauffman, E. G., 1981, Ecological reappraisal of the German Posidonienschiefer (Toarcian) and the stagnant basin model: *In* J. Gray, A. J. Boucot, and W. B. N. Berry (eds.), Communities of the past: Hutchinson Ross, Stroudsburg, Pa., p. 311–381.

Keast, A. (ed.), 1981, Ecological biogeography of Australia, v. I–III: Dr. W. Junk, Publishers, the Hague, The Netherlands, 2142 p.

Keen, A. M., 1937, An abridged check list and bibliography of west North American marine mollusca: Stanford Univ. Press, Stanford, Calif., 87 p.

Keigwin, L. D., Jr., 1979, Late Cenozoic stable isotope stratigraphy and paleoceanography of DSDP sites from the east equatorial and central north Pacific Ocean: Earth Planet. Sci. Lett., *45:* 361–382.

Kennedy, W. J., J. D. Taylor, and A. Hall, 1969, Environmental and biological controls on bivalve shell mineralogy: Biol. Rev., *44:* 499–530.

Kennett, J. P., 1976, Phenotypic variation in some Recent and Late Cenozoic planktonic foraminifera: *In* R. H. Hedley and C. G. Adams (eds.), Foraminifera, v. 2: Academic Press, New York, p. 111–170.

Kennett, J. P., 1982, Marine geology: Prentice-Hall, Englewood Cliffs, N.J., 813 p.

Kennett, J. P. and N. J. Shackleton, 1975, Laurentide ice sheet melt water recorded in the Gulf of Mexico deep-sea cores: Science, *188:* 147–150.

Kennish, M. J. and R. K. Olsson, 1975, Effects of thermal discharges on the microstructural growth of *Mercenaria mercenaria:* Environ. Geol., *1:* 41–64.

Kershaw, S., 1984, Patterns of stromatoporoid growth in level bottom communities: Palaeontology, *27:* 113–130.

Keupp, H., 1977, Ultrafazies und Genese der solnhofener Plattenkalke (Oberer Malm, Sudliche Frankenalb): Abh. Naturhist. Ges. Nürnberg, *37,* 128 p.

Kidwell, S. M., 1982, Time scales of fossil accumulation: Patterns from Miocene benthic assemblages: Proc. 3rd North Am. Paleontol. Conv., *1:* 295–300.

Kidwell, S. M., 1988, Taphonomic comparison of passive and active continental margins: Neogene shell beds of the Atlantic coastal plain and northern Gulf of California: Palaeogeogr. Palaeoclimatol. Palaeoecol., *63:* 201–223.

Kidwell, S. M., F. T. Fürsich, and T. Aigner, 1986, Conceptual framework for the analysis and classification of fossil concentrations: Palaios, *1:* 228–238.

Kidwell, S. M. and D. Jablonski, 1983, Taphonomic feedback: ecological consequences of shell accumulation: *In* M. J. S. Tevesz and P. L. McCall (eds.), Biotic interactions in Recent and fossil benthic communities: Plenum Press, New York, p. 195–248.

Kier, P. M., 1972, Upper Miocene echinoids from the Yorktown Formation of Virginia and their environmental significance: Smithsonian Contrib. Paleobiol., *13,* 41 p.

Kier, P. M., 1974, Evolutionary trends and their functional significance in the post-Paleozoic echinoids: Paleontol. Soc. Mem., *5* [J. Paleontol., *48*(3), suppl.], 96 p.

Killingley, J. S. and M. A. Rex, 1985, Mode of larval development in deep-sea gastropods indicated by oxygen-18 values of their carbonate shells: Deep Sea Res., *32:* 809–818.

Kinne, O., 1971, Environmental factors, salinity, animals, invertebrates: *In* O. Kinne (ed.), Marine ecology: a comprehensive, integrated treatise on life in oceans and coastal waters, v. 1, pt. 2: Wiley, New York, p. 821–1083.

Kinsman, D. J. J. and H. D. Holland, 1969, The co-precipitation of cations with $CaCO_3$. IV. The co-precipitation of Sr^{2+} with aragonite between 16° and 96° C: Geochim. Cosmochim. Acta, *33:* 1–17.

Kitano, Y., N. Kanamori, and S. Yoshoika, 1976, Influence of chemical species on crystal type of calcium carbonate: *In* N. Watabe and K. M. Wilbur (eds.), The mechanisms of mineralization in the invertebrates and plants: Univ. South Carolina Press, Columbia, S.C., p. 191–202.

Kitchell, J. A., J. F. Kitchell, D. L. Clark, and L. Dangeard, 1978, Deep-sea foraging behavior: Its bathymetric potential in the fossil record: Science, *200:* 1289–1291.

Knutson, D. W., R. W. Buddemeier, and S. V. Smith, 1972, Coral chronometers: Seasonal growth bands in reef corals: Science, *177:* 270–272.

Kolodny, Y. and M. Raab, 1988, Oxygen isotopes in phosphatic fish remains from Israel: Paleothermometry of tropical Cretaceous and Tertiary shelf waters: Palaeogeogr. Palaeoclimatol. Palaeoecol., *64:* 59–67.

Kontrovitz, M. and J. H. Myers, Jr., 1988, Ostracode eyes as paleoenvironmental indicators: Physical limits of vision in some podocopids: Geology, *16:* 293–295.

Krantz, D. E., D. F. Williams, and D. S. Jones, 1987, Ecological and paleoenvironmental information using stable isotope profiles from living and fossil molluscs: Palaeogeogr. Palaeoclimatol. Palaeoecol., *58:* 249–266.

Krantz, D. E., A. T. Kronick, and D. F. Williams, 1988, A model for interpreting continental-shelf hydrographic processes from the stable isotope and cadmium: calcium profiles of scallop shells: Palaeogeogr. Palaeoclimatol. Palaeoecol., *64:* 123–140.

Kranz, P. M., 1974, The anastrophic burial of bivalves and its palaeoecological significance: J. Geol., *82:* 237–265.

Krauskopf, K. B., 1967, Introduction to geochemistry: McGraw-Hill, New York, 721 p.

Kroopnick, P. M., S. V. Margolis, and C. S. Wong, 1977, ^{13}C variations in marine carbonate sediments as indicators of the CO_2 balance between the atmosphere and oceans: *In* N. R. Anderson and A. Malahoff (eds.), The fate of fossil fuel CO_2 in the oceans: Plenum Press, New York, p. 295–321.

Kummel, B., 1948, Environmental significance of dwarfed cephalopods: J. Sediment. Petrol., *18:* 61–64.

Kurtén, B., 1954, Population dynamics: A new method in paleontology: J. Paleontol., *28:* 286–292.

Kurtén, B., 1964, Population structure in paleoecology: *In* J. Imbrie and N. D. Newell (eds.), Approaches to paleoecology: Wiley, New York, p. 91–106.

LaBarbera, M., 1977, Brachiopod orientation to water movement. 1. Theory, laboratory behavior, and field orientation: Paleobiology, *3:* 270–287.

Ladd, H. S. (ed.), 1957, Treatise on marine ecology and paleoecology, v. 2, Paleoecology: Geol. Soc. Am. Mem., *67*(2), 1077p.

Lagaaij, R. and P. L. Cook, 1973, Some Tertiary to Recent Bryozoa: *In* A. Hallam (ed.), Atlas of palaeobiogeography: Elsevier, Amsterdam, p. 489–498.

Lagaaij, R. and Y. V. Gautier, 1965, Bryozoan assemblages from marine sediments of the Rhône delta, France: Micropaleontology, *11:* 39–58.

Landman, N. H., D. M. Rye, and K. L. Skelton, 1983, Early ontogeny of *Endrophoceras* compared to recent *Nautilus* and Mesozoic ammonites: Evidence from shell morphology and light stable isotopes: Paleobiology, *9:* 269–279.

Landman, N. H., E. R. M. Druffel, J. K. Cochran, D. J. Donahue, and A. J. T. Jull, 1988, Bomb-produced radiocarbon in the shell of the chambered nautilus: Rate of growth and age at maturity: Earth Planet. Sci. Lett., *89:* 28–34.

Lane, N. G., 1973, Paleontology and paleoecology of the Crawfordsville fossil site (Upper Osagian: Indiana): Univ. Calif. Publ. Geol. Sci., *99,* 141 p.

Lasker, H., 1976, Effects of differential preservation on the measurement of taxonomic diversity: Paleobiology, *2:* 84–93.

Latimer, W. M., 1952, Oxidation potentials, 2nd ed.: Prentice-Hall, Englewood Cliffs, N.J., 392 p.

Lawrence, D. R., 1968, Taphonomy and information losses in fossil communities: Geol. Soc. Am. Bull., *79:* 1315–1330.

Lawrence, D. R., 1971a, The nature and structure of paleoecology: J. Paleontol., *45:* 593–607.

Lawrence, D. R., 1971b, Shell orientation in recent and fossil oyster communties from the Carolinas: J. Paleontol., *45:* 347–349.

Lawrence, J., 1987, A functional biology of echinoderms: Johns Hopkins Univ. Press, Baltimore, 340 p.

Leclerc, A. J. and L. Labeyrie, 1987, Temperature dependence of the oxygen isotopic fractionation between diatom silica and water: Earth Planet. Sci. Lett., *84:* 69–74.

Lehmann, U., 1981, The ammonites: Their life and their world: Cambridge Univ. Press, Cambridge, 246 p.

Leigh, E. G., Jr., 1971, Adaptation and diversity: Freeman, Cooper, San Francisco, 288 p.

Leutenegger, S., 1984, Symbiosis in benthic foraminifera: Specificity and host adaptations: J. Foraminiferal Res., *14:* 16–35.

Lever, J., 1958, Quantitative beach research I. The "left-right-phenomenon": sorting of Lamellibranch valves on sandy beaches: Basteria, *22:* 21–68.

Levins, R., 1968, Evolution in changing environments: Princeton Univ. Press, Princeton, N.J., 120 p.

Levinson, S. A., 1961, Identification of fossil ostracodes in thin section: *In* R. C. Moore (ed.), Treatise on invertebrate paleontology, pt. Q, Arthropoda 3: Geol, Soc. Am., Boulder, Colo., p. 70–73.

Levinton, J. S., 1970, The paleoecological significance of opportunistic species: Lethaia, *3:* 69–78.

Levinton, J. S. and R. K. Bambach, 1970, Some ecological aspects of bivalve mortality patterns: Am. J. Sci., *268:* 97–112.

Lewy, Z. and C. Saintleben, 1979, Functional morphology and paleoecological significance of the conchiolin layeres in corbulid pelecypods: Lethaia, *12:* 341–351.

Li, Y. H., T. Takahashi, and W. S. Broecker, 1969, Degree of saturation of $CaCO_3$ in the oceans: J. Geophys. Res., *74:* 5507–5525.

Linsley, R. M., 1978a, Shell form and the evolution of gastropods: Am. Sci. *66:* 432–441.

Linsley, R. M., 1978b, Locomotion rates and shell form in the Gastropoda: Malacologia, *17:* 193–206.

Linsley, R. M., E. L. Yochelson, and D. M. Rohr, 1978, A reinterpretation of the mode of life of some Paleozoic frilled gastropods: Lethaia, *11:* 105–112.

Livingstone, D. A., 1963, Chemical composition of rivers and lakes: U.S. Geol. Surv. Prof. Pap., *440-G,* 64 p.

Lloyd, M. and R. J. Ghelardi, 1964, A table for calculating the "equitability" component of species diversity: J. Anim. Ecol., *33:* 217–225.

Lockley, M. G., 1986, The paleobiological and paleoenvironmental importance of dinosaur footprints: Palaios, *1:* 37–47.

Loffler, H. and D. Danielopol (eds.), 1977, Aspects of ecology and zoogeography of Recent and fossil Ostracoda: Dr. W. Junk, Publishers, The Hague, The Netherlands, 521 p.

Logan, B. W., P. Hoffman, and C. D. Gebelein, 1974, Algal mats, cryptalgal fabrics, and structures, Hamelin Pool, Western Australia: *In* B. W. Logan et al. (eds.), Evolution and diagenesis of Quaternary carbonate sequences, Shark Bay, Western Australia: Am. Assoc. Pet. Geol. Mem., *22:* 140–194.

Longinelli, A. and S. Nuti, 1973, Revised phosphate–water isotopic temperature scale: Earth Planet. Sci. Lett., *19:* 373–376.

Lorens, R. B. and M. L. Bender, 1980, The impact of solution chemistry on *Mytilus edulis* calcite and aragonite: Geochim. Cosmochim. Acta, *44:* 1265–1278.

Lorenz, K. Z., 1974, Analogy as a source of knowledge: Science, *185:* 229–234.

Loutit, T. S. and J. P. Kennett, 1979, Application of carbon isotope stratigraphy to late Miocene shallow marine sediments, New Zealand: Science, *204:* 1196–1199.

Lowe, J. J. and M. J. C. Walker, 1984, Reconstructing Quaternary environments: Longman Group, Harlow, Essex, England, 389 p.

Lowenstam, H. A., 1950, Niagaran reefs of the Great Lakes area: J. Geol., *58:* 430–487.

Lowenstam, H. A., 1954a, Factors affecting the aragonite calcite ratios in carbonate-secreting marine organisms: J. Geol., *62:* 284–322.

Lowenstam, H. A., 1954b, Environmental relations of modification compositions of certain carbonate-secreting marine invertebrates: Natl. Acad. Sci. Proc., *40:* 39–48.

Lowenstam, H. A., 1957, Niagaran reefs of the Great Lakes area: *In* H. S. Ladd (ed.),

Treatise of marine ecology and paleoecology, v. 2: Geol. Soc. Am. Mem., *67*(2): 215–248.

Lowenstam, H. A., 1961, Mineralogy, O^{18}/O^{16} ratios, and strontium and magnesium contents of Recent and fossil brachiopods and their bearing on the history of the oceans: J. Geol., *69:* 241–260.

Lowenstam, H. A., 1962, Magnetite in denticle capping in Recent chitons (Polyplacophora): Geol. Soc. Am. Bull., *73:* 435–438.

Lowenstam, H. A., 1964a, Sr/Ca ratio of skeletal aragonites from the recent marine biota at Palau and from fossil gastropods: *In* H. Craig et al. (eds.), Isotopic and cosmic chemistry: North-Holland, Amsterdam, p. 114–132.

Lowenstam, H. A., 1964b, Paleotemperatures of the Permian and Cretaceous periods: *In* A. E. M. Nairn, (ed.), Problems in paleoclimatology: Wiley-Interscience, New York, p. 227–248.

Lowenstam, H. A., 1974, Impact of life on chemical and physical processes: *In* E. D. Goldberg (ed.), The sea, v. 5: Wiley-Interscience, New York, p. 715–796.

Lowenstam, H. A., 1981, Minerals formed by organisms: Science, *211:* 1126–1131.

Lowenstam, H. A. and S. Epstein, 1954, Paleotemperatures of the post-Aptian Cretaceous as determined by the oxygen isotope method: J. Geol., *62:* 207–248.

Lowenstam, H. A. and D. McConnell, 1968, Biologic precipitation of fluorite: Science, *162:* 1496–1498.

Lowenstam, H. A. and S. Weiner, 1989, On biomineralization: Oxford Univ. Press, Oxford, 324 p.

Lutz, R. A. and D. C. Rhoads, 1977, Anaerobiosis and a theory of growth line formation: Science, *198:* 1222–1227.

Luz, B., Y. Kolodny, and J. Kovach, 1984, Oxygen isotope variations in phosphate of biogenic apatites. III. Conodonts: Earth Planet. Sci. Lett., *69:* 255–262.

Ma, T. Y. H., 1958, The relation of growth rate of reef corals to surface temperature of sea water as basis for study of causes of diastrophisms instigating evolution of life: Res. Past Clim. Continental Drift, *14:* 1–60.

McAlester, A. L. and D. C. Rhoads, 1967, Bivalves as bathymetric indicators: Mar. Geol. *5:* 383–388.

MacArthur, R. H., 1972, Geographical ecology: Patterns in the distribution of species: Harper & Row, New York, 269 p.

MacArthur, R. H. and J. H. Connell, 1966, The biology of populations: Wiley, New York, 200 p.

MacArthur, R. H. and E. O. Wilson, 1963, An equilibrium theory of insular zoogeography: Evolution, *17:* 373–387.

MacArthur, R. H. and E. O. Wilson, 1967, The theory of island biogeography: Princeton Univ. Press, Princeton, N.J., 203 p.

McCall, P. L. and M. J. S. Tevesz, (eds.), 1982, Animal–sediment relations: Plenum Press, New York, 336 p.

McCarthy, B., 1977, Selective preservation of mollusc shells in a Permian beach environment, Sydney Basin, Australia: Neues Jahrb. Geol. Palaeontol. Monatsh., *8:* 466–474.

MacClintock, C., 1967, Shell structure of patelloid and bellerophontoid gastropods (Mollusca): Peabody Mus. Nat. Hist. Yale Univ. Bull., *22,* 140 p.

McConnaughey, T., 1989a, ^{13}C and ^{18}O isotopic disequilibrium in biological carbonates. I. Patterns: Geochim. Cosmochim. Acta, *53:* 151–162.

McConnaughey, T., 1989b, ^{13}C and ^{18}O isotopic disequilibrium in biological carbonates. II. In vitro simulation of kinetic isotope effects: Geochim. Cosmochim. Acta, *53:* 163–171.

McCoy, E. D. and K. L. Heck, 1983, Centers of origin revisited: Paleobiology, *9:* 17–19.

McCrea, J. M., 1950, On the isotopic chemistry of carbonates and a paleotemperature scale: J. Chem. Phys., *18:* 849–857.

MacDonald, G. J. F., 1956, Experimental determination of calcite–aragonite equilibrium relations at elevated temperatures and pressures: Am. Mineral., *41:* 744–756.

MacGeachy, J. K. and C. W. Stearn, 1976, Boring by macro-organisms in the coral *Montastrea annularis* on Barbados reefs: Int. Rev. Gesamten Hydrobiol., *61:* 715–745.

McGhee, G. R., Jr., 1980, Shell geometry and stability strategies in the biconvex Brachiopoda: Neues Jahrb. Geol. Palaeontol. Monatsh., *1980:* 155–184.

McIntyre, A., 1967, Coccoliths as paleoclimatic indicators of Pleistocene glaciations: Science, *158:* 1314–1317.

McIntyre, A., A. W. H. Bé, and M. B. Roche, 1970, Modern Pacific coccolithophorida: A paleontological thermometer: Trans. N.Y. Acad. Sci., *32:* 720–731.

McKenna, M. C., 1980, Early history and biogeography of South America's extinct land mammals: *In* R. L. Ciochon and A. B. Chiarelli (eds.), Evolutionary biology of the New World Monkeys and continental drift: Plenum Press, New York, p. 43–77.

McKerrow, W. S. (ed.), 1978, The ecology of fossils: Gerald Duckworth & Co., London, 383 p.

McKerrow, W. S. and L. R. M. Cocks, 1976, Progressive faunal migration across the Iapetus Ocean: Nature, *263:* 304–306.

McKinney, F. K., 1977, Functional interpretation of lyre-shaped bryozoans: Paleobiology, *3:* 90–97.

McKinney, M. L. and L. G. Zachos, 1986, Echinoids in biostratigraphy and paleoenvironmental and reconstructions: A cluster analysis from the Eocene Gulf Coast (Ocala Limestone): Palaios, *1:* 420–423.

Machel, H.-G., 1983, Cathodoluminescence in calcite and dolomite and its chemical interpretation: Geosci. Can., *12:* 139–147.

Macurda, D. B., Jr. and D. L. Meyer, 1983, Sealilies and feather stars: Am. Sci. *71:* 354–365.

Magurran, A. E., 1988, Ecological diversity and its measurement: Princeton Univ. Press, Princeton, N.J., 179 p.

Majewske, O. P., 1969, Recognition of invertebrate fossil fragments in rocks and thin sections: E. J. Brill, Leiden, The Netherlands, 101 p.

Malmgren, B. A. and J. P. Kennett, 1978, Test size variation in *Globigerina bulloides* in response to Quaternary palaeoceanographic changes: Nature, *275:* 123–124.

Mancini, E. A., 1978a, Origin of micromorph faunas in the geologic record: J. Paleontol., *52:* 311–322.

Mancini, E. A., 1978b, Origin of the Grayson micromorph fauna (Upper Cretaceous) of north-central Texas: J. Paleontol., *52:* 1294–1314.

Mangerud, J., E. Sonstegaard, and H. P. Sejrup, 1979, Correlation of the Eemian (interglacial) stage and the deep-sea oxygen-isotpe stratigraphy: Nature, *277:* 189–192.

Manker, J. P. and B. D. Carter, 1987, Paleoecology and paleogeography of an extensive rhodolith facies from the lower Oligocene of south Georgia and north Florida: Palaios, *2:* 181–188.

Marshall, L. G. et al., 1982, Mammalian evolution and the Great American Interchange: Science, *215:* 1351–1357.

Martinsson, A., 1970, Taphonomy of trace fossils: Geol. J. Spec. Issue *3:* 323–330.

Matthews, R. K. and R. Z. Poore, 1980, Tertiary ^{18}O record and glacio-eustatic sea-level fluctuations: Geology, *8:* 501–504.

May, R. M. and J. Seger, 1986, Ideas in ecology: Am. Sci. *74:* 256–267.

Mayor, A. G., 1924, Causes which produce stable conditions in the depth of the floors of Pacific fringing reef-flats: Pap. Dep. Mar. Biol., Carnegie Inst., Wash., *19:* 27–36.

Mayr, E., 1982, Adaptation and selection: Biol. Zentralbl., *101:* 161–174.

Mayr, E., 1983, How to carry out the adaptationist program?: Am. Nat., *121:* 324–334.

Menard, H. W. and A. J. Boucot, 1951, Experiments on the movement of shells by water: Am. J. Sci., *249:* 131–151.

Menzies, R. J., R. Y. George, and G. T. Rowe, 1973, Abyssal environment and ecology of the world oceans: Wiley-Interscience, New York, 488 p.

Merrill, G. K. and M. D. Martin, 1976, Environmental control on conodont distribution in the Bued and Matoon Formations (Pennsylvanian, Missourian) northern Illinois: *In* C. R. Barnes (ed.), Conodont Paleoecology: Geol. Assoc. Can. Spec. Pap., *15*, 243–271.

Meyer, D. L., 1980, Ecology and biogeography of living classes: *In* T. W. Broadhead and J. A. Waters (eds.), Echinoderms: Notes for a short course: Univ. Tenn., Dep. Geol. Sci., Stud. Geol., *3*, 1–14.

Meyer, D. L. and N. G. Lane, 1976, The feeding behavior of some Paleozoic crinoids and Recent basketstars: J. Paleontol., *50:* 472–480.

Meyer, D. L. and K. B. Meyer, 1986, Biostratinomy of recent crinoids (Echinodermata) at Lizard Island, Great Barrier Reef, Australia: Palaios, *1:* 294–302.

Middlemiss, F. A., P. F. Rawson, and G. Newall, 1971, Faunal provinces in space and time: Geol. J. Spec. Issue, *4*, 236 p.

Miller, K. G., 1982, Cenozoic benthic foraminifera case histories of paleoceanographic and sea-level changes: *In* M. A. Buzas and B. K. Sen Gupta (eds.), Foraminifera: Notes for a short course: Univ. Tenn., Dep. Geol. Sci., Stud. Geol., *6:* 106–126.

Miller, W., III, 1986, Community replacement in estuarine Pleistocene deposits of eastern North Carolina: Tulane Stud. Geol. Paleontol., *19:* 97–122.

Miller, M. F., and C. W. Byers, 1984, Abundant and diverse early Paleozoic infauna indicated by the stratigraphic record: Geology, *12:* 40–43.

Miller, K. G., G. S. Mountain, and B. E. Tucholke, 1985, Oligocene glacio-eustasy and erosion on the margins of the North Atlantic: Geology, *13:* 10–13.

Miller, K. G., R. G. Fairbanks, and G. S. Mountain, 1987, Tertiary oxygen isotope synthesis, sea level history, and continental margin erosion: Paleoceanography, *2:* 1–19.

Miller, K. B., C. E. Brett, and K. M. Parsons, 1988, The paleoecologic significance of storm-generated disturbance within a middle Devonian muddy epeiric sea: Palaios, 3: 35–52.

Milliman, J. D., 1974. Marine carbonates: Springer-Verlag, New York, 375 p.

Mix, A. C. and N. G. Pisias, 1988, Oxygen isotope analyses and deep-sea temperature changes: Implications for rates of oceanic mixing: Nature, 331: 249–251.

Möbius, K., 1877. Die Auster und die Austernwirtschaft: Hempel and Parry, Berlin, 126 p.

Monty, C., 1977. Evolving concepts on the nature and the ecological significance of stromatolites: In E. Flügel (ed.), Fossil algae: Recent results and developments, Springer-Verlag, West Berlin, p. 15–35.

Monty, C. (ed.), 1981. Phanerozoic stromatolites: Springer-Verlag, West Berlin, 249 p.

Mook, W. G., 1971. Paleotemperatures and chlorinities from stable carbon and oxygen isotopes in shell carbonate: Palaeogeogr. Palaeoclimatol. Palaeoecol., 9: 245–263.

Moran, P. J., R. E. Reichelt, and R. H. Bradbury, 1986, An assessment of the geological evidence for previous *Acanthaster* outbreaks: Coral Reefs, 4: 235–238.

Morton, J. E., 1979. Molluscs. 5th ed.: Hutchinson Univ. Library, London, 254 p.

Mucci, A., 1987. Influence of temperature on the composition of magnesian calcite overgrowths precipitated from seawater: Geochim. Cosmochim. Acta, 51: 1977–1984.

Muhs, D. R. and T. K. Kyser, 1987. Stable isotope composition of fossil mollusks from southern California: Evidence for a cool last interglacial ocean: Geology, 15: 119–122.

Muller, C. H., 1958, Science and philosophy of the community concept: Am. Sci. 46: 294–322.

Munthe, J. and M. C. Coombs, 1979, Miocene dome-skulled chalicotheres (Mammalia, Perisodactyla) from the western United States: A preliminary discussion of a bizarre structure: J. Paleontol., 53: 77–91.

Murray, B. G., 1979, Population dynamics: Alternate models: Academic Press, New York, 212 p.

Murray, J. W., 1987, Benthic foraminiferal assemblages: Criteria for the distinction of temperature and sub-tropical carbonate environments: In M. B. Hart (ed.), Micropaleontology of carbonate environments: Ellis Horwood, Chichester, West Sussex, England, p. 9–20.

Mutvei, H., 1983, Flexible nacre in the nautiloid *Isorthoceras*, with remarks on the evolution of cephalopod nacre: Lethaia, 16: 233–240.

Mutvei, H. and R. A. Reyment, 1973, Buoyancy control and siphuncle function of ammonoids: Palaeontology, 16: 623–636.

Myers, A. C., 1979. Summer and winter burrows of a mantis shrimp, *Squilla empusa*, in Narragansett Bay, Rhode Island (U.S.A.): Estuarine Coastal Mar. Sci. 8: 87–98.

Nagle, J. S., 1967. Wave and current orientation of shells: J. Sediment. Petrol., 37: 1124–1138.

Natland, M. L., 1933. The temperature and depth distribution of some recent and fossil foraminifera in the southern California region: Bull. Scripps Inst. Oceanogr., Tech. Ser., 3: 225–230.

Neale, J. W. (ed.), 1969, The taxonomy, morphology and ecology of recent ostracoda: Oliver & Boyd, Edinburgh, 553 p.

Nelson, P. C., 1975, Community structure and evaluation of trophic analysis in paleoecology: Stone City Formation (Middle Eocene-Texas): unpublished M.S. thesis, Texas A & M Univ., 168 p.

Neuman, R. B., 1984, Geology and paleobiology of islands in the Ordovician Iapetus Ocean: Review and implications: Geol. Soc. Am. Bull., 95: 1188–1201.

Newell, R., 1965, The role of detritus in the nutrition of two marine deposit feeders, the prosobranch Hydrobia ulvae and the bivalve Macoma balthica: Zool. Soc. London Proc., 144: 25–45.

Newell, N. D., J. Imbrie, E. G. Purdy, and D. L. Thurber, 1959, Organism communities and bottom facies, Great Bahama Bank: Bull. Am. Mus. Nat. Hist., 117: 183–228.

Nichols, D., 1959, Changes in the chalk heart urchin Micraster interpreted in relation to living forms: R. Soc. London, Philos. Trans., B242: 347–437.

Nichols, D., 1972, Echinoderms, 4th ed.: Hutchinson Univ. Library, London, 200 p.

Nicol, D., 1962, The biotic development of some Niagaran reefs: An example of an ecological succession or sere: J. Paleontol., 36: 172–176.

Nicol, D., 1964, An essay on size of marine pelecypods: J. Paleontol., 38: 968–974.

Nicol, D., 1967, Some characteristics of cold-water marine pelecypods: J. Paleontol., 41: 1330–1340.

Nier, A. O., 1950, A redetermination of the relative abundances of the isotopes of carbon, nitrogen, oxygen, argon, and potassium: Phys. Rev., 77: 789–793.

Noble, J. P. A. and A. Logan, 1981, Size-frequency distributions and taphonomy of brachiopods: A Recent model: Palaeogeogr. Palaeoclimatol. Palaeoecol., 36: 87–105.

Noe-Nygaard, N. and F. Surlyk, 1988, Washover fan and brackish bay sedimentation in the Berriasian–Valanginian of Bornholm, Denmark: Sedimentology, 35: 197–217.

Noe-Nygaard, N., F. Surlyk, and S. Piasecki, 1987, Bivalve mass mortality caused by toxic dinoflagellate blooms in a Berriasian–Valanginian Lagoon, Bornholm, Denmark: Palaios, 2: 263–273.

Nomland, J. O., 1916, Fauna from the Lower Pliocene at Jacalitos Creek and Waltham Canyon, Fresno County, California: Univ. Calif. Publ. Bull. Dep. Geol., 9: 199–214.

Norris, R. D., 1986, Taphonomic gradients in shelf fossil assemblage: Pliocene Purisima Formation, California: Palaios, 1: 256–270.

Odum, E. P., 1969, The strategy of ecosystem development: Science, 164: 262–270.

Odum, E. P., 1971, Fundamentals of ecology, 3rd ed.: Saunders, Philadelphia, 574 p.

Ohno, T., 1985, Experimentelle Analysen zur Rhythmik des Schalenwachstums einiger Bivalven und ihrer paläobiologischen Bedeutung: Palaeontographica, 189A: 63–123.

Oliver, W. A., Jr. and A. G. Coates, 1987, Phylum Cnidaria: In R. S. Boardman, A. H. Cheetham, and A. J. Rowell (eds.), Fossil invertebrates: Blackwell Scientific, Boston, p. 140–193.

Olsen, C. R., 1978, Sedimentation rates: In R. W. Fairbridge and J. Bourgeois (eds.), The encyclopedia of sedimentology: Dowden, Hutchinson & Ross, Stroudsburg, Pa., p. 687–692.

Olsen, P. E., C. L. Remington, B. Cornet, and K. S. Thomson, 1978, Cyclic change in Late Triassic lacustrine communities: Science, *201:* 729–733.

Olson, E. C. and K. Bolles, 1975, Permo-Carboniferous fresh water burrows: Fieldiana Geol., *33:* 271–290.

Olson, E. C. and P. P. Vaughn, 1970, The changes of terrestrial vertebrates and climates during the Permian of North America: Forma Functio, *3:* 113–138.

O'Neil, J. R., R. N. Clayton, and T. K. Mayeda, 1969, Oxygen isotope fractionation in divalent metal carbonates: J. Chem. Phys. *51:* 5547–5558.

Paasche, E., 1968, Biology and physiology of coccolithophorids: Ann. Rev. Microbiol., *22:* 71–76.

Palmer, A. A., 1986, Cenozoic radiolarians as indicators of neritic versus oceanic conditions in continental-margin deposits: U.S. Mid-Atlantic coastal plain: Palaios, *1:* 122–132.

Pannella, G., 1972, Paleontological evidence on the earth's rotational history since Early Precambrian: Astrophys. Space Sci., *16:* 212–237.

Pannella, G. and C. MacClintock, 1968, Biological and environmental rhythms reflected in molluscan shell growth: J. Paleontol., *42*(5, II): 64–80.

Pannella, G., C. MacClintock, and M. N. Thompson, 1968, Paleontologic evidence of variations in length of synodic month since Late Cambrian: Science, *162:* 792–796.

Papp, A., H. Zapfe, F. Bachmayer, and A. F. Tauber, 1947, Lebensspuren mariner Krebse: K. Akad. Wiss. Vienna Math. Naturwiss. Kl. Sitzungsber., *155:* 281–317.

Park, R., 1976, A note on the significance of lamination in stromatolites: Sedimentology, *23:* 379–393.

Parker, R. H., 1956, Macro-invertebrate assemblages as indicators of sedimentary environments in east Mississippi Delta region: Bull. Am. Assoc. Pet. Geol., *40:* 295–376.

Pearse, J. S. and V. B. Pearse, 1975, Growth zones in the echinoid skeleton: Ann. Zool., *15:* 731–753.

Pearse, V., J. Pearse, R. Buchsbaum, and M. Buchsbaum, 1987, Living Invertebrates: Blackwell Scientific, Boston, 848 p.

Pearse, A. S. and G. Gunter, 1957, Salinity: *In* J. W. Hedgpeth (ed.), Treatise on marine ecology and paleoecology, v. 1: Geol. Soc. Am. Mem. *67*(1): 129–158.

Peel, J. S., 1975, A new Silurian gastropod from Wisconsin and the ecology of uncoiling in Palaeozoic gastropods: Bull. Geol. Soc. Den. *24:* 211–221.

Peet, R. K., 1974, The measurement of species diversity: Annu. Rev. Ecol. Syst., *5:* 285–307.

Pemberton, S. G. and R. W. Frey, 1985, The *Glossifungites* ichnofacies: Modern examples from the Georgia coast, U.S.A.: *In* H. A. Curran (ed.), Biogenic structures: Their use in interpreting depositional environments: Soc. Econ. Paleontol. Mineral. Spec. Publ., 35: 237–259.

Perkins, R. D. and C. I. Tsentas, 1976, Microbial infestation of carbonate substrates planted on the St. Croix shelf, West Indies: Geol. Soc. Am. Bull., *87:* 1615–1628.

Peryt, T. M. and T. S. Piatkowski, 1977, Stromatolites from the Zechstein Limestone (Upper Permian) of Poland: *In* E. Flügel (ed.), Fossil algae: Recent results and developments: Springer-Verlag, West Berlin, p. 124–135.

Petersen, C. G. J., 1915, On the animal communities of the sea bottom in the Skagerrak, the Christiana Fjord, and the Danish waters: Rep. Dan. Biol. Sta., *23:* 3–28.

Peterson, C. H., 1977, The paleoecological significance of undetected short-term temporal variability: J. Paleontol., *51:* 976–981.

Peterson, C. H., 1985, Patterns of lagoonal bivalve mortality after heavy sedimentation and their paleoecological significance: Paleobiology, *11:* 139–153.

Peterson, C. H. and W. G. Ambrose, Jr., 1985, Potential habitat dependence in deposition rate of presumptive annual lines in shells of the bivalve *Protothaca staminea:* Lethaia, *18:* 257–260.

Philcox, M. E., 1971, Growth form and role of colonial coelenterates in reefs of the Gower Formation (Silurian), Iowa: J. Paleontol., *45:* 338–346.

Pianka, E. R., 1978, Evolutionary ecology, 2nd ed.: Harper & Row, New York, 397 p.

Pielou, E. C., 1979, Biogeography: Wiley-Interscience, New York, 351 p.

Pielou, E. C., 1983, Spatial and temporal change in biogeography: Gradual or abrupt?: *In* R. W. Sims, J. H. Price, and P. E. S. Whalley (eds.), Evolution, time and space: The emergence of the bioshpere: Academic Press, London, p. 29–56.

Pilkey, O. H. and J. Hower, 1960, The effect of environment on the concentration of skeletal magnesium and strontium in *Dendraster:* J. Geol., *68:* 203–216.

Playford, P. E., A. E. Cockbain, E. C. Druce, and J. L. Wray, 1976, Devonian stromatolites from the Canning Basin, Western Australia: *In* M. R. Walter (ed.), Stromatolites: Elsevier, Amsterdam, p. 543–564.

Plotnick, R. E., 1986, Taphonomy of a modern shrimp: Implications for the arthropod fossil record: Palaios, *1:* 286–293.

Pojeta, J., Jr., B. Bunnegard, J. S. Peel, and M. Gordon, Jr., 1987, Phyllum mollusca: *In* R. S. Boardman, A. H. Cheetham, and A. J. Rowell (eds.), Fossil invertebrates: Blackwell Scientific, Boston, p. 220–444.

Pokorny, V., 1978, Ostracodes: *In* B. U. Haq, and A. Boersma (eds.), Introduction to marine micropaleontology: Elsevier, New York, p. 109–149.

Popp, B. N., T. F. Anderson, and P. A. Sandberg, 1986a, Brachiopods as indicators of original isotopic compositions in some Paleozoic limestones: Geol. Soc. Am. Bull., *97:* 1262–1269.

Popp, B. N., F. A. Podosek, J. C. Brannon, T. F. Anderson, and J. Pier, 1986b, [87]Sr/[86]Sr ratios in Permo-Carboniferous sea water from the analyses of well-preserved brachiopod shells: Geochim. Cosmochim. Acta, *50:* 1321–1328.

Powell, E. N., 1977, Particle size selection and sediment reworking in a funnel feeder, *Leptosynapta tenuis* (Holothuroidea, Synaptidae): Int. Rev. Gesamten Hydrobiol, *62:* 385–408.

Powell, E. N. and R. J. Stanton, Jr., 1985, Estimating biomass and energy flow of molluscs in palaeocommunities: Palaeontology, *28:* 1–35.

Powell, E. N., R. J. Stanton, Jr., D. Davies, and A. Logan, 1986, Effect of a large larval settlement and catastrophic mortality on the ecologic record of the community in the death assemblage: Estuarine Coastal Shelf Sci., *23:* 513–525.

Prell, W. L., 1982, Oxygen and carbon isotope stratigraphy for the Quaternary of hole 502B: Evidence for two modes of isotopic variability: *In* W. L. Prell, J. V. Gardner, et al. (eds.), Initial reports of the Deep Sea Drilling Project, v. 68: U.S. Govt. Printing Office, Washington, D.C., p. 455–464.

Prell, W. L., J. Imbrie, D. G. Martinson, J. J. Morley, N. G. Pisias, N. J. Shackleton, and H. F. Streeter, 1986, Graphic correlation of oxygen isotope stratigraphy application to the late Quaternary: Paleoceanography, *1:* 137–162.

Prentice, M. L. and R. K. Matthews, 1988, Cenozoic ice-volume history: Development of a composite oxygen isotope record: Geology, *16:* 963–966.

Pryor, W. A., 1975, Biogenic sedimentation and alteration of argillaceous sediments in shallow marine enviroments: Geol. Soc. Am. Bull., *86:* 1244–1254.

Purchon, R. D., 1968, The biology of the Mollusca: Pergamon Press, Elmsford, N.Y., 560 p.

Purchon, R. D., 1977, The biology of the Mollusca, 2nd ed.: Pergamon Press, Oxford, 560 p.

Radtke, R. L., 1987, Age and growth information available from the otoliths of the Hawaiian snapper, *Pristipomoides filamentosus:* Coral Reefs, *6:* 19–25.

Railsback, L. B. and T. F. Anderson, 1987, Control of Triassic seawater chemistry and temperature on the evolution of post-Paleozoic aragonite secreting faunas: Geology, *15:* 1002–1005.

Raup, D. M., 1956, *Dendraster:* A problem in taxonomy: J. Paleontol., *30:* 685–694.

Raup, D. M., 1966a, The exoskeleton: *In* R. A. Boolootian (ed.), Physiology of Echinodermata: Wiley-Interscience, New York, p. 379–395.

Raup, D. M., 1966b, Geometric analysis of shell coiling: General problems: J. Paleontol., *40:* 1178–1190.

Raup, D. M., 1968, Theoretical morphology of echinoid growth: Paleontol. Soc. Mem., *2* [J. Paleontol., *42*(5), suppl.]: 50–63.

Raup, D. M., 1972, Approaches to morphologic analysis: *In* T. J. M. Schopf (ed.), Models in paleobiology: Freeman, Cooper, San Francisco, p. 28–44.

Raup, D. M., 1977, Probabilistic models in evolutionary paleobiology: Am. Sci. *65:* 50–57.

Raup, D. M. and S. M. Stanley, 1978, Principles of paleontology, 2nd ed.: W. H. Freeman, San Francisco, 481 p.

Raymond, A., 1988, The paleoecology of a coal-ball deposit from the Middle Pennsylvania of Iowa dominated by cordaitalean gymnosperms: Rev. Palaeobot. Palynol., *53:* 233–250.

Reeh, N., H. H. Thomsen, and H. B. Clausen, 1987, The Greenland ice-sheet margin: A mine of ice for paleo-environmental studies: Palaeogeogr. Palaeoclimatol. Palaeoecol., *58:* 229–234.

Reif, W. E., 1984, Preadaptation and the change of function: A discussion: Neues Jahrb. Geol. Palaeontol. Monatsh., *1984:* 90–94.

Reyment, R. A., 1971, Introduction to quantitative paleoecology: Elsevier, New York, 226 p.

Rhoads, D., 1975, The paleoecological and environmental significance of trace fossils: *In* R. W. Frey (ed.), The study of trace fossils: Springer-Verlag, New York, p. 147–160.

Rhoads, D. C. and J. W. Morse, 1971, Evolutionary and environmental significance of oxygen-deficient marine basins: Lethaia, *4:* 413–428.

Rhoads, D. C. and G. Pannella. 1970. The use of molluscan shell growth patterns in ecology and paleoecology: Lethaia, *3:* 143–161.

Rhoads, D. C. and D. J. Stanley. 1965. Biogenic graded bedding: J. Sediment. Petrol., *35:* 956–963.

Rhoads, D. C. and D. K. Young. 1970. The influence of deposit-feeding organisms on bottom-sediment stability and community trophic structure: J. Mar. Res., *28:* 150–178.

Rhoads, D. C., I. G. Speden, and K. M. Waage. 1972. Trophic group analysis of Upper Cretaceous (Maestrichtian) bivalve assemblages from South Dakota: Bull. Am. Assoc. Pet. Geol., *56:* 1100–1113.

Richards, F. A.. 1957. Oxygen in the oceans: *In* J. W. Hedgpeth (ed.), Treatise on marine ecology and paleoecology, v. 1: Geol. Soc. Am. Mem. *67*(1): 185–238.

Richards, R. P. and R. K. Bambach. 1975. Population dynamics of some Paleozoic brachiopods and their paleoecological significance: J. Paleontol., *49:* 775–798.

Richardson, J. R., 1986, Brachiopods: Sci. Am., *255:* 100–106.

Richter, R.. 1929. Gründung and Aufgaben der Forschungsstelle für Meeresgeologie "Senckenberg" in Wilhelmshaven: Nat. Mus. *59:* 1–30.

Rider, J. and R. Cowan, 1977, Adaptive architectural trends in incrusting Ectoprocta: Lethaia, *10:* 29–41.

Rigby, J. K.. 1987. Phylum porifera: *In* R. S. Boardman, A. H. Cheetham, and A. J. Rowell (eds.), Fossil invertebrates: Blackwell Scientific, Boston, p. 116–139.

Rigby, J. K. and C. W. Stearn (eds.), 1983, Sponges and spongiomorphs: Notes for a short course: Univ. Tenn., Dep. Geol. Sci., Stud. Geol., *7,* 220 p.

Risk, M. J., S. E. Pagani, and R. J. Elias, 1987, Another internal clock: Preliminary estimates of growth rates based on cycles of algal boring activity: Palaios, *2:* 323–331.

Robinson, R. A. and C. Teichert (eds.), 1979, Treatise on invertebrate paleontology, pt. A, Introduction: Geol. Soc. Am., Boulder, Colo., and Univ. of Kansas Press, Lawrence, Kans., 569 p.

Rollins, H. B., M. Carothers, and J. Donahue, 1979, Transgression, regression and fossil community succession: Lethaia, *12:* 89–104.

Romanek, C. S., D. S. Jones, D. F. Williams, D. E. Krantz, and R. Radtke, 1987, Stable isotopic investigation of physiological and environmental changes recorded in shell carbonate from the giant clam *Tridacna maxima:* Mar. Biol., *94:* 385–393.

Romer, A. S., 1961. Palaeozoological evidence of climate: Vertebrates: *In* A. E. M. Nairn (ed.), Descriptive paleoclimatology: Wiley-Interscience, New York, p. 183–206.

Root, R. B., 1967. The niche exploitation pattern of the blue-gray gnatcatcher: Ecol. Monogr., *97:* 317–350.

Rosen, B. R.. 1977. The depth distribution of recent hermatypic corals and its palaeontological significance: Bur. Rech. Geol. Mineral. Mem. *89:* 507–517.

Rosenberg, G. D.. 1982. Growth rhythms in the brachiopod *Rafinesquina alternata* from the late Ordovician of southeastern Indiana: Paleobiology, *8:* 389–401.

Rosenberg, G. D. and S. K. Runcorn (eds.), 1975, Growth rhythms and the history of the earth's rotation: Wiley-Interscience, New York, 559 p.

Ross, C. A. (ed.), 1974, Paleogeographic provinces and provinciality: Soc. Econ. Paleontol. Mineral. Spec. Publ., *21,* 233 p.

Ross, J. R. P., 1976, Body wall ultrastructure of living cyclostrome ectoprocts: J. Paleontol., *50:* 350–353.

Ross, J. R. P. (ed.), 1987, Bryozoa: Present and past: Western Washington Univ., Bellingham, Wash., 333 p.

Ross, C. A. and J. R. P. Ross, 1985, Carboniferous and Early Permian biogeography: Geology, *13:* 27–30.

Rowell, A. J. and R. E. Grant, 1987, Phylum brachiopoda: *In* R. S. Boardman, A. H. Cheetham, and A. J. Rowell (eds.), Fossil invertebrates: Blackwell Scientific, Boston, p. 445–496.

Rudwick, M. J. S., 1964a, The function of zigzag deflections in the commissure of fossil brachiopods: Palaeontology, *7:* 135–171.

Rudwick, M. J. S., 1964b, The inference of function from structure in fossils: Br. J. Philol. Sci., *15:* 27–40.

Rudwick, M. J. S., 1970, Living and fossil brachiopods: Hutchinson Univ. Library, London, 199 p.

Runcorn, S. K., 1975, Paleontological and astronomical observations on the rotational history of the earth and moon: *In* G. D. Rosenberg and S. K. Runcorn (eds.), Growth rhythms and the history of the earth's rotation: Wiley-Interscience, New York, p. 285–291.

Ryland, J. S., 1970, Bryozoans: Hutchinson Univ. Library, London, 175 p.

Sackett, W. M. and R. P. Thompson, 1963, Isotopic organic carbon composition of recent continental derived clastic sediments of eastern Gulf Coast, Gulf of Mexico: Bull. Am. Assoc. Pet. Geol., *47:* 525–528.

Sanchez, J. D., J. P. Bradbury, B. F. Bohor, and D. A. Coates, 1987, Diatoms and tonsteins as paleoenvironmental and paleodepositional indicators in a Miocene coal bed, Costa Rica: Palaios, *2:* 158–164.

Sandberg, P. A., 1977, Ultrastructure, mineralogy, and development of bryozoan skeletons: *In* R. M. Woollacott and R. L. Zimmer (eds.), Biology of bryozoans: Academic Press, New York, p. 143–181.

Sandberg, P. A., 1984, Recognition criteria for calcitized skeletal and non-skeletal aragonites: Paleontogr. Am. *54:* 272–279.

Sanders, H. L., 1956, Oceanography of Long Island Sound, 1952–1954. X. The biology of marine bottom communities: Bingham Oceanogr. Coll. Bull., *15:* 345–414.

Sanders, H. L., 1968, Marine benthic diversity: A comparative study: Am. Nat., *102:* 243–282.

Sanders, H. L., P. C. Mangelsdorf, Jr., and G. R. Hampson, 1965, Salinity and faunal distribution in the Pocasset River, Massachusetts: Limnol. Oceanogr., *10* (suppl.): R216–R228.

Saunders, W. B., 1984, *Nautilus* growth and longevity: Evidence from marked and recaptured animals: Science, *224:* 990–992.

Saunders, W. B. and P. D. Ward, 1979, Nautiloid growth and lunar dynamics: Lethaia, *12:* 172.

Saunders, W. B. and D. A. Wehman, 1977, Shell strength of *Nautilus* as a depth limiting factor: Paleobiology, *3:* 83–89.

Savazzi, E., 1989, Burrowing mechanisms and sculptures in Recent gastropods: Lethaia, *22:* 31–48.

Savilov, H. L., 1957, Biological aspect of the bottom fauna groupings of the North Okhotsk Sea: In B. N. Nikitin (ed.), Trans. Inst. Oceanol. Marine Biol., USSR, Acad. Sci. Press, 20: 67–136. (Published in the United States by Amer. Inst. Biol. Sci., Washington, D.C.).

Savin, S. M., R. G. Douglas and F. G. Stehli, 1975, Tertiary marine paleotemperatures: Geol. Soc. Am. Bull., 86: 1499–1510.

Savrda, C. E. and D. J. Bottjer, 1986, Trace fossil model for reconstruction of paleo-oxygenation in bottom waters: Geology, 14: 3–6.

Savrda, C. E., D. J. Bottjer, and D. S. Gorsline, 1984, Development of a comprehensive oxygen-deficient marine biofacies model: Evidence from Santa Monica, San Pedro, and Santa Barbara Basins, California continental borderland: Am. Assoc. Petrol. Geologists Bull., 68: 1179–1192.

Schäfer, W., 1962, Aktuo-Paläontologie nach Studien in der Nordsee: Kramer Verlagsbuchhandlung, Frankfurt, West Germany, 666 p.

Schäfer, W., 1972, Ecology and paleoecology of marine environments: Univ. Chicago Press, Chicago, 568 p.

Scheihing, M. H. and H. W. Pfefferkorn, 1984, The taphonomy of land plants in the Orinoco Delta: A model for the incorporation of plant parts in clastic sediments of Upper Carboniferous age in Euramerica: Rev. Palaeobot. Palynol., 41: 205–240.

Schidlowski, M., 1988, A 3800-million-year isotopic record of life from carbon in sedimentary rocks: Nature, 333: 313–318.

Schifano, G., 1984, Environmental, biological and mineralogical controls on strontium incorporation into skeletal carbonates in some intertidal gastropod species: Palaeogeogr. Palaeoclimatol. Palaeoecol., 46: 303–312.

Schindel, D. E., 1980, Microstratigraphic sampling and the limits of paleontologic resolution: Paleobiology, 6: 408–426.

Schlanger, S. O. and M. B. Cita (eds.), 1982, Nature and origin of Cretaceous carbon-rich facies: Academic Press, New York, 229 p.

Schmalfuss, H., 1981, Structure, patterns and function of cuticular terraces in trilobites: Lethaia, 14: 331–341.

Schmidt, W. J., 1924, Die Bausteine des Tierkörpers in polarisiertem Licht: F. Cohen, Bonn, Germany, 528 p.

Schmidt, K. P., 1954, Faunal realms, regions, and provinces: Q. Rev. Biol., 29: 322–331.

Schopf, T. J. M., 1969, Paleoecology of ectoprocts (bryozoans): J. Paleontol., 43: 234–244.

Schopf, T. J. M., 1978, Fossilization potential of an intertidal fauna: Friday Harbor, Washington: Paleobiology, 4: 261–270.

Schrader, A., 1971, Fecal pellets: Role in sedimentation of pelagic diatoms: Science, 174: 55–57.

Schroeder, J. H., E. J. Dwonik, and J. J. Papike, 1969, Primary protodolomite in echinoid skeletons: Geol. Soc. Am. Bull., 80: 1613–1616.

Scott, G. H., 1963, Uniformitarianism, the uniformity of nature, and paleoecology: N. Z. J. Geol. Geophys., 6: 510–527.

Scott, R. W., 1978, Approaches to trophic analysis of paleocommunities: Lethaia, 11: 1–14.

Scrutton, C. T., 1964, Periodicity in Devonian coral growth: Palaeontology, 7: 552–558.

Segerstråle, S. G., 1957, Baltic Sea: *In* J. W. Hedgpeth (ed.), Treatise on marine ecology and paleoecology, v. 1, Ecology: Geol. Soc. Am. Mem., *67*(1): 751–800.

Seidel, M. R., 1979, The osteoderms of the American alligator and their functional significance: Herpetologica, *35:* 375–380.

Seilacher, A., 1957, An-aktualistisches Wattenmeer?: Palaeontol. Z., *31:* 198–206.

Seilacher, A., 1960, Strömungsanzeichen im Hunsrückschiefer: Hess. Landesamtes Bodenforsch. Notizbl., *88:* 88–106.

Seilacher, A., 1964, Sedimentological classification and nomenclature of trace fossils: Sedimentology, *3:* 253–256.

Seilacher, A., 1967, Bathymetry of trace fossils: Mar. Geol., *5:* 413–428.

Seilacher, A., 1970, Arbeitskonzept zur Konstruktionsmorphologie: Lethaia, *3:* 393–396.

Seilacher, A., 1975, Mechanische Simulation und funktionelle Evolution des Ammonitenseptums: Palaeontol. Z., *49:* 268–286.

Seilacher, A., 1977a, Evolution of trace fossil communities: *In* A. Hallam (ed.), Patterns of evolution as illustrated by the fossil record: Elsevier, Amsterdam, p. 359–376.

Seilacher, A., 1977b, Pattern analysis of *Paleodictyon* and related trace fossils: Geol. J. Spec. Issue, *9:* 289–334.

Seilacher, A., 1978, Use of trace fossils for recognizing depositional environments: *In* P. B. Basan (ed.), Soc. Econ. Paleontol. Mineral. Short Course Notes, *5:* 185–201.

Seilacher, A., 1979, Constructional morphology of sand dollars: Paleobiology, *5:* 191–221.

Seilacher, A., 1984, Constructional morphology of bivalves: Evolutionary pathways in primary versus secondary soft-bottom dwellers: Palaeontology, *27:* 207–237.

Seilacher, A., 1985, Bivalve morphology and function: Univ. Tenn., Dep. Geol. Sci., Stud. Geol., *13:* 88–101.

Selley, R. C., 1976, An introduction to sedimentology: Academic Press, New York, 408 p.

Severin, K. P., 1983, Test morphology of benthic foraminifera as a discriminator of biofacies. Mar. Micropaleontol., *8:* 65–76.

Shackleton, N. J., 1967, Oxygen isotope analyses and Pleistocene temperatures reassessed: Nature, *215:* 15–17.

Shackelton, N. J., 1973, Attainment of isotopic equilibrium between ocean water and the benthonic foraminifera genus *Uvigerina:* Isotopic changes in the ocean during the last glacial age: Colloq. Int. C. N. R. S., *219:* 203–209.

Shackleton, N. J. and J. P. Kennett, 1975, Paleotemperature history of the Cenozoic and the initiation of Antarctic glaciation: Oxygen and carbon isotope analysis in DSDP sites 277, 279, and 281: *In* J. P. Kennett, R. E. Houtz, et al. (eds.), Initial reports of the Deep Sea Drilling Project, v. 29: U. S. Government Printing Office, Washington D.C., p. 743–755.

Shackleton, N. J. and R. K. Matthews, 1977, Oxygen isotope stratigraphy of Late Pleistocene coral terraces in Barbados: Nature, *268:* 618–620.

Shackleton, N. J. and N. D. Opdyke, 1973, Oxygen isotope and paleomagnetic stratigraphy of equatorial Pacific core V28-238: Oxygen isotope temperatures and ice volumes in a 10^5 and 10^6 year scale: Quat. Res., *3:* 39–55.

Shackleton, N. J. and N. D. Opdyke, 1976, Oxygen isotope and paleomagnetic stratigraphy of Pacific core V28-239 Late Pliocene to latest Pleistocene: *In* R. M. Cline and J. D. Hayes (eds.), Investigation of Late Quaternary paleoceanography and paleoclimatology: Geol. Soc. Am. Mem., *145:* 449–464.

Shaver, R. H., 1974, Silurian reefs of northern Indiana: Reef and interreef macro-faunas: Am. Assoc. Pet. Geol. Bull., *58:* 934–956.

Shaver, R. H. and J. A. Sunderman (1989). Silurian seascapes: Water depth, clinothems, reef geometry, and other motifs—A critical review of the Silurian reef model: Geol. Soc. Am. Bull., *101:* 939–951.

Sheehan, P. M. and D. R. J. Schiefelbein, 1984, The trace fossil *Thalassinoides* from the Upper Ordovician of the eastern Great Basin: Deep burrowing in the early Paleozoic: J. Paleontol., *58:* 440–447.

Sheldon, A. L., 1969, Equitability indices: Dependence on the species count: Ecology, *50:* 466–467.

Shelford, V. E., 1963, The ecology of North America: Univ. Illinois Press, Urbana, Ill., 610 p.

Shen, G. T. and E. A. Boyle, 1987, Lead in corals: reconstruction of historical industrial fluxes to the surface ocean: Earth and Planet. Sci. Ltrs., *82:* 289–304.

Shen, G. T., E. A. Boyle, and D. W. Lea, 1987, Cadmium in corals as a tracer of historical upwelling and industrial fallout: Nature, *328:* 794–796.

Shimoyama, S., 1985, Size-frequency distribution of living populations and dead shell assemblages in a marine intertidal sand snail, *Umbonium* (*Suchium*) *moniliferum* (Lamarck), and their palaeoecological significance: Palaeogeogr. Palaeoclimatol. Palaeoecol., *49:* 327–353.

Shipman, P., 1981, Life history of a fossil: Harvard Univ. Press, Cambridge, Mass., 222 p.

Simberloff, D., 1983, When is an island community in equilibrium?: Science, *220:* 1275–1277.

Simkiss, K., 1964, Phosphates as crystal poisons of calcification: Biol. Rev., *39:* 487–505.

Simonsen, R. (ed.), 1972, First symposium on Recent and fossil marine diatoms: Nova Hedwigia, *39:* 1–294.

Simpson, G. G., 1940, Mammals and land bridges: J. Wash. Acad. Sci., *30:* 137–163.

Simpson, G. G., 1960, Notes on the measurement of faunal resemblance: Am. J. Sci., *258:* 300–311.

Sims, R. W., J. H. Price, and P. E. S. Whalley (eds.), 1983, Evolution, time and space: The emergence of the biosphere: Academic Press, London, 493 p.

Skougstad, M. W. and C. A. Horr, 1963, Occurrence and distribution of strontium in natural water: U.S. Geol. Surv. Water Supply Pap., *1496-D,* 97 p.

Smith, A. B., 1984, Echinoid palaeobiology: Allen & Unwin, London, 190 p.

Smith, R. K., 1987, Fossilization potential in modern shallow-water benthic for-aminiferal assemblages: J. Foraminiferal Res., *17:* 117–122.

Smith, P. L. and H. W. Tipper, 1986, Plate tectonics and paleobiogeography: Early Jurassic (Pliensbachian) endemism and diversity: Palaios, *1:* 399–412.

Smith, S. V., R. W. Buddemeier, R. C. Redalje, and J. G. Houck, 1979, Strontium–calcium thermometry in coral skeletons: Science, *204:* 404–407

Sokal, R. R. and F. J. Rohlf, 1969, Biometry: The principles and practice of statistics in biological research: W. H. Freeman, San Francisco, 778 p.

Solbrig, O. T. and G. H. Orians, 1977, The adaptive characteristics of desert plants: Am. Sci., *65:* 412–421.

Sommer, M. A. and D. M. Rye, 1978, Oxygen and carbon isotope internal thermometry using benthic calcite and aragonite foraminifera pairs: *In* R. E. Zartman (ed.), Short papers 4th international conference geochronology, cosmochemistry, isotope geology: U.S. Geol. Surv., Open-File Rep., *78-701:* 408–410.

Sorauf, J. E., 1971, Microstructure in the exoskeleton of some Rugosa (Coelenterata): J. Paleontol., *45:* 23–32.

Sorauf, J. E., 1972, Skeletal microstructure and microarchitecture in Scleractinia (Coelenterata): Palaeontology, *15:* 88–107.

Sorauf, J. E., 1974, Observations on microstructure and biomineralization in coelenterates: Biomineralization, *7:* 37–55.

Sørensen, M., 1984, Growth and mortality in two Pleistocene bathyal micromorphic bivalves, Lethaia, *17:* 197–210.

Southward, A. J., 1964, The relationship between temperature and rhythmic cirral activity in some Cirripedia considered in connection with their geographic distribution: Helgol. Wiss. Meeresunters., *10:* 391–403.

Spaeth, C., J. Hoefs, and V. Vetter, 1971, Some aspects of isotopic composition of belemnites and related paleotemperatures: Geol. Soc. Am. Bull., *82:* 3139–3150.

Spicer, R. A. and A. G. Greer, 1986, Plant taphonomy in fluvial and lacustrine systems: Univ. Tenn., Dept. Geol. Sci., Stud. Geol., *15:* 10–26.

Staff, G. M., E. N. Powell, R. J. Stanton, Jr., and H. Cummins, 1985, Biomass: Is it a useful tool in paleocommunity reconstruction?: Lethaia, *18:* 209–232.

Staff, G. M., R. J. Stanton, Jr., E. N. Powell, and H. Cummins, 1986, Time-averaging, taphonomy, and their impact on paleocommunity reconstruction: Death assemblages in Texas Bays: Geol. Soc. Am. Bull., *97:* 428–443.

Stanley, S. M., 1968, Post-Paleozoic adaptive radiation of infaunal bivalve mollusks: A consequence of mantle fusion and siphon formation: J. Paleontol., *42:* 214–229.

Stanley, S. M., 1970, Relation of shell form to life habits of the Bivalvia (Mollusca): Geol. Soc. Am. Mem., *125:* 296 p.

Stanley, S. M., 1972, Functional morphology and evolution of byssally attached bivalve mollusks: J. Paleontol., *46:* 165–212.

Stanton, R. J., Jr., 1967, The effects of provenance and basin-edge topography on sedimentation in the basal Castaic Formation (Upper Miocene, marine), Los Angeles County, California: Calif. Div. Mines Geol. Spec. Rep., *92:* 21–31.

Stanton, R. J., Jr., 1976, The relationship of fossil communities to the original communities of living organisms: *In* R. W. Scott and R. R. West (eds.), Structure and classification of paleocommunities: Dowden, Hutchinson & Ross, Stroudsburg, Pa., p. 107–142.

Stanton, R. J., Jr., and J. R. Dodd, 1970, Paleoecologic techniques: Comparison of faunal and geochemical analyses of Pliocene paleoenvironments, Kettleman Hills, California: J. Paleontol., *44:* 1092–1121.

Stanton, R. J., Jr., and J. R. Dodd, 1972, Pliocene cyclic sedimentation in the Kettleman Hills, California: *In* E. W. Rennie, Jr., (ed.), Geology and oil fields, west side central San Joaquin Valley: Pacific Section Amer. Assoc. Petrol. Geol. Guidebook 1972, p. 50–58.

Stanton, R. J., Jr., and J. R. Dodd, 1976, The application of trophic structure of fossil communities in paleoenvironmental reconstruction: Lethaia, *9:* 327–342.

Stanton, R. J., Jr., and I. Evans, 1971, Environmental controls of benthic macrofaunal patterns in the Gulf of Mexico adjacent to the Mississippi Delta: Trans. Gulf Coast Assoc. Geol. Soc., *21:* 371–378.

Stanton, R. J. Jr., and I. Evans, 1972, Community structure and sampling requirements in paleoecology: J. Paleontol., *46:* 845–858.

Stanton, R. J., Jr., and P. C. Nelson, 1980, Reconstruction of the trophic web in paleontology: Community structure in the Stone City Formation (Middle Eocene, Texas): J. Paleontol., *54:* 118–135.

Stanton, R. J. Jr., and J. E. Warme, 1971, Stop 1: Stone City Bluff: *In* B. F. Perkins (ed.), Trace fossils: A field guide: Louisiana State Univ., School of Geosci. Misc. Publ., *71-1:* 3–10.

Stanton, R. J., Jr., J. R. Dodd, and R. R. Alexander, 1979, Eccentricity in the clypeasteroid echinoid *Dendraster:* Environmental significance and application in Pliocene paleoecology: Lethaia, *12:* 75–87.

Stanton, R. J., Jr., E. N. Powell, and P. L. Nelson, 1981, The role of carnivorous gastropods in the trophic analysis of a fossil community: Malacologia, *20:* 451–469.

Stearn, C. W., 1982, The shapes of Paleozoic and modern reef-builders: A critical review: Paleobiology, *8:* 228–241.

Stearn, C. W., 1984, Growth forms and macrostructural elements of the coralline sponges: Paleontogr. Am., *54:* 315–325.

Stearn, C. W. and A. J. Male, 1987, Skeletal microstructure of Paleozoic stromatoporoids and its mineralogical implications: Palaios, *2:* 76–84.

Stehli, F. G., 1956, Shell mineralogy in Paleozoic invertebrates: Science, *123:* 1031–1032.

Stehli, F. G., 1968, Taxonomic diversity gradients in pole location: The Recent model: *In* E. T. Drake (ed.), Evolution and environment: Yale Univ. Press, New Haven, Conn., p. 163–277.

Stehli, F. G. and C. E. Helsley, 1963, Paleontologic technique for defining ancient pole positions: Science, *142:* 1057–1059.

Stehli, F. G. and S. D. Webb, (eds.), 1985, The great American biotic interchange: Plenum Press, New York, 532 p.

Stein, R. S., 1975, Dynamic analysis of *Pteranodon ingens:* A reptilian adaptation to flight: J. Paleontol., *49:* 534–548.

Stenzel, H. B., 1935, A new formation in the Claiborne Group: Univ. Tex. Bull., *3501:* 267–279.

Stenzel, H. B., 1963, Aragonite and calcite as constituents of adult oyster shells: Science, *142:* 232–233.

Stephenson, W., W. T. Williams, and S. D. Cook, 1972, Computer analyses of Petersen's original data on bottom communities: Ecol. Monogr., *42:* 387–415.

Stitt, J. H., 1976, Functional morphology and life habits of the Late Cambrian trilobite *Stenopilus pronus* Raymond: J. Paleontol., *50:* 561–576.

Stommel, H., 1958, The abyssal circulation: Deep-Sea Res., *5:* 80–82.

Strauch, F., 1968, Determination of Cenozoic sea-temperatures using *Hiatella arctica* (Linne): Palaeogeogr. Palaeoclimatol. Palaeoecol., *5:* 213–233.

Strong, D. R., Jr., D. Simberloff, L. G. Abele, and A. B. Thistle, 1984, Ecological communities: Conceptual issues and the evidence: Princeton Univ. Press, Princeton, N.J., 613 p.

Surlyk, F., 1972, Morphological adaptations and population structures of the Danish chalk brachiopods (Maastrichtian, Upper Cretaceous): K. Dan. Vidensk. Selsk. Biol. Skr. *19:* 1–57.

Sverdrup, H. V., M. W. Johnson, and R. H. Fleming, 1942, The oceans: Their physics, chemistry, and general biology: Prentice-Hall, Englewood Cliffs, N.J., 1087 p.

Swart, P. K., 1983, Carbon and oxygen isotope fractionation in scleractinian corals: A review: Earth Sci. Rev., *19:* 51–80.

Swinchatt, J. P., 1969, Algal borings: A possible depth indicator in carbonate rocks and sediments: Geol. Soc. Am. Bull., *80:* 1391–1396.

Sylvester-Bradley, P. C., 1971, Dynamic factors in animal palaeogeography: *In* F. A. Middlemiss, P. F. Rawson, and G. Newall (eds.), Faunal provinces in space and time: Geol. J. Spec. Issue *4:* 1–18.

Tan, F. C. and J. D. Hudson, 1974, Isotopic studies on the paleoecology and diagenesis of the Great Estuarine Series (Jurassic) of Scotland: Scott. J. Geol., *10:* 91–128.

Tanabe, K., 1988, Age and growth rate determinations of an intertidal bivalve, *Phacosoma japonicum,* using internal shell increments: Lethaia, *21:* 231–241.

Tanabe, K. and E. Arimura, 1987, Ecology of four infaunal bivalve species in the Recent intertidal zone, Shikoku, Japan: Palaeogeogr. Palaeoclimatol. Palaeoecol., *60:* 219–230.

Tarutani, T., R. N. Clayton, and T. K. Mayeda, 1969, The effects of polymorphism and magnesium substitution on oxygen isotope fractionation between calcium carbonate and water: Geochim. Cosmochim. Acta. *33:* 987–996.

Tavener-Smith, R. and A. Williams, 1972, The secretion and structure of the skeleton of living and fossil bryozoa: Philos. Trans. R. Soc. London, *B264:* 97–159.

Taylor, J. D. and M. Layman, 1972, The mechanical properties of bivalve (Mollusca) shell structures: Palaeontology, *15:* 73–87.

Taylor, B. E. and P. D. Ward, 1983, Stable isotope studies of *Nautilus macromphalus* Sowerby (New Caledonia) and *Nautilus pompilius* L (Fiji): Palaeogeogr. Palaeoclimatol. Palaeoecol., *41:* 1–16.

Taylor, J. D., W. J. Kennedy, and A. Hall, 1969, The shell structure and mineralogy of the Bivalvia, introduction. Nuculacea–Trigonacea: Bull. Br. Mus. Nat. Hist. Zool., suppl. 3, 125 p.

Termier, H. and G. Termier, 1975, Rôle des ésponges hypercalcifiées en paléoécologie et en paléobiogéographie: Bull. Soc. Geol. Fr., *17*(7): 803–819.

Thayer, C. W., 1975a, Morphologic adaptations of benthic invertebrates to soft substrata: J. Mar. Res., *33:* 177–189.

Thayer, C. W., 1975b, Strength of pedicle attachment in articulate brachiopods: Ecological and paleoecological significance: Paleobiology, *1:* 388–399.

Thayer, C. W., 1975c, Size-frequency and population structure of brachiopods: Palaeogeogr. Palaeoclimatol. Palaeoecol., *17:* 139–148.

Thayer, C. W., 1977, Recruitment, growth, and mortality of a living articulate brachiopod, with implications for the interpretation of survivorship curves: Paleobiology, *3:* 98–109.

Thayer, C. W., 1983, Sediment-mediated biological disturbance and the evolution of marine benthos: *In* M. J. S. Tevesz and P. L. McCall (eds.), Biotic interactions in Recent and fossil benthic communities, Plenum Press, New York, p. 479–625.

Thompson, J. B., H. J. Mullins, C. R. Newton, and T. L. Vercoutere, 1985, Alternative biofacies model for dysaerobic communities: Lethaia, *18:* 167–179.

Thomson, K. S., 1976, On the heterocercal tail in sharks: Paleobiology, *2:* 19–38.

Tipper, J. C., 1975, Lower Silurian animal communities: Three case histories: Lethaia, *8:* 287–299.

Toggweiler, J. R. and S. Trumbore, 1985, Bomb-test ^{90}Sr in Pacific and Indian Ocean surface water as recorded by banded corals: Earth Planet. Sci. Lett., *74:* 306–314.

Toots, H., 1965, Random orientation of fossils and its significance: Wyo. Univ. Contrib. Geol., *4:* 59–62.

Tourtelot, H. A. and R. O. Rye, 1969, Distribution of oxygen and carbon isotopes in fossils of Late Cretaceous age, western interior region of North America: Geol. Soc. Am. Bull., *80:* 1903–1922.

Towe, K. M., 1972, Invertebrate shell structure and the organic matrix concept: Biominer. Res. Rep. *4:* 1–14.

Towe, K. M. and R. Cifelli, 1967, Wall ultrastructure in the calcareous foraminifera: Crystallographic aspects and a model for calcification: J. Paleontol., *41:* 742–762.

Travis, D. F., 1960, Matrix and mineral deposition in skeletal structures of decapod crustacea: *In* R. F. Sognnaes (ed.), Calcification in biological systems: Am. Assoc. Adv. Sci. Publ., *64,* 57–116.

Trewin, N. H. and W. Welsh, 1976, Formation and composition of a graded estuarine shell bed: Palaeogeogr. Palaeoclimatol. Palaeoecol., *19:* 219–230.

Trueman, A. E., 1940, The ammonite body-chamber, with special reference to the buoyancy and mode of life of the living ammonite: Q. J. Geol. Soc. London, *96:* 339–383.

Turner, J. V., 1982, Kinetic fractionation of carbon-13 during calcium carbonate precipitation: Geochim. Cosmochim. Acta, *46:* 1183–1191.

Turner, J. V., P. Fritz, P. F. Karrow, and B. G. Warner, 1983, Isotopic and geochemical composition of marl lake waters and implications for radiocarbon dating of marl lake sediments: Can. J. Earth Sci., *20:* 599–615.

Turpaeva, E. P., 1957, Food interrelationships of dominant species in marine benthic biocoenoses: *In* B. N. Nikitin (ed.), Trans. Inst. Oceanology, Marine Biol. USSR Acad. Sci. Press *20:* 137–148. (Translated and published in the United States by Amer. Inst. Biol. Sci., Washington, D.C., 1959.)

Underwood, A. J., 1978, The detection of non-random patterns of distribution of species along a gradient: Oecologia, *36:* 317–326.

Urey, H. C., 1947, The thermodynamic properties of isotopic substances: J. Chem. Soc., *1947:* 562–581.

Urey, H. C., H. A. Lowenstam, S. Epstein, and C. R. McKinney, 1951, Measurement of paleotemperatures and temperatures of the Upper Cretaceous of England, Denmark, and southeastern United States: Geol. Soc. Am. Bull., *62:* 399–416.

Valentine, J. W., 1961, Paleoecological molluscan geography of the Californian Pleistocene: Univ. Calif. Publ. Geol. Sci., *34:* 309–442.

Valentine, J. W., 1963, Biogeographic units as biostratigraphic units: Am. Assoc. Pet. Geol. Bull., *47:* 457–466.

Valentine, J. W., 1966, Numerical analysis of marine molluscan ranges on the extratropical northeastern Pacific shelf: Limnol. Oceanogr., *11:* 198–211.

Valentine, J. W., 1971a, Plate tectonics and shallow marine diversity and endemism: An actualistic model: Syst. Zool., *20:* 253–264.

Valentine, J. W., 1971b, Resource supply and species diversity patterns: Lethaia, *4:* 51–61.

Valentine, J. W., 1973a, Evolutionary paleoecology of the marine biosphere: Prentice-Hall, Englewood Cliffs, N.J., 511 p.

Valentine, J. W., 1973b, Phanerozoic taxonomic diversity: A test of alternate models: Science, *180:* 1078–1079.

Valentine, J. W., 1984, Neogene marine climate trends: Implications for biogeography and evolution of the shallow-era biota: Geology, *12:* 647–650.

van der Spoel, S., 1983, Patterns in plankton distribution and their relation to speciation: The dawn of pelagic biogeography: *In* R. W. Sims, J. H. Price, and P. E. S. Whalley (eds.), Evolution, time and space: The emergence of the biosphere: Academic Press, London, p. 291–334.

Van Hinte, J. E., M. B. Cita, and C. H. Van der Weijden, 1987, Extant and ancient anoxic basin conditions in the eastern Mediterranean: Mar. Geol., *7*(1/4): 281 p.

Veizer, J., 1982, Chemical diagenesis of carbonates: Theory and application of trace element technique: *In* M. A. Arthur (ed.), Stable isotopes in sedimentary geology: Soc. Econ. Paleontol. Mineral. Short Course Notes, *10:* 3.1–3.100.

Veizer, J., P. Fritz, and B. Jones, 1986, Geochemistry of brachiopods: Oxygen and carbon isotopic records of Paleozoic oceans: Geochim. Cosmochim. Acta, *50:* 1679–1696.

Vermeij, G. J., 1978, Biogeography and adaptation patterns of marine life: Harvard Univ. Press, Cambridge, Mass., 332 p.

Vinogradov, A. P., 1953, The elementary chemical composition of marine organisms: Sears Found. Mar. Res. Mem., *2:* 647 p.

Vogel, S., 1988, How organisms use flow induced pressures: Am. Sci., *76:* 28–34.

Voigt, E., 1979, The preservation of slightly or non-calcified fossil Bryozoa (Ctenostomata and Cheilostomata) by bioimmuration: *In* G. P. Larwood and M. B. Abbott, (eds.), Advances in Bryozoology: Academic Press, London, p. 541–564.

von Arx, W. S., 1962, An introduction to physical oceanography: Addison-Wesley, Reading, Mass., 422 p.

Voorhies, M. R., 1969, Taphonomy and population dynamics of an Early Pliocene vertebrate fauna, Knox County, Nebraska: Wyo. Univ. Contrib. Geol. Spec. Pap., *1*, 69 p.

Waage, K. M., 1964, Origin of repeated fossiliferous concretion layers in the Fox Hills Formation: Kansas Geol. Surv. Bull., *169, 2:* 541–563.

Waage, K. M., 1968, The type Fox Hills Formation, Cretaceous (Maestrichtian), South Dakota. 1. Stratigraphy and paleoenvironments: Peabody Mus. Nat. Hist. Yale Univ. Bull., *27:* 1–171.

Walker, K. R., 1972, Trophic analysis: A method for studying the function of ancient communities: J. Paleontol., *46:* 82–93.

Walker, K. R. and L. P. Alberstadt, 1975, Ecological succession as an aspect of structure in fossil communities: Paleobiology, *1:* 238–257.

Walker, K. R. and R. K. Bambach, 1971, The significance of fossil assemblages from fine-grained sediments: Time-averaged communities: Geol. Soc. Am. Abstr. Programs, *3:* 783–784.

Walker, K. R. and W. W. Diehl, 1986, The effect of synsedimentary substrate modification on the composition of paleo-communities: Paleoecologic succession revisited: Palaios, *1:* 65–74.

Walker, K. R. and L. F. Laporte, 1970, Congruent fossil communities from Ordovician and Devonian carbonates of New York: J. Paleontol., *44:* 928–944.

Walker, K. R. and W. C. Parker, 1976, Population structure of a pioneer and a later stage species in an Ordovician ecological succession: Paleobiology, *2:* 191–201.

Wallace, A. R., 1876, The geographical distribution of animals: Macmillan Press, London, 2 vols., 503 and 607 p.

Walter, M. R. (ed.), 1976, Stromatolites: Elsevier Scientific, Amsterdam, 790 p.

Walter, M. R., 1977, Interpreting stromatolites: Am. Sci. *65:* 563–571.

Ward, P. D., L. Greenwald, and F. Rougerie, 1980, Shell implosion depth for living *Nautilus macromphalus* and shell strength of extinct cephalopods: Lethaia, *13:* 182.

Ward, P. D. and E. G. Westermann, 1985, Cephalopod paleoecology: Univ. Tenn., Dep. Geol. Sci., Stud. Geol., *13:* 215–229.

Warme, J. E., 1967, Graded bedding in the Recent sediments of Mugu Lagoon, California: J. Sediment. Petrol., *37:* 540–547.

Warme, J. E., 1975, Borings as trace fossils, and the processes of marine bioerosion: *In* R. W. Frey (ed.), The study of trace fossils: Springer-Verlag, New York, p. 181–227.

Warme, J. E., 1977, Carbonate borers: Their role in reef ecology and preservation: Am. Assoc. Pet. Geol. Stud. Geol. *4:* 261–279.

Warme, J. E. and R. J. Stanton, Jr., 1971, Stop 2: Rockdale railroad cut: *In* B. F. Perkins (ed.), Trace fossils, a field guide: Louisiana State Univ., School Geosci. Misc. Publ., *71-1;* 11–15.

Watkins, R., W. B. N. Berry, and A. J. Boucot, 1973, Why "communities"?: Geology, *1:* 55–58.

Webb, T., 1985, Holocene palynology and climate: *In* A. D. Hecht (ed.), Paleoclimate analysis and modeling: Wiley-Intersciences, New York, p. 163–195.

Weber, J. N. and E. W. White, 1977, Caribbean reef corals *Montastrea annularis* and *Montastrea cavernosa:* Long-term growth data as determined by skeletal x-radiography: Am. Assoc. Pet. Geol. Stud. Geol., *4:* 171–179.

Weber, J. N. and P. M. Woodhead, 1970, Carbon and oxygen isotope fractionation in the skeletal carbonate of reef-building corals: Chem. Geol., *6:* 93–117.

Weber, J. N., E. W. White, and P. H. Weber, 1975, Correlation of density banding in reef coral skeletons with environmental parameters: The basis for interpretation of chronological records preserved in the coralla of corals: Paleobiology, *1:* 137–149.

Wells, J. W., 1957, Corals: *In* J. W. Hedgpeth (ed.), Treatise on marine ecology and paleoecology, v. 1, Ecology: Geol. Soc. Am. Mem., *67*(1): 1087–1104.

Wells, J. W., 1963, Coral growth and geochronometry: Nature, *197:* 948–950.

Wells, H. W., 1961, The fauna of oyster beds, with special reference to the salinity factor: Ecol. Monogr., *31:* 239–266.

West, R. M. (ed.), 1977, Paleontology and plate tectonics with special reference to the history of the Atlantic Ocean: Milwaukee Pub. Mus. Spec. Publ. Biol. Geol., *2*, 109 p.

Wetmore, K. L., 1987, Correlations between test strength, morphology and habitat in some benthic foraminifera from the coast of Washington: J. Foraminiferal Res., *17:* 1–13.

Wetzel, A. and T. Aigner, 1986, Stratigraphic completeness: Tiered trace fossils provide a measuring stick: Geology, *14:* 234–237.

Weyl, P. K., 1978, Micropaleontology and ocean surface climate: Science, *202:* 475–481.

Weymouth, F. W., 1923, The life-history and growth of the Pismo Clam (*Tivela stultorum* Mawe): State Calif. Fish Game Comm. Fish Bull., *7,* 120 p.

Whittaker, R. H., 1972, Evolution and measurement of species diversity: Taxon, *21:* 213–251.

Wilbur, K. M., 1964, Shell formation and regeneration: *In* K. M. Wilbur and C. M. Yonge (eds.), Physiology of Mollusca: Academic Press, New York, p. 243–282.

Wilbur, K. M., 1976, Recent studies in invertebrate mineralization: *In* N. Watabe and K. M. Wilbur (eds.), The mechanisms of mineralization in the invertebrates and plants: Univ. South Carolina Press, Columbia, S.C., p. 79–108.

Wilbur, K. M. (ed.), 1983–1985, The Mollusks, v. 1–10: Academic Press, Orlando, Fla.

Williams, A., 1968, Evolution of the shell structure of articulate brachiopods: Spec. Pap. Palaeontol., *2:* 1–55.

Williams, A. and A. D. Wright, 1970, Shell structure of the Craniacea and other calcareous inarticulate Brachiopoda: Spec. Pap. Palaeontol., *7:* 1–51.

Williams, D. F., M. A. Arthur, D. S. Jones, and N. Healy-Williams, 1982, Seasonality and mean annual sea surface temperatures from isotopic and sclerochronological records: Nature, *296:* 432–434.

Wilson, E. O., 1975a, Sociobiology: Harvard Univ. Press–Belknap Press, Cambridge, Mass., 697 p.

Wilson, J. L., 1975b, Carbonate facies in geologic history: Springer-Verlag, New York, 471 p.

Wilson, M. A., 1982, Origin of brachiopod–bryozoan assemblages in an Upper Carboniferous limestone: Importance of physical and ecological controls: Lethaia, *15:* 263–273.

Wolfe, J. A., 1978, A paleobotanical interpretation of Tertiary climates in the northern hemisphere: Am. Sci., *66:* 694–703.

Woodring, W. P., R. Stewart, and R. W. Richards, 1940, Geology of the Kettleman Hills oilfield, California: Stratigraphy, paleontology, and structure: U.S. Geol. Surv. Prof. Pap., *195:* 1–170.

Woodruff, F., S. M. Savin, and R. G. Douglas, 1981, Miocene stable isotope record: A detailed deep Pacific Ocean study and its implications: Science, *212:* 665–668.

Woodward, S. P., 1856, A manual of the Mollusca: A treatise on Recent and fossil shells: Virtue Brothers, London, 542 p.

Woollacott, R. M. and R. L. Zimmer, 1977, Biology of bryozoans: Academic Press, New York, 566 p.

Wray, J. L., 1971, Ecology and geologic distribution: *In* R. Ginsburg, R. Rezak, and J. L. Wray (eds.), Geology of calcareous algae (notes for a short course): Comparative Sedim. Lab., University of Miami, p. 5.1–5.6.

Wray, J. L., 1977, Calcareous algae: Elsevier, Amsterdam, 185 p.

Wright, R. P., 1974, Jurassic bivalves from Wyoming and South Dakota: A Study of feeding relationships: J. Paleontol., *48:* 425–433.

Wright, V. P., 1985, Seasonal banding in the alga *Solenopora jurassica* from the middle Jurassic of Gloucestershire, England: J. Paleontol., *59:* 721–732.

Wright, J., J. F. Miller, and W. T. Holser, 1987a, Conodont chemostratigraphy across the Cambrian–Ordovician boundary: western U.S.A. and southeast China: *In* R. L. Austin, (ed.), Conodonts: Investigative techniques and applications: Ellis Horwood, Chichester, West Sussex, England, p. 246–283.

Wright, J., H. Schrader, and W. T. Holser, 1987b, Paleoredox variations in ancient oceans recorded by rare earth elements in fossil apatite: Geochim. Cosmochim. Acta, *51:* 631–644.

Wüst, G., W. Brogmus, and E. Noodt, 1954, Die zonale Verteilung von Salzgehalt, Niederschlag, Verdunstung, Temperatur und Dichte an der Oberfläche der Ozeane: Kiel. Meeresforsch., *10:* 137–164.

Yancey, T. and P. Wilde, 1970, Faunal communities on the central California shelf near San Francisco: A sedimentary environmental study: Calif. Univ. Hydraul. Eng. Lab. Tech. Rep. *HEL-2-29,* 65 p.

Yonge, C. M. and T. E. Thompson, 1976, Living marine mollusks: William Collins Sons, London.

Ziegler, A. M., L. R. M. Cocks, and R. K. Bambach, 1968, The composition and structure of Lower Silurian marine communities: Lethaia, *1:* 1–27.

Zolotarev, V. N., 1974, Magnesium and strontium in the shell calcite of some modern pelecypods: Geochem. Int., *11:* 347–353.

Index